The Great Power Competition Volume 3

Adib Farhadi · Ronald P. Sanders · Anthony Masys
Editors

The Great Power Competition Volume 3

Cyberspace: The Fifth Domain

Springer

Editors
Adib Farhadi
College of Arts and Sciences
University of South Florida
Tampa, FL, USA

Ronald P. Sanders
Florida Center for Cybersecurity
University of South Florida
Tampa, FL, USA

Anthony Masys
College of Public Health
University of South Florida
Tampa, FL, USA

ISBN 978-3-031-04585-1 ISBN 978-3-031-04586-8 (eBook)
https://doi.org/10.1007/978-3-031-04586-8

This Springer imprint is published by the registered company Springer Nature Switzerland AG
The registered company address is: Gewerbestrasse 11, 6330 Cham, Switzerland

Foreword

Cybersecurity has been an important topic of debate in the United States for over 30 years but has only recently gained bipartisan traction as a national strategic focus, especially in terms of its impact on our national security. In that regard, *cyberspace* has emerged as a domain of warfare in its own right (along with land, sea, air, and space domains), and as such, it has become a key component of our national security strategy and our military's "way of war."

Unfortunately, our Nation's digital dependence has enabled other nation-states—those that aspire to Great Power status like Russia and China, as well as other regional geopolitical rivals—and their non-state proxies to gain asymmetric strategic advantage over the US, as they have unlawfully and covertly accessed our Nation's data, systems, and networks—including elements of our Defense Industrial Base and our critical infrastructure—for purposes of exploitation, theft of intellectual property, espionage (economic and otherwise), and, at least potentially, sabotage...as a part of and/or precursor to more kinetic forms of warfare.

That is what this book is about. Entitled *Cyberspace: The Fifth Domain*, this work examines the various cybersecurity threats facing the United States and its allies and provides realistic response options that range from the tactical to the strategic. Based in large part on the third Great Power Competition Conference hosted by the University of South Florida, US Central Command, and the National Defense University's Near East South Asia Center for Strategic Studies, it elaborates on many of the topics and ideas raised during that dynamic event, held in April 2021.

The authors—all of whom come from a national security background—address critical cyberspace topics, including the risk associated with cybersecurity threats, adversary capabilities and intentions, and the pros and cons of various U.S. responses. Further, the authors consider a U.S. offensive cyberspace posture and explore strategic military and technological cyberoperations to combat (or deter) foreign cyberspace activities.

Through expert examination of the cyberspace threats the United States faces today, *Cyberspace: The Fifth Domain* serves to contribute to this important discussion. Accordingly, I hope that our many readers—including military leaders, scholars, and policymakers—find that the following pages challenge your thinking and enlighten your perspective on the often-subtle influence cyberspace operations have on our national security, both defensively and offensively.

HON J. Michael "Mike" McConnell, VADM, USN, Retired
Executive Director, The Florida Center for Cybersecurity
Former Director of U.S. National Intelligence
Former Director of the National Security Agency
Chair Emeritus, National Intelligence University Foundation
Tampa, USA

Acknowledgments

The editors would like to extend their sincerest gratitude to the entire University of South Florida and National Defense University Near East South Asia Center for Strategic Strategies for their work on the Great Power Competition Conferences Series and the resulting edited volume.

They would like to express their appreciation to Dr. Eric Eisenberg, USF Dean of College of Arts and Science, for his continued and enthusiastic support for the initiative.

A most heartful thanks to Kathleen Whitaker, Arman Mahmoudian, Dr. Dianna Donnelly, Andrew Roberts, Sarah T. White and the contributing authors for making this book possible.

Dr. Adib Farhadi would like to thank his wife, Elaha, children, Adam, Sophia, and mother, Maliha, for their support, patience, and love, not only in regard to this project but in everything that life brings.

Overview

Influence Operations

Alternate Reality—The Use of Disinformation to Normalize Extremism

By Elie Alhajjar

Disinformation is becoming abundant in our civic society, and it poses a national cyberthreat to our democracy. While some conspiracy theories are born spontaneously (in the heat of the moment), others are premeditated by near-peer nation-states such as Russia, China, and their proxies. In this chapter, it is argued that our enemies are using game-like cognitive tactics to reach a large number of people within the United States and radicalize their beliefs. By turning random events and unrelated pieces of information into an alternative reality in which the vulnerable population is getting immersed, a new cognitive state is being established that can basically justify any extreme thought one might have.

The Commercialization of Influence Operations

By Sean Ryan, Ian Conway, and Kathleen Cassedy

This research establishes a clear threat to U.S. national security through the information domain. Influence exercised through this domain impacts domestic stability and internal political processes in the United States. Scenarios presented demonstrate deliberate activities that focused on influencing the U.S. and allies. Scenarios demonstrated actual intent by Russia, China, and regional actors like Iran, to impact the perceptions of target audiences in U.S., Europe, and Africa working by, with, and through commercial platforms to both obfuscate attribution and corrupt objectivity in reporting. Going further, this research connects activities by Great Power competitors, namely, Russia and China, to the subversion of commercial enterprises for the primary purpose of influencing media outlets that provide information to the American (and global) public, and to lever the global business environment to their advantage. Exploiting commercial relationships built legally through overt and clandestine corporate ventures is a primary vehicle for exerting influence in Great Power

Competition. Loopholes in antiquated U.S. laws and policies allow such exploitation to go unchecked and often unobserved. High net worth individuals, multinational corporations, sovereign wealth funds, and NGOs are tied to political support leading to subversion and corruption through the direct use of finance and digital information. Finally, recommendations are made to tighten loopholes in U.S. policies and statutes to mitigate foreign exploitation of information aimed at distracting and disrupting the United States in Great Power Competition for global influence.

The Future of Cyber-Enabled Influence Operations: Emergent Technologies, Disinformation, and the Destruction of Democracy

By Joe Littell

Nation-states have been embracing online influence campaigns through disinformation at breakneck speeds. Countries such as China and Russia have completely revamped their military doctrine to information-first platforms (Cunningham, 2020) to compete with the United States and the West. The Chinese principle of "Three Warfares" and Russian Hybrid Warfare have been used and tested across the spectrum of operations ranging from competition to active conflict. With the COVID-19 pandemic limiting most means of face-to-face interpersonal communication, many other nations have transitioned to online tools to influence audiences both domestically and abroad (Strick, 2020) to create favorable environments for their geopolitical goals and national objectives. This chapter focuses on the landscape that allows nations like China and Russia to attack democratic institutions and discourse within the United States, the strategies and tactics employed in these campaigns, and the emergent technologies that will enable these nations to gain an advantage with key populations within their spheres of influence or to create a disadvantage to their competitors within their spheres of influence. Advancements in machine learning through generative adversarial networks (Creswell et al., 2018) that create deepfakes (Whittaker; Letheren and Mulcahy, 2021) and attention-based transformers (Devlin et al., 2018) that create realistic speech patterns and interaction will continue to plague online discussion and information spread, attempting to cause further partisan divisions and decline of U.S. stature on the world stage and democracy as a whole.

Countering Influence Operations

The Need to Inoculate Military Servicemembers Against Information Threats: The Case for Digital Literacy Training for the Force

By Peter W. Singer and Eric Johnson

Every minute of every day, men and women in uniform are attacked by a weapon that threatens them, their services, and the nation. Yet the U.S. military has not trained them to prepare for this onslaught. It is time for this to change.

Over the last several years, misinformation and deliberately spread disinformation, pushed by both foreign and domestic sources, have proliferated online. They have shaped not just what people read and believe, but also how they act. This "weaponization of social media" has created a formidable challenge in nearly every policy area, from aiding the forces of terrorism and extremism to being a tool of great power competition to damaging the vitality of our democracy.

This challenge is not just to our wider national security, but also to the military itself. Every day, millions of service members at every rank use social media. In so doing, they regularly are targeted by and engage with the viral spread of false information online. The resulting effects on them and the military affect operational security, force reputation, and even the physical health of service members.

Regional Cyber Issues

Countering Violent Extremism in Central Asia and South Asia: Islamophobia and Cyber-Radicalization in the Digital Era

By Adib Farhadi

Widespread political and economic uncertainty following the COVID-19 pandemic, along with increased access to social media and digital messaging in rural areas, has rendered vulnerable populations in Central Asia and South Asia (CASA) even more susceptible to misinformation and population targeting by violent extremists. More, studies show that violent radicalization is inextricably linked to Islamophobia—which is on the rise alongside an endless stream of digital news reporting. Violent extremists capitalize on publicized Islamophobic events to spread misinformation and lure disenfranchised recruits. Compounding these issues in the CASA region is a severely debilitated Afghanistan, ripe for the proliferation of violent extremist activity that will reach far beyond its borders. To mitigate the proliferation of violent extremism in Central Asia and South Asia, Islamophobia must be addressed in earnest at home and abroad. The need for more effective mitigation of Islamophobia and violent Islamic radicalization in the region remains particularly acute during this period of intense insecurity in Afghanistan.

Cyber and Great Power Competition in the Western Hemisphere

By Alexander Crowther, Fabiana Perera, and Brian Fonseca

Great Power Competition is happening in every geographic region and across most domains. The People's Republic of China, Russian Federation, the United States, and other actors are engaged globally, regionally, and even locally. Not everyone competes everywhere, but there is no country in the world today that is not on a field of competition. The United States own neighborhood, the western hemisphere, has seen competition increase as the People's Republic of China, Iran, North Korea, and Russian Federation have all sought to engage to contest historical US hegemony over the region.

Competition in the western hemisphere and in other regions is happening across all three major domains—land, sea, and air. It is also happening in the cyberdomain. This chapter discusses great power competition in the cyberdomain in the western hemisphere. It summarizes the interests and activities of US near-peers in the western hemisphere and presents an overview of their operations in cyberspace. The chapter argues that each of the US near-peers is pursuing different tools for different aims within the western hemisphere. However, in all cases, advances from US near-competitors in the cyberdomain are facilitated by weak governance of the defense sector, and weak capabilities in the cyberdomain in the target countries.

Cyberspace Leadership

The Cyber Pandemic that Could Redefine the Great Power Competition: Preparing the Defense Industrial Base

By Adib Farhadi, Ian Galloway, and Ayman Bekdash

The Fourth Industrial Revolution (4IR) stands poised to transform the geopolitics and geoeconomics of the Great Power Competition (GPC) as digital and cyberworlds permeate societies, governments, and nation-states. The United States must reconceptualize its public-private approach to cybersecurity, starting with the Defense Industrial Base (DIB). As policymakers continue to grapple with the COVID-19 pandemic, the World Economic Forum (WEF) warns that "we should prepare for a COVID-like global cyber pandemic that will spread faster and further than a biological virus, with an equal or greater economic impact" (Davis and Pipkaite 2020). Despite this warning, the new cybersecurity regulations being rolled out under what is commonly known as Cybersecurity Maturity Model Certification (CMMC) are inadequate to protect the DIB from future cyberthreats. Underpinned by the conceptual framework of stakeholder capitalism, this chapter posits that cybersecurity regulations must be reimagined to foster greater process and systems agility, transparency, and trust between the government and the private sector.

Cyber Leadership in the Era of the Great Power Competition

By Garrett Potts

In an effort to consider what the U.S. can do to mitigate threats to the relatively new battlefield of cyberspace, this chapter calls readers' attention to how critical infrastructure and software breaches often happen. Next, the chapter engages with questions regarding who can help us to learn from the breaches further. An answer is provided in light of the ideal type of the "Cyber Leader"—defined herein as someone who demonstrates craft-expertise in the practice of cybersecurity to promote the internal goods of privacy and security. A partial account of what virtues cyber leaders require is also sketched before chapter conclusions are drawn.

Deterrence in Cyberspace

Cybersecurity and Strategic Deterrence: Changing Adversary's Risk Versus Reward Calculations

By Hon J. Mike McConnell, VADM, USN, RET, and Mark Grzegorzewski

This chapter is an extension of the April 16, 2021 panel discussion at the University of South Florida-Cyber Florida Great Power Competition conference series titled "Cybersecurity: The Fifth Domain." Frank Cilluffo moderated the panel "Cybersecurity and Strategic Deterrence" which included the following panelists: HON Mike McConnell; VADM, USN; RET., HON Michael Chertoff; and LTGEN Dennis Crall, USMC.

SolarWinds, Microsoft Exchange Server, JBS, Colonial Pipeline, Kaseya…cyberattacks, cyberintrusions, and exploitation breaches keep mounting, not only from cybercriminals but from nation-states, most notably Russia and China. While cyberattacks against U.S. companies and U.S. Government organizations are not new, the scale and frequency are increasing at an alarming rate. With the U.S. becoming increasing "digitally dependent," this problem has reached strategic proportions. Why is this the case? How can the most military-capable country in the world be so exposed to cyberattacks and not forcefully respond? We proffer the answer can be found in General (GEN) Paul Nakasone's 2018 Senate confirmation as Commander of USCYBERCOMMAND. Responding to Sen. Dan Sullivan's line of inquiry as to why the United States is the "cyber punching bag of the world," GEN Nakasone responded, "I would say right now they do not think that much will happen to them." He added, "they don't fear us" and "the longer that we have inactivity, the longer that our adversaries are able to establish their own norms." In essence, the U.S. is not deterring its adversaries in cyberspace since they do not fear a consequence. Part of their risk-reward calculation surely assesses that although the U.S. may have some of the most exquisite cybercapabilities, it also has many more unpatched cybersecurity vulnerabilities and consequently lacks cyberspace resiliency. Thus, adversarial states assess that the reward of carrying out an action through cyberspace far exceeds any potential cost.

While there are certainly factors constraining how the U.S. can respond, it should not mean the U.S. government (USG) could not respond beyond its current efforts. The cliché that the U.S. lives in the largest glass house and is therefore reluctant to throw stones can only hold for so long. It is well past time the USG fortify its glass house, through cyberresiliency, so it can make its adversaries fear consequences for their actions. In this chapter, we propose a way in which the United States can respond, namely, through cyberresiliency (i.e., deterrence by denial), to build toward establishing a more effective overall deterrence posture.

In what follows, we extend the important insights shared during the "Cybersecurity and Strategic Deterrence" panel. First, we address the risk inherent in the U.S.' large attack surface. This overview provides context to our argument that increased cyberresiliency could bolster overall U.S. national security. Second, we present our thesis to the reader. Our central premise is that the U.S. needs to strengthen its cyberresiliency to deter more effectively, and we demonstrate this claim by illuminating the scope of the problem and providing specific recommendations for changing the risk versus reward calculations of U.S. adversaries. We also address challenges to our recommendations. We then layout what is meant by deterrence and cyberdeterrence, so that the reader can have a deeper appreciation of its complexity and understand how cyberresiliency is just one part of a larger calculation. Finally, we critique the nuclear deterrence-cyberdeterrence analogy before moving to a summation of our chapter.

Cyberspace Strategy and Doctrine

Great Power Competition: Critical Infrastructure

By Morey Haber

The tensions between adversarial nations have escalated far beyond the physical build up troops and machines of war. It has long been predicted that the next conflict would use the Internet as a new battle field to cripple critical infrastructure and disrupt the financial and business operations in the theater of conflict. Whether a cyberattack occurs first, like in the Operations of Desert Storm in 1990, crippling the Iraq power grid or through Stuxnet to interfere with Uranium production in Iran, the fact remains a well-coordinated cyberattack can be an effective weapon during a nation-state mission or as a precursor to war. There is no doubt that the opposing forces are considering cyberwarfare as a part of their offensive and defensive strategy. While the world has generally condoned chemical and biological weapons that could cause mass causalities outside of traditional war fighters, cyberattacks against critical infrastructure could cause the poisoning of civilians, disruptions in the food supply chain, and even the ability to provide life-saving health services to civilians. To that end, politicians, the military, and all realms of technology professionals must consider critical infrastructure as a target and the threats and mitigation strategies are something we will explore in this chapter.

Examining Systemic Risk in the Cyber Landscape

By Dr. Anthony Masys

Cyberattacks and incursions have certainly emerged as a national security issue. Globally we are seeing the effects of such attacks not only on the financial domain but also in health care, government, and critical infrastructure (Masys, 2014, 2021a). Understanding the extent of the impact of cyberincursions and attacks requires understanding the systemic cyberrisks "…of risks spreading across interdependent systems" (Welburn et al., 2021). The World Economic Forum (2016) defines Systemic cyberrisk as "…the risk that a cyber event (attack(s) or other adverse event(s)) at an individual component of a critical infrastructure ecosystem will cause significant delay, denial, breakdown, disruption or loss, such that services are impacted not only in the originating component but consequences also cascade into related (logically and/or geographically) ecosystem components, resulting in significant adverse effects to public health or safety, economic security or national security." Lucas et al (2018) argue that "…systemic risk refers to a potential collapse of a system of potentially global importance and criticality to services that humans urgently need. This dimension of a large potential threat within a complex web of interacting elements distinguishes systemic from other types of risk." As such traditional risk management approaches are not sufficient for dealing with them IRGC (2018:5). This chapter explores systemic risk across the cyber landscape through the non-traditional security lens (Masys, 2021b) and presents applications of systems thinking, scenario planning, and High-Reliability Security Organizations to support systemic risk awareness and management.

When the Levee Breaks: A Global Trend of Cyber-Physical and Cyber-Operational Attacks Against Critical Infrastructure and Future Implications on the Great Power Competition

By Steve Sin, and Rhyner Washburn

Cyber-physical systems (CPSs) refer to systems that connect computers, communication channels, and physical devices. They lie at the heart of today's critical infrastructure. CPSs are currently one of the most targeted systems of adversarial actors operating in the cyberdomain. Cyber-physical and cyber-operational attacks on critical infrastructures via attacks on CPSs have the potential to damage physical infrastructure assets and have widespread consequences for national security as well as society. We analyzed 427 publicly reported cyber-physical and cyber-operational attacks conducted against critical infrastructures globally between January 1, 1992 and July 9, 2021. We find that of the attacks that can be attributed to an actor type, state actors (including state-affiliated and state-supported actors) were found to be the predominant actors that conduct cyber-physical attacks while state and non-state actors occupied approximately the same ratio of attacks for cyber-operational attacks. We also find espionage to be the most statistically significant motivation for the state actors to conduct cyber-physical and/or cyber operational attacks. Additionally, we find the rivalry between the attacker and the target to be the most statistically significant international security-relevant variable. Finally, we provide an assessment of the implications of cyber-physical and cyber-operational attacks on critical infrastructure in the contexts of current and future irregular warfare and great power competition.

Training and Talent

The Cyber-Grand Strategy Gap

By Jacob Shively

This chapter finds that America's central cyberstrategic challenge is a massively skewed risk/reward calculation that favors peer competitors. Leading proposals to address this imbalance are technical and operational, but such solutions are inadequate for an unbalanced strategic environment. This study applies a grand strategy framework to discussions of national security and cyberarticulated by senior officials, military commanders, and other experts at the April 2021 Great Power Competition conference. For attackers, the consequences of being seen as a cyberthreat by the United States are relatively minor compared to the rewards of hacking US systems, stealing intelligence and intellectual property, conducting information warfare, and developing capabilities to quickly devastate critical infrastructure. In short, the benefits of violating US cybersystems are specific and valuable. The costs are diffuse and, attackers seem to agree, manageable. These are classic conditions favoring offensive behavior. In response, professionals at the conference recommended operational and technical solutions. These focused on versions of deterrence, namely, collective defense and defending forward. Here, collective defense refers to cooperation and collaboration among private US actors and the US government. It is designed to convince attackers to redirect their efforts because US systems are resilient and difficult to penetrate. Defending forward refers to US agencies launching persistent, offensive attacks that keep adversary resources distracted, and adversary operators worried about US reprisals. Viewed in the context of grand strategy, these are necessary but not sufficient policy solutions. US planners need to bridge the gap between a current strategic context that favors challengers and a future in which that balance favors US defenders. As in prior eras of technological change, policymakers likely need to pair technical and operational prowess with statecraft and other diplomatic tools.

The War for Cyber Talent: Can the US Win It?

By Ronald Sanders

Our Nation has become increasingly digitally dependent, and, in so doing, it has also become more vulnerable in cyberspace. A significant part of that vulnerability concerns people. For example, people—users like us—are still the principal cause of cybersecurity breaches; over three-quarters of them the result of some socially engineered attack vector, with the remaining breaches more technical (and technically sophisticated) in nature, and as a consequence, people must also stand guard against those incursions. The latter is the focus of this chapter…the people who are responsible for protecting our data and our networks, including those who engage in what are arguably deterrent offensive cyberoperations. In that regard, our Nation's cyberworkforce has become a major factor in its ability to protect our critical—and increasingly vulnerable—information and communications technology (ICT) infrastructure…as well as to project power in the cyberspace domain to influence the behavior of other aspiring "Great Powers" like Russia and China, their proxies, and other regional geopolitical rivals. However, this is not a war in the benign, labor market sense. One involving the competition to talent, although that is part of it. No, the US is in a real "people" war in the cyberspace domain, with real consequences for national security. To that end, the chapter will begin with a brief introduction that seeks to define the cybersecurity talent gap that is the basis of that war, and then it will then discuss the size of that gap, and/or how the US can attempt to measure it. The chapter will then discuss how the US can begin to close that gap, especially with respect to the pipeline that ultimately produces cyber talent that may be deployed in the US national interest, beginning with efforts to attract more young people to cyberspace-related academic disciplines in elementary and secondary schools, post-secondary vocational schools, and colleges and universities (the most obvious "raw" source of cyber talent), before eventually finding their way into cyberspace-related professions. In so doing, the chapter compares and contrasts how our Great Power competitors develop such talent and offer some suggestions on how the US's "supply side" cyberpipeline may be improved. The chapter will then narrow its focus to the Nation's largest employer of such talent: The US government, including its armed forces and its civilian and contractor workforce. And as we will see, the Federal government is also potentially the largest producer and supplier of that talent, both directly and indirectly, and the chapter will close with a description of the US government's current efforts in that regard, as well as with some recommendations on how those efforts may be accelerated to help close the national security cyber talent gap. For that gap—between the US and its global geopolitical rivals, particularly those that aspire to Great Power status and otherwise—is real and growing.

Leveraging Talent to Dominate in Cyber War—An Army Perspective

By Colonel Chad Bates and Major Charlene Rose

DISCLAIMER: The views expressed in this work are those of the author(s) and do not reflect the official policy or position of the United States Military Academy, the Department of the Army, or the Department of Defense.

In order to dominate in a persistent engagement with the nation's adversaries in cyberwarfare, the United States Government must posture a mature and experienced force through a strategic, nationwide partnership for a whole-of-society defense against adversarial threats. In order to recruit and maintain such a high-performing workforce requires an innovative talent management plan that will prevail in an expanding war on acquiring these highly skilled cyberexperts. This plan will assist the United States Cyber Command (USCYBERCOM) to effectively enforce the nation's cybersecurity objectives in the growing great power competition within cyberspace. In this domain, adversarial cyberstrategies range from malicious and sporadic, to calculating and cohesive. Requiring an agile national U.S. cybersecurity posture to engage against these threats through innovative technological solutions, maximizing cyberhuman capital, and continuously evaluating adversarial cyberstrategies. The ingenuity of the human brain is critical in meeting the nation's strategic goals in the cyberdomain. While also leveraging technological advancements in order to enable the pursuit for dominance in this great power competition. Across the continuum of engagement within cyberspace, the country must leverage its cyber talent across all sectors in order to more effectively meet cybersecurity demands through streamlined lateral linkages as a whole-of-society defense. Nationwide partnerships will support the response rate necessary for the evolving cyberdomain, leading to an agile cyber-workforce reinforced by the matrix of U.S. cyberagencies and organizations. The cyberposture strategy must consider talent recruitment, development, and retention to successfully meet talent management goals. This paper will review the U.S. Army Cyber Command's perspective of these key talent management components as it relates to a national, whole-of-society cybersecurity concept.

Contents

Editors and Contributors

About the Editors

Dr. Adib Farhadi is an Assistant Professor, Faculty Director of Executive Education, and lead editor of the Great Power Competition book and conference series at the University of South Florida. His research is at the intersection of religion, politics, economics, and conflict, with a particular focus on the Central and South Asia (CASA) Region. He is the author of *Countering Violent Extremism by Winning Hearts and Minds* and a frequent presenter on the topics of Strategic Competition, Countering Violent Extremism (CVE), and Strategic Negotiations & Communication. Formerly, Dr. Farhadi served in senior positions for Afghanistan and has extensively advised the United States government and various other international organizations. He earned his Ph.D. in Economics from the University of Canberra, M.A. from New York University, and B.S. from East Carolina University.

Dr. Ronald P. Sanders is a Fellow of the National Academy of Public Administration and a decorated, 20-year member of the US government's Senior Executive Service, he currently serves as Staff Director for the Florida Center for Cybersecurity, where he helps to coordinate cybersecurity research, education, and outreach efforts among Florida's 12 State University System (SUS) institutions. From 2017 to 2020, he served as the Director of the University of South Florida's School of Public Affairs, where he oversaw various graduate education and research programs in public administration and urban and regional planning, as well as its Florida Institute of Government. In addition, until his unintentionally well-publicized resignation in October 2020, he chaired the US Federal Salary Council (a Presidential appointment), which makes recommendations on pay raises for the Federal government's 2M+ white-collar employees. He also served on the Advisory Board of the National Security Agency and Vice Chair of the National Intelligence University Foundation.

In addition to his USF appointments, he has served in senior leadership positions in both public and private sectors. For example, he was Vice President of the consulting firm Booz Allen Hamilton and the firm's very first Fellow (2010 to 2017);

the first Associate Director of National Intelligence for Human Capital (2005–2010), where he earned one of Harvard University's *Innovations in American Government Awards* and the National Intelligence Distinguished Service Medal; the first Associate Director of the US Office of Personnel Management for Policy (2002 to 2005); the Internal Revenue Service's first Chief Human Resource Officer (1998 to 2002); and the US Defense Department's Director of Civilian Personnel (1991 to 1998). He has also served on the faculties of Syracuse University's Maxwell School of Citizenship and Public Affairs (from 1995 to 1996) and the George Washington University's School of Business and Public Management (from 1997 to 1998) and directed research centers at both institutions, and he has co-authored four other books, including two published by the Brookings Institution and one by NAPA, as well as numerous articles and monographs in both academic and professional outlets.

Dr. Anthony Masys is an Associate Professor and Director of Global Disaster Management, Humanitarian Assistance and Homeland Security. A former Senior Air Force Officer, he has a BSc in Physics and MSc in Underwater Acoustics and Oceanography from the Royal Military College of Canada and a Ph.D. from the University of Leicester. He is Editor in Chief for Springer Publishing book series: Advanced Sciences and Technologies for Security Applications and holds various advisory board positions with academic journals and books series. He is an internationally recognized author, speaker, and facilitator and has held workshops on security, visual thinking, design thinking, and systems thinking in Europe, Canada, South America, West Africa, and Asia. He has published extensively in the domains of physics and the social sciences. He supports the University of Leicester (U.K.) as an Associate Tutor in their Distance MSc Program on Risk Crisis and Disaster Management.

Contributors

Alhajjar Elie United States Military Academy, West Point, NY, USA

Bates Colonel Chad U.S. Army War College, Carlisle Barracks, PA, USA

Bekdash Ayman DGC International, McLean, VA, USA

Cassedy Kathleen West Liberty University, West Liberty, WV, USA

Conway Ian West Liberty University, West Liberty, WV, USA

Crowther Alexander Florida International University, Miami, FL, USA

Farhadi Adib University of South Florida, Tampa, FL, USA

Fonseca Brian Florida International University, Miami, FL, USA

Galloway Ian DGC International, McLean, VA, USA

Grzegorzewski Mark St. Petersburg College, St. Petersburg, Tampa, FL, USA

Haber Morey BeyondTrust, Johns Creek, USA

Johnson Eric New America, Washington, DC, USA

Littell Joe Army Cyber Institute at the West Point, United States Military Academy, West Point, NY, USA

Masys Anthony University of South Florida, Tampa, FL, USA

McConnell Hon J. Mike University of South Florida, Tampa, FL, USA

Perera Fabiana National Defense University, Washington, D.C, USA

Potts Garrett University of South Florida, Tampa, FL, USA

Rose Major Charlene United States Military Academy, West Point, NY, USA

Ryan Sean West Liberty University, West Liberty, WV, USA

Sanders Ronald University of South Florida, Tampa, FL, USA

Shively Jacob University of West Florida, Pensacola, FL, USA

Sin Steve University of Maryland, College Park, MD, USA

Singer Peter W. Arizona State University, Tempe, AZ, USA

VADM, USN, RET University of South Florida, Tampa, FL, USA

Washburn Rhyner University of Maryland, College Park, MD, USA

Cyberspace Strategy and Doctrine

Great Power Competition: Critical Infrastructure

Morey Haber

Abstract The tensions between adversarial nations have escalated far beyond the physical build up troops and machines of war. It has long been predicted that the next conflict would use the Internet as a new battle field to cripple critical infrastructure and disrupt the financial and business operations in the theater of conflict. Whether a cyber attack occurs first, like in the Operations of Desert Storm in 1990, crippling the Iraq power grid or through Stuxnet to interfere with Uranium production in Iran, the fact remains a well coordinated cyber attack can be an effective weapon during a nation state mission or as a precursor to war. The is no doubt that the opposing forces are considering cyber warfare as a part of their offensive and defensive strategy. While the world has generally condoned chemical and biological weapons that could cause mass causalities outside of traditional war fighters, cyber attacks against critical infrastructure could cause the poisoning of civilians, disruptions in the food supply chain, and even the ability to provide life saving health services to civilians. To that end, politicians, the military, and all realms of technology professionals must consider critical infrastructure as a target and the threats and mitigation strategies are something we will explore in this chapter.

Keywords Cyber warfare · Great power competition · Cyber attacks · Critical infrastructure security · Industrial control systems

Cyber Warfare

A true cyber war could be nearly as devastating as conventional bombs and weapons in terms of human life and destruction to infrastructure and the economy. If you think that this is farfetched, consider these real world examples of the escalation of **sophisticated cyberattacks** targeting critical infrastructure covering the last

---

M. Haber (✉)
BeyondTrust, Johns Creek G30097, USA
e-mail: mhaber@beyondtrust.com

A. Farhadi et al. (eds.), *The Great Power Competition Volume 3*,
https://doi.org/10.1007/978-3-031-04586-8_1

15 years (recent and distant past) and the steps necessary to conduct a successful attack[1].

Oldsmar Water Treatment: In February 2021, a threat actor successfully hacked into the computer systems of a water treatment plant in Oldsmar, Florida. During their exploitation, the hacker changed the plant's sodium hydroxide levels temporary to a dangerous level. While the change was immediately identified and fail safes would have prevented the poisoning of the local population, the attack demonstrates the risk to critical infrastructure and the potential to cause harm via the manipulation of industrial control systems and critical software.[2]

Colonial Pipeline: In May 2021, a ransomware attack on the largest pipeline in the United States successfully proved that treat actors could impact the general public and United States economy by disabling the distribution of fossil fuel along the east coast of the United States. What started as a leaked password on the dark web turned into a full blown attack that affected primary and secondary systems in the organization and crippled the company until a ransom was paid to the ransomware operators. The attack demonstrated the potential dangers of cyber warfare to the general public as prices soared and fuel supplies ran dry for consumers for several weeks.[3]

JBS: In June 2021, the world's largest meat processor paid a staggering ransomware to threat actors that forced the shutdown of all of the organizations poultry and beef processing plants. The interruption of the food supply change scared the consumer market and temporarily drove prices up in the fear of a nationwide shortage. Luckily, JBS was able to recover relatively quickly and the worst fears where not realized. The attack however demonstrated the technology dependance for the food supply chain and how badly the attack could be on livestock, food processing, and the distribution of food to consumers.[4]

Triton: In December 2017, an unidentified power generation facility, believed to be in Saudi Arabia, was compromised when the Triconex industrial safety control system made by Schneider Electric SE was exploited allegedly by a state-sponsored attack.[5] The malware, known as "Triton", exploited a vulnerability in computers running the Microsoft Windows operating system, and potentially allowed the **threat actors** nearly full control of the safety systems for the power plant.[6]

UBS: In 2006, a disgruntled system administrator planted a "logic bomb" in 1,500 computers that, had it been successful, would have driven UBS's stock price into the ground.[7]

---

1 Smith [1].

2 Lyngaas [2].

3 Turton [3].

4 Collier [4].

5 Seals [5].

6 Smith [1].

7 https://www.justice.gov/archive/criminal/cybercrime/press-releases/2002/duronioIndict.htm

Fannie Mae: In late 2008, a contract engineer launched a logic-bomb attack after he was terminated. Had his attack not been thwarted, it could have wiped data off of more than 4,000 servers that may not have been recoverable.

Wand Corp: And in another 2008 attack, a laid-off tech support worker initiated a partially successful attack that crashed 25 computers and cost thousands of dollars to clean up and restore normal operations.

And the research and threats continue to demonstrate the potential threats:

GPS Manipulation: According to a study conducted by The Center for Advanced Defense Studies, the Russian Government is actively hacking the global navigation satellite system to confuse thousands of ships and airplanes regarding their current location. As an example of this potential threat reported by the UK Space Agency, Britain's entire critical infrastructure relies on the GNSS and GPS for operations.[8]

GHIDRA[9]: Ghidra is a software reverse engineering framework developed by National Security Agencies Research Directorate to support ongoing United States government cybersecurity missions. The tool helps analyze malicious code and malware, and can give **cybersecurity professionals** [10]a better understanding of potential vulnerabilities in their networks and other **advanced file-based threats**.[11] The tool was released to the public on April 4th, 2019.

Medical Devices[12]: The earliest attacks on medical devices can be linked back to the late researcher, **Barnaby Jack**[13] (a former colleague of mine at eEye Digital Security). As early as 2011, demonstrations on the **vulnerabilities in medical devices**[14]were published and raised awareness of the potential risks when IT meets implanted medical equipment. This threat has become vastly more prevalent as mhealth and IoT have exploded in popularity. Some IT-enabled medical devices still have default, **embedded passwords**,[15] which is a particularly risky practice.

HOPLIGHT[16]: According to the **US-CERT**,[17] this malware was created by the North Korean government and is designed to inventory a host, collect targeted information, and relay that data to a list of addresses on the Internet. This default payload contains a reconnaissance tool, but future versions are speculated to use this information for deeper, targeted attacks based on attributes discovered on a host system.

---

[8] Edwards [6].

[9] Khandelwal [7].

[10] Stoneff [8].

[11] BeyondTrust [9].

[12] Khandelwal [10].

[13] Wikipedia [11].

[14] Haber [12].

[15] Miller [13].

[16] Seals [14].

[17] Cyber Security and Infrastructure Security Agency [15].

With these recent attacks in mind, tools being publicly disclosed, and research on cyber attacks continuing, nations of the world have the technology means to commit to a cyber war without even firing a single bullet or dropping a single bomb. The results could be equally as devasting even if their physical military (adversary) would be no match for the target nation (potentially the United States). Some of this is already playing out in the Russia and Ukrainian conflict we are all monitoring today.

Critical Infrastructure Security

It is clear that we are dealing with much more than **ransomware** [18]given the sophistication and sources of these attacks. In summary, these cyberattacks and malware have the capability to:

- To penetrate critical infrastructure (Triton) and inflict wide-ranging damage—from service disruption to catastrophic systems failure, and even extreme damage that could cost lives and require massive physical repairs.
- Misdirect commercial navigation (GPS) that could cause collisions in dependent services, like shipping, aviation, and civil rescue.
- Reverse engineer attacks for countermeasures and new attack vectors (GHIDRA) that could cause an all out war after a cyber attack once the source is known.
- Target critical individuals with medical history (medical devices) for assassination.
- Provide reconnaissance and surveillance of foreign activities (HOPLIGHT) to reveal sensitive information and plan for future attacks, both physical and cyber, across any industry.
- Impact resources that could devastate a company, governments, or critical infrastructure.
- If we consider that these are only a fraction of what nation states wield in their cyber war arsenal, we should all take note. Therefore, the best way to predict the future is to rediscover the past. Some of the earliest cyber attacks help paint this picture even from domestic threats; with domestic threats being a key concept to remember.
- If you're like most people, you're thinking to yourself, "I do not work for a big company, so this isn't something I need to worry about." Only recently has the general population begin to realize the scope and impact of these threats. Perhaps you're in an organization that has little connected devices as a part of their digital transformation strategy. Surely, you won't be targeted, right? Stop a minute and think about how many things are attached and computerized in your home, office, and business. Control systems are everywhere. The fire suppression systems, security cameras, alarms and battery backups could all be compromised: Any

[18] BeyondTrust [16].

device with an IP address is fair game and this impacts the critical infrastructure that supports are daily lives. Therefore, a domestic attack would disrupt everyone.

Logic Bombs

As a primary threat to critical infrastructure consider logic bombs. A logic bomb is a piece of malware that disrupts industrial control systems by placing false logic into the workflow to confuse systems and cause disruption. As early as November 2009, an episode of "60 Minutes on CBS" showed a test attack, called Operation Aurora[19] (conduct in 2007), that hacked into the control system of a power generator and caused it to self destruct.

Logic bombs have been suspected in several cyberespionage attacks. The **hack into the U.S. electrical grid**[20] in 2018, and the Bugat trojan that hit financial wire transfer services all had characteristics similar to logic bombs.

In these incidents, the intent wasn't to cause immediate damage like ransomware—although the power grid hack may have left behind malicious payloads that could cause future physical damage. Instead, these attacks were designed to steal either money, information, or provide reconnaissance for a future critical infrastructure attacks.

Attacks against power grids represent one of the scariest aspects of these new attacks—they are coordinated mostly by state-sponsored entities. The list of known state-sponsored cyberattacks is itself a long one, with China's alleged attacks on Google and the U.S. power grid only the most recent examples. Others include Russian[21] attacks on Georgia and Estonia, and U.S. cyberwar efforts in Iraq and Serbia, and even organizations like Anonymous introducing themselves as a biased attacker against Russian assets.

The United States isn't solely a victim when it comes to cyberwarfare. Our government has admitted to conducting previous cyberattacks. Two involved Iraq. One targeted the cell phones of insurgents involved with roadside improvised explosive devices, while another was intended to take down the Iraqi banking system during the invasion. The banking attack never got beyond the planning stages for fear of it spreading beyond Iraq's borders and affecting allies. A third rumored penetration supposedly kept the power on in Baghdad during the invasion. These demonstrate the potential capabilities of cyber attacks as an additional theater operations during a military conflict.

---

[19] https://www.muckrock.com/news/archives/2016/nov/14/aurora-generator-test-homeland-security/

[20] Atherton [17].

[21] https://fortune.com/2022/04/11/anonymous-cyber-war-russia-ukraine/

The Threat

Critical infrastructure attacks inspires fear-based headlines, but what does this have to do with the security of our (of any) nation? Quite a lot actually. Cyberextortion is nothing new (especially ransomware), and even if the organization you work for is not a high-value target like a government agency, a utility, or a even social media, all it takes is one disgruntled employee to wreak havoc on your business with a cyber attack.

It is therefore easy to conclude attacks are inevitable and the time to start thinking about them is now. Most U.S. industries are already under regulatory pressure to modernize and safeguard their IT security systems. Most large organizations have a compliance program in place too. And, recent presidential executive orders are designed to push modernization and security along to adapt to these modern threats. We will never be able to stop 100% of attacks but how we respond or limit their success will be crucial in our ability to survive a cyber war domestically or from abroad.

Critical Infrastructure Defined

These threats require concise definitions. Critical infrastructure is defined as physical or electronic information technology systems and assets that are so vital to the United States (or other governments) that if incapacitated or destroyed would have a debilitating impact on the physical or economic security or public health or safety. A countries critical infrastructure provides the essential services that underpin the culture and society for the general population.

With this in mind, critical infrastructure and supporting software used across private industry and the federal government should have rigorous security measures applied to ensure safe and secure use. Compromise, misuse, or disruption of critical infrastructure can jeopardize the safety and well being of citizens and cause financial damage, loss of trust, loss of life and widespread societal consequences. Not only must the critical infrastructure itself be protected, but the providing entity, its customers, partners, and the public must be protected from any potential misuse or rogue activity posed by the software itself like in the cases of SolarWinds Orion and Kaseya supply chain attacks.

With a wave of industry-shaking cyberattacks kicking off the first half of 2021 that we previously explored, the U.S. government has mobilized efforts and marshalled minds and resources to improve supply chain cyber resilience and protect critical infrastructure. Recently, President Biden issued the **Executive Order (EO) on Improving the Nation's Cybersecurity**.[22] The remainder of this chapter will discuss the definitions and best practices to address Sect. 4 of Biden's EO: Enhancing

[22] White Hours [18].

Software Supply Chain Security that is dire for critical infrastructure. To start, lets establish what critical software is, and then breakdown several fundamental cybersecurity categories that all critical infrastructure should implement to ensure a resilient implementation that can thwart a cyber attack.

Critical Software

To help ensure critical infrastructure align the appropriate security rigor for protection of critical software and help ensure the continuity and integrity of their missions, a basic requirement is to define what constitutes "critical software." The factors by which software packages are deemed "critical" can vary, but include understanding the types of data it accesses, the level of systems it controls, and/or the open attack surface that it carries. While some software may always be classified as critical, classification can fluctuate based on the data being processed at any given time. Therefore, in today's enterprise-level environments, "critical software" for critical infrastructure is defined as:

- Application software, which relates to the software's access to, maintenance of, or transmission or storage of sensitive data.
- System Software, categorized as software that controls and manages networked devices, including client, server, or network hardware including those used in operational technology and industrial control systems.
- And include this superset of definitions provided by **NIST**[23]:

- Software that is designed to run with elevated privilege or manage privileges;
- Software that has direct or privileged access to networking or computing resources;
- Software that is designed to control access to data or operational technology;
- Software that performs a function critical to trust;
- Or software that operates outside of normal trust boundaries with privileged access.
- This superset includes operating systems, web browsers, hypervisors, endpoint security tools, identity and access management applications, network monitoring tools and other products, according to NIST that could also be present in a critical infrastructure environment.
- With these in mind, the critical software can be furthered expanded to include key components for including commercial software and industrial control systems. In the next few sections, we will explore a break down of each of them.

[23] Freguson [19].

Application Software

The first category of critical software, application software, is installed using a variety of network architectural techniques or on an individual system. Software within this category receives its critical designation because it is considered sensitive and/or classified under privacy laws or government designations based on its mission and usage.

For example, applications that maintain, transmit, or access data governed under the Health Insurance Portability and Privacy Act (HIPAA) should be designated "critical", considering the type of data and the potential for massive damage resulting from a breach. Similarly, Personally Identifiable Information (PII) is defined by **OMB Memorandum M 071,616** [24]as "information that can be used to distinguish or trace an individual's identity, either alone or when combined with other personal or identifying information that is linked or linkable to a specific individual." This data, as dictated by OMB, would create a potential risk to any users operating critical infrastructure. In this case, not only is the harm substantive in cost, physical and electronic disruption, but it also creates a potential life-threatening risk if obtained and exploited by a threat actor.

As an example, in April of 2015, operators at the Office of Personnel Management **discovered a breach**[25] in personnel management software, including "SF-86" Forms, which contained the extensive background information for cleared civilian personnel working in the federal government. Fingerprints, financial information, known friends and relatives, and job assignments are just a sample of information included in the breach. Multiple sources cited the trove of information as a significant value to the international intelligence community, and a potential significant risk to those employees who might be serving abroad or undercover. This is a clear example of combining sensitive and/or classified data with PII to form a potential for harm. If any of these people are operating critical infrastructure as well, regardless if for military or civilian operators, the results, if leveraged, could be catastrophic.

System Software

The second category, system software, provides management for networks, databases, client management systems, industrial control systems, and other utilities as necessary for the operation of critical infrastructure. The critical status of system software is determined by the impact to the runtime or mission if that software and data were no longer functioning, compromised, or properly secured from access or leakage.

A particularly sensitive subset of system software includes authoritative directories, which provide the base of authentication for most modern networks. A breach of

[24] Office of Budget and Management [20].

[25] Zetter and Greenberg [21].

this software allows threat actors to access almost everything of value on a network and may allow them to regain access in the future. These directories may be the most significant critical piece of software In an environment. This is why most organizations go to great lengths to protect their Identity Services governing identities, accounts, credentials, roles, group membership, and secrets within an environment.

Database servers are the backbone of modern critical infrastructure and operational software. Often containing tens of thousands of data entries and supporting the runtime and historical performance of a critical infrastructure environment, databases provide the essential storage for maintaining a mission. Without them, or if compromised, the state of critical infrastructure from one time to another is not known nor are any key settings needed for appropriate operations. To a threat actor, compromising a database provides a near complete ownership and dominance in an environment.

In order for a threat actor to typically obtain access to a database and other resources, they need to compromise the system software or exploit an application based vulnerability. Access to this software will grant lateral movement or allow unmitigated access to sensitive data. Even as NIST (National Institute of Standards and Technology) and CISA (Cybersecurity and Infrastructure Security Agency) begin to recognize identity as the perimeter (Zero Trust Architectures), network hardware and the underlying management software still cannot be understated in their importance. These services provide dependent functions for other critical software. Any unauthorized use or breach of this software by a threat actor could render multiple functions inoperable or allow continued access to an environment as a persistent presence.

Finally, software that is considered "public facing", defined by **Sect. 508**[26] as "[software] made available by the agency to members of the general public", provides an extraordinarily vulnerable attack vector into an environment, specifically if the underlying data services also serve internal software or components of the network. These software packages should take precedence as "critical" because a breach of them could open multiple attack vectors into a network, and create a larger attack surface for threat actors' to probe for additional opportunities into critical infrastructure. This is an undisputable form of critical system software that should always be considered in an organization's risk profile.

Best Practices

A robust strategy should include the concepts from NIST's definition of critical software merged with the superset of resources needed to operate critical infrastructure:

[26] Section 508 [22].

- Systems are hardened, are assessed for vulnerabilities, and security patches or other mitigating controls are applied in a timely manner. This includes adhering to best practices like configuration management for the entire process.
- Privileged access management (PAM) and application control for the secure storage of privileged credentials and appropriate access to applications.
- Zero trust architecture, with an emphasis on least privilege, following guidance in NIST -800–207.

Each of the three areas above overlap, and in each of these, the Principle of Least Privilege (PoLP) is a strong underlying theme. Why is that? Almost every attack (malware, external hacker, insider, etc.) today requires privilege—either to execute and initially compromise an environment, and/or to move laterally within an organization's systems. The exception to this is based on exploits that can perform privileged escalation which has become a rare attack vector[27] (less than 10% in 2021) for modern threats. Therefore, The more privileges a piece of software or application, user, account, or process has, the greater the potential for abuse, exploit, insider threat, or error. Today, many planes of privileges are being created at tremendous scale due to digital transformation, cloud expansion, virtualized environments, and IoT. Remote workers pose many compounding risks as well as they operate critical infrastructure from home verses operations pre-pandemic. All of these add risk.

To that end, applying the **PoLP**[28] is one of the most powerful and well-recognized ways of securing crtitical software from attack and preventing misconfigurations, while also protecting infrastructure from rogue, compromised, or misused applications. A least privilege security posture can not only outright eliminate many types of malware and other attacks from executing and gaining a foothold, it can also essentially maroon attackers who do gain a foothold by sharply reducing potential for privilege escalation and lateral movement. After all, **lateral movement**[29] between resources is the difference between compromising a single asset and potentially traversing throughout an environment to establish a persistent presence on any compromised system. After all **70% of attacks today**[30] are reported to involve some form of lateral movement. And today, it's the scale and disruption wrought by society-shaking breaches (SolarWinds Orion, Colonial Pipeline, JBS, Verkada, etc.) that should concern everyone operating critical infrastructure.

As best practices go, lets look in detail at the several recommendations for security critical software present in critical infrastructure:

Systems Hardening,Vulnerability Management (VM), Patch Management (PM), Privileged Access Management (PAM), and Configuration Management (CM).

The first strategy, **systems hardening,**[31] provides a baseline for secure software and configuration. In rushing to support a large remote workforce during the early

---

[27] https://www.beyondtrust.com/resources/whitepapers/microsoft-vulnerability-report

[28] Miller [23].

[29] Haber [24].

[30] VMware [25].

[31] BeyondTrust [26].

days of the COVID-19 pandemic, agencies and enterprises relaxed their hardening policies. In March 2020, **internet-facing RDP**[32] **(Remote Desktop Protocol) ports increased by 50%**[33] in an ill-advised attempt to support these WFH (work from home) initiatives. Over the following 12 months, **52% of ransomware attacks**[34] leveraged publicly accessible RDP servers to gain initial access. These breaches are a direct result of ignoring basic system and network hardening tenets. RDP should never be enabled directly on the Internet as a part of basic system hardening.

In addition. systems should be regularly scanned and have automated ways to remove unnecessary, or block-listed software and applications. Organizations must automate the process of privilege discovery, auditing, and onboarding across their environment to encompass the activities of humans, applications, and machines to ensure these are not leveraged to gain access to the installed critical software. These controls should be applied prior to the device or endpoint receiving access to an environment. This is a part of system hardening and commonly implemented using Configuration Management solutions.

Finally, hardening activities should be rigorously enforced across the device life-cycle. The proliferation of BYOD (Bring Your Own Device)and work-from-home amongst users that interact with critical infrastructure increases the necessity for the continual scan and hardening of such devices, especially if they periodically leave and re-enter an environment and interact with critical infrastructure. Changes at home can negatively impact an environment if the device is modified in one location and physically (or electronically) connects to another.

Unfortunately, attackers leveraging known vulnerabilities is not new. However, this practice continues to give attackers a high rate of success in compromising their targets and is a common component of modern cyber attack chains. In late 2020, **the FBI and CISA warned**[35] that APT hackers are targeting government networks, critical infrastructure, and election organizations with chained (multiple exploits used as a single attack) vulnerability cyberattacks. These attacks were stringing together older vulnerabilities (exploits) to achieve a foothold and continue using lateral movement with additional bundle exploits in order continue propagation.

A vulnerability management strategy where software, application, and other vulnerabilities are continuously and systematically discovered, assessed, prioritized, and addressed, is critical to reducing the attack surface and preventing attackers from gaining a foothold or escalating an attack. Addressing vulnerabilities typically entails: patching or another mitigation (i.e. a configuration change or use of cybersecurity tools), while accepting a vulnerability could include actions that range from doing nothing to falsely relying on cyber insurance. In the end, mitigations and acceptance do not eliminate the risk. Only remediating the vulnerability using

---

[32] https://www.beyondtrust.com/blog/entry/what-is-rdp-how-do-you-secure-or-replace-it

[33] Roccia [27].

[34] Skulkin [28].

[35] Health IT Security [29].

a Patch Management solution can truly block a potential exploit if other security controls faulter or your scoring of risk is inappropriate.

In the end, prioritization is a key component of vulnerability management and essential to effectively address the sheer multitude of vulnerabilities that will exist across any environment hosting critical infrastructure. It is important to note that patching itself is not always a riskless activity—for instance, it could cause incompatibilities, software disruption, or have its own flaws that magnify the situation. That's why information technology and security teams need automated tools that can help them make smart vulnerability management decisions and prioritize which risks to address first. And with any change, test, test, test, before installing something new or patching critical software.

Privileged Access Management (PAM)

Privileged access management (PAM)[36] and identity access security must be applied across all resources in an environment hosting critical infrastructure. Organizations must automate the process of privilege discovery, auditing, and onboarding across their environment and beyond the perimeter, to encompass the activities of humans (employees, vendors, on-premise or remote), applications, and machines.

In implementing PAM, organizations should strive to manage privileged credentials. Management of these credentials and their associated secrets also encompasses reducing the number of identities with privileged access, the frequency of privileged access, and the duration of privileged access. This approach not only drastically condenses the attack surface, but also vastly minimizes threat windows.

Credential security best practices should be consistently applied—such as enforcing unique, complex passwords that are not easily human readable. This helps prevent password re-use, improper documentation of passwords, verbal transmission of passwords, and many other attack vectors. As much as possible, organizations should also eliminate shared accounts, so there is clear oversight over what a specific identity did with a privileged account. Any **hardcoded credentials should be eliminated**[37] and replaced with API calls or dynamic secrets, such as with DevOps and CI/CD pipeline toolsets.

As proof, in 2020, 87% of critical vulnerabilities in Internet Explorer and Edge **would have been mitigated by removing admin rights.**[38] These are common browser-based tools for managing resources in critical infrastructure. A similarly influential risk-reducing power of least-privilege has also been demonstrated across important third-party applications, with a quick search of the National Vulnerability

[36] BeyondTrust [30].

[37] Miller [31].

[38] BeyondTrust [32].

Database, **published by NIST**,[39] revealing a significant number of registered vulnerabilities mitigated by appropriate Elevated Privilege Management. Eliminating admin rights translates into a significant risk reduction across many platforms, even absent an appropriate patching process in place. This alone can mitigate many risks in critical infrastructure as well as the risks in a typical corporate environment.

In addition, organizations should remove local admin rights on endpoints that manage critical infrastructure and manage all root and admin access rights to network based assets using architectures that implement PAWs (Privileged Access Workstations). Moreover, by defaulting all users to standard privileges, while enabling elevated privileges for applications to perform specific tasks **malware such as samples deployed by Darkside**,[40] the group who perpetrated the devastating attack on Colonial Pipeline, could be minimized or fully mitigated. This fact alone demonstrates the important of managing administrative rights in critical infrastructure to thwart an attack and implementing PoLP.

Privileged access, except in few exceptional and unavoidable cases such as in certain automated workflows, should also only be provisioned **Just-In-Time (JIT)**[41] when certain contextual parameters are sufficiently met, and subsequently removed or expired immediately upon completion of the activity or change in appropriate behavior. Context should also dynamically consider real-time vulnerability and threat data about an application, asset, or user so privileged access may be further restricted, denied, or launch workflows requiring additional approvals or monitoring.

As an example, **The SolarWinds Orion supply chain attack** [42]was particularly devastating because the Orion application needed unrestricted "god" like privileges in order to operate within an environment on every asset from the same application. This included unrestricted administrative privileges and allow listing in common security tools such as anti-virus and intrusion detection. Since the Orion application itself was compromised, threat actors leveraged this unrestricted privileged access throughout victims' environments using the application. This attack demonstrates why it's essential for organization to identify and address over-privileged users and applications, and wherever possible and **implement least privilege application management**.[43] However, since many legacy applications (i.e., SolarWinds) may not work without these high levels of privilege, critical infrastructure should implement more layers of mitigation or discontinue the software's use in the environment. And, if access is needed, use a just in time model to secure privileged access outside of appropriate usage.This is another justification for PAM within critical infrastructure.

In order to further protect critical infrastructure further, organizations should ensure only authorized applications execute approved functions or communications. This type of control would also prevent unsigned binaries from executing on the system, as used in Darkside malware attacks. This entails exercising granular control

[39] National Vulnerability Database [33].

[40] Maude [34].

[41] Beyond Trust [35].

[42] Haber [36].

[43] Haber [36].

over access and privilege elevation for all applications, commands, files, and scripts to prevent or mitigate errors, and reduce the attack surface. Allow lists, block lists, and trust but verify lists should also be applied in tandem with endpoint privilege management to efficiently control application privileged execution. Trusted operating system applications (Powershell, Wscript, Csript, etc.) are often leveraged as part of the attack chain (i.e., in Darkside ransomware attacks). These are typically referred to as Living of the Land (LotL) attacks since they use native trusted applications against the user. This allows threat actors to wield and abuse an organization's legitimate tools in a malicious manner that typically circumvent most detection techniques. In fact, fileless attacks as they are also often called and **surged 888% in 2020**,[44] and were often part of ransomware campaigns. By applying context-rich privileged access security controls, such as content and application control rules and control over launch of child processes, organizations can **protect trusted applications from misuse and privileged escalation**.[45]

Adhering to the security principle of "Individual Accountability" where privileged accounts are not shared, rather a privileged account is restricted to one user at a time, gives organizations clear oversight into what actions specific users and their associated identities are performing with the elevated access. Organizations should apply **privilege separation** within critical infrastructure,[46] which involves defining and delineating employee, application, and system roles and tasks so that access is only granted to specific, discrete parts of systems or data, as necessary, and each account can only perform a specific, narrow range of actions. Separation of privileges helps contain intruders close to the point of compromise and restricts lateral movement. This is fundamentally key for protecting critical infrastructure from lateral movement and large scale infiltration by a threat actor Thus, if one account is compromised, the range of privileges it affords the attacker is limited and the damage can be contained without jeopardizing the mission of critical infrastructure.

As an example, the **Verkada Breach** provides a striking illustration of how a breach scope can get to mind-boggling size when separation of privilege and separation of duties are not implemented.[47] In 2020, Verkada network exposed the live feeds of 150,000 security cameras used in jails, hospitals, and even companies like Tesla. The exposure revealed live feeds from some incredibly sensitive environments including women's health clinics, psychiatric facilities, and even police departments. Many controls (removing admin rights, etc.) detailed in this section would have helped prevent or mitigate this breach, as would separation of privilege. No one account should have control over so many different customer accounts and have such high-level privileges across so many systems. It is a glaring example of unchecked privileged access controls and segregation. Rather than having one **superuser account**[48]perform all of an IT admin's duties, implement separation of

---

[44] Internet Security Report [37].

[45] Pitfield [38].

[46] Matt [39].

[47] Haber [40].

[48] Miller [40].

privileges and duties across different accounts, with each account requiring unique login credentials and used only for a specific set of functions and tasks. Some practical ways critical infrastructure can learn from this separation of privileges include:

- Separating various administrative account functions from each other
- Separating administrative and standard account capabilities
- Separating auditing and logging capabilities within administrative accounts
- Separating software functions like read, edit, write, execute, etc.

Finally, organizations should strive to provide secure application access independent of Virtual Private Networks (VPN), and only enabling remote employees, contractors, and vendors the applications and systems they need verses broad network access. This is a part of secure identity access management. This will increase protection against malicious insiders, external actors, and potential human errors. Organizations protecting critical infrastructure should also **proxy access to control planes** and other critical software to segment and isolate remote access traffic to on premise resources,[49] hybrid environments, and when cloud solutions have been deployed as a part of critical software digital transformation. While this is not directly related to PAM, privileged access typically occurs using remote access technology and when not properly secured becomes yet another attack vector.

Zero Trust Architecture

As a part of a successful Zero Trust strategy, organizations must consider how segmenting access can be applied to help enforce least privilege and enable continuous identity authentication. Segmentation effectively helps organizations, in broad strokes, protect cloud deployments and legacy enclave systems found in critical infrastructure from being the victim of an expanding risk surface. It is an essential strategy for protecting sensitive software, applications, data, users, and environments from the dissolving perimeter that has been accelerating due to the work from home requirements needed to mitigate the personal risks of COVID-19.

To start, every zero trust based attempted access to an asset is properly vetted using identity and asset controls with location awareness, identity authentication, asset security, etc. and assumes every attempt at access is a threat actor until verified. This allows for an identity-centric approach to network segmentation. In this model, the network is not segmented and controls like firewalls or intrusion prevention are not the primary method to protect an asset. Access is continuously monitored to identify inappropriate behavior and respond automatically if necessary. Application-level micro-segmentation should be implemented to prevent users from discovering applications they are not authorized to access creating isolation between individual resources themselves.

[49] Beyond Trust [41].

One of the core tenants of zero trust architecture is the continuous and dynamic monitoring of identities that access an asset (network, application, operating system, etc.). Organizations should monitor this architecture and manage any session involving privileged activity, whether by human, application, or machine. Monitoring should include performing screen recording, command logging, scripts executed, API monitoring, and screen outputs. Anomalous activity should be flagged, with the ability to pinpoint, pause, and terminate suspicious sessions in real-time based on policy. Privilege analytics should also be performed to spot potential issues and to optimize a least privilege posture, such as removing unused privileges. This can be done using a rules engine or even advanced artificial intelligence technologies trained for appropriate behavior.

As a matter of reference, NIST defines Zero Trust as "an evolving set of cybersecurity paradigms that move defenses from static, network-based perimeters to focus on users, assets, and resources." Zero Trust is increasingly relevant for industrial control systems, as technologies and have blurred or dissolved the idea of a network-zoned perimeter being operated by users in a specific geolocation. The seismic shift to remote working has also accelerated the demise and created a higher risk for critical infrastructure that can be solved by zero trust.

NIST's guidelines also provide a clear playbook for organizations seeking guidance on how to adopt Zero Trust principles. Many organizations have begun to embrace Zero Trust frameworks and are building these into their security strategies as they hardened critical infrastructure. The basic tenants follow the security best practices we have already outlined in this chapter and add a few that are specific to industrial environments:

- Enforcing the philosophy of least privilege for remote access sessions used in a zero trust architecture.
- Treating managed devices with the same level of trust as an unmanaged device—which is zero. Nothing should be assumed as trusted.
- Providing application access independent of network access. This follows the model of perimeter-less defense.
- Recording all activities performed using remote access and disabling functionality such as copy/paste and file transfers/downloads.
- Enabling API security to protect the integrity of data being sent from IoT devices to back-end systems.
- Establish an enclave mode (zoned network) for legacy technology that can be adapted to zero trust using segmentation.
- Prevent any and all access to a critical infrastructure resource from adjacent resources. In other words, clearly define all communication paths and monitor and block any lateral movement.
- Applying these disciplines can help achieve Zero Trust objectives and ensures all access is appropriate, managed, and documented, regardless of how the perimeter has been redefined.

Critical Infrastructure Technology, Standards, and Models

In order to take the security and best practices previously discussed and apply them to critical infrastructure, a basic understanding of the Purdue Reference Model is required for resources that typically make up critical infrastructure.

The Purdue Enterprise Reference Architecture (PERA) model (commonly referred to as just the Purdue model) was adopted by ISA-99 and used as a concept model for Industrial Control Systems (ICS) and their appropriate network segmentation. It is an industry adopted reference model that shows the interconnections and interdependencies of all the main components of a typical ICS. After all, most critical infrastructure is based on ICS, IIOT, and OT and typically uses Commercial Of The Shelf (COTS) technology for management and network communications. This is illustrated in Fig. 1 TBD.

The Purdue model divides an ICS environment into three zones and five levels.

Level 0: The physical process - Defines the actual physical processes controlled in a critical infrastructure environment. These are typically mechanical, robotic, or runtime sensors that provide physical operation.
Level 1: Intelligent devices—These resources provide sensing and manipulation of the physical processes themselves. These include devices that have software and logic to perform operations like sensors, analyzers, actuators, and related instrumentation.
Level 2: Control systems—These resources are responsible for supervising, monitoring and controlling the physical processes. These are real-time controls and

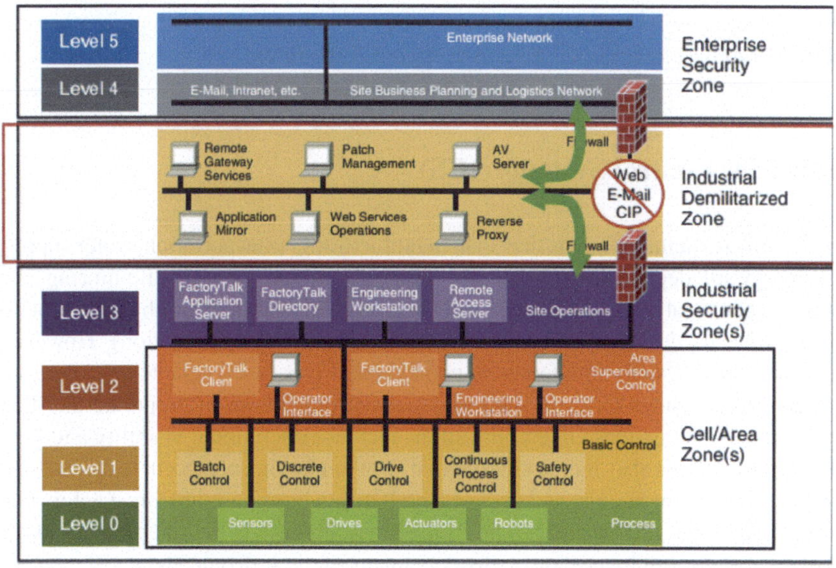

Fig. 1 Purdue Model for ICS

software, human–machine interface (HMI) resources; supervisory and data acquisition (SCADA) software, and typically based on COTS operating systems and devices.

Level 3: Manufacturing operations systems—Supporting resources for managing production workflow are included in this level to produce the desired products or sustain a mission like power generation. Operational concepts like batch management; manufacturing execution and operations management systems, laboratory, maintenance. and plant performance management systems are designated to operate in this level.

Level 4: Business logistics systems—Every production environment needs supporting technology to manage the business-related activities for the environment. Typically, ERP is the primary system; establishes the basic plant production schedule, material use, shipping and inventory levels and other consumables used to support the mission.

Level 5: Enterprise Network—The remaining business processes utilize the network for communications. In a tradition environment, this is secure with a perimeter and firewalls. In a zero trust architecture, this provides the proxy, gateway, or policy administrator for brokering and monitoring access and authentication.

As we consider the Purdue model for critical infrastructure, we can easily identify the critical software at each layer. Then, if we apply the security disciplines previously discussed, we can build a model for securing critical software at each level and thus the entire infrastructure from cyber attacks. In order to do this, we need a good understanding of the environment, what level each resource resides in, and a plan to monitor, manage, and maintain each resource. They key is knowing how attacks are occurring and how an organization should respond and mitigate the risk. The lessons learned from past attacks will help us prepare for the future.

Industrial Control Systems (ICS)

Critical Infrastructure systems that span manufacturing, transportation, water supply, and energy all depend heavily on information systems for their monitoring and control. Historically, Industrial Control Systems (ICS) relied on physical separation (Purdue model) as the primary means for security (segmentation). However, modern control system architectures, management processes, remote workers, and cost control measures have resulted in increased integration of corporate and ICS environments. While these interconnections increase operational visibility and flexible control, it can also increase risks that previously did not occur with isolated ICS. Through an interconnected network, an ICS can be exposed to threat actors that have already exploited or compromised the environment via the Internet, corporate network, remote employees, or by insiders misusing their privileges.

Table 1 Cyber Risks and ICS-CERT Recommendations

Cyber Risk	ICS-CERT recommendation
Secure passwords	Remove, disable, or rename any default system accounts wherever possible
Strong password management	Establish and implement policies requiring the use of strong passwords
Reduce risks of brute force attacks	Implement account lockout policies to reduce the risk from brute forcing attempts
Minimize network exposure	This activity includes the implementation of firewalls and network segmentation. This can reduce the attack surface for bad actors and reduce the risks of lateral movement within a compromised environment
Secure remote access	This activity includes deployment and appropriately updating remote access solutions, such as VPN, if required
Third-party vendors	Monitor the creation of administrator level accounts by third-party vendors
Vulnerability management	Apply patches in the ICS environment, when possible, to mitigate known vulnerabilities
Threat detection	ICS-Cert recommends that organizations monitor for suspect activities and to report their findings to ICS-CERT for incident response support and correlation with other similar incidents

The ICS-CERT[50] (Industrial Control Systems Cyber Emergency Response Team) provides ICS-CERT alerts[51] to assist organizations on how to monitor for these threats, and provides actionable mitigation strategies to prevent exploitation of ICS assets. In addition, in order to manage threats overall, ICS-CERT encourages sound security practices using a "defense-in-depth" approach, including, but not limited to, the following defensive in-depth measures and managing privileges in Table 1.

As you can see, the recommendations follow the security recommendations covered in the beginning of this chapter. While ICS represents a specific set of recommendations for critical infrastructure, the benefits for any implementation can be utilized in almost any environment:

- Discover all managed and unmanaged assets across your interconnected corporate and ICS infrastructure across all zones and levels. This will help you create an architecture that follows the best practices of the Purdue Model.
- Automatically discover and inventory privileged accounts used internally by employees and contractors and externally by all third-party vendors. This will help you identify privileged risks and design a PoLP plan.

[50] Reference [42].

[51] Reference [43].

- Provide central control by securely storing all credentials and SSH keys in a secure database that is hardened, encrypted, and strictly monitored for access. This is a fundamental strategy for Privileged Access Management (PAM).
- Reduce the risk of lost or stolen credentials by systematically rotating passwords for all managed systems based on time and employee retention. This helps secrets remain obfuscated from all vendors, contractors, and employees until they are actually needed.
- Implement secure vendor enclaves to isolate ICS and vendor devices to reduce the risks of malware and attack. Consider using zero trust architectures for these environments.
- Verify that no default passwords exist on any managed system or device. This is a vulnerability and configuration basic recommendation that should always be followed for everything from SNMP (Simple Network Management Protocol) to interactive remote access.
- Manage all managed devices automatically and store a unique password per each device. Passwords should never be reused to prevent lateral movement.
- Automatically rotate each device's password based on age, after each remote vendor session, or an identified risk to minimize leakage.
- Provide a complete workflow for device access, including an approval process for when remote vendor or employee access is required as a part of change and configuration management.
- Record all or select remote sessions with playback to document and review what occurs when a device is accessed and approve appropriate behavior. This is core tenant of zero trust to monitor appropriate behavior and take action when something inappropriate occurs.
- Provide detailed reporting of all privileged credentials used regardless of location and potentially block access from inappropriate geolocations based on your critical infrastructure. For example, no one in Russia should be accessing power generation in the United States.
- Based on these recommendations, and the security guidance provided by ICS-CERT, ICS devices can be securely managed against the threats that typically compromise critical infrastructure.

Industrial Internet of Things (IIoT)

Industrial Internet of Things (IIoT) devices introduce a unique set of risks based on privileges and asset attack vectors to critical infrastructure. By definition, IIoT devices are single-purpose assets with embedded operating systems to perform a specific function. They possess unique characteristics, including the capability to interact with a physical environment, localized role-based access, and potentially a web server to provide interactive or API based functionality. IIoT devices include everything from network-based cameras, digital video recorders, thermostats, thermocouples, valve controls, lighting, to motion sensors that may be used in critical infrastructure to maintain security, operations, and environmental conditions. The list of network-based IIoT devices is growing every single day and differ from tradition Internet of Thing (IoT) devices based on their ability to handle harsh environments, hardened physical security, specified function, and various forms of tamper protection. As an example, these devices could be a biometric door lock but differ from consumer versions based stronger mounting hardware or certifications based on military standards (MIL-SPEC). While these types of devices have existed for years, they have only recently been grouped and labeled IIoT based on their mass adoption for industry and, more importantly, their mass identification of security risks and privileged attack vectors. Therefore, as IIoT devices become more commonplace, there is a need to ensure that they do not represent an unnecessary security risk to critical infrastructure. Unfortunately, it has already been proven that many of these devices are insecure by design, have unresolvable flaws (unmatchable firmware or software), and can be leveraged to compromise an entire environment with something as simple as a default credential regardless of level they may be designated in the Purdue model. Therefore, for any IIoT deployment, consider these seven recommendations to mitigate security risks of these devices when used in critical infrastructure.

- **Segment Networks**: Using basic capabilities in modern network routers and switches, all IIoT devices should be networked using separate wireless networks and VLANs, and segmented on a level within the Purdue model. All communications from IIoT networks should be explicitly blocked from critical servers, databases, and workstations that should not communicate directly with other devices or between levels—including the Internet. This helps ensure that, even if an IIoT device is compromised, it cannot directly be leveraged to negatively impact the remainder of the critical infrastructure. If possible, all IoT network communications should be monitored to other trusted networks to identify any anomalous behavior.
- **Manage All Credentials**: Almost all IIoT devices ship with default passwords for initial configuration or have default passwords that can be reset with physical access. End users should change all usernames AND passwords on these devices to complex passwords and unique usernames, and consider changing at least the passwords periodically if they are exposed to end users during normal operations.
- **Limit Connectivity**: Never place IIoT devices of any type directly on the Internet with public IP addresses or on level 5 with broad user access. It is just a matter of time before they will be compromised or subject to a DDOS attack. IIoT devices are based on very simple networking technology and not robust enough to thwart all the potential IP traffic that contains malicious code or attack vectors.
- **Identify Shadow IT**: Shadow IT is another buzzword for rogue devices and unsanctioned assets. Make sure any IIoT devices placed on your network are approved and adhere to the security considerations outlined previously. Shadow IT based on IIoT could easily violate many of your security policies and introduce a risk. Standard network discovery tools can find these rogue devices and help place them under proper management.

- **Vulnerability Management Service-level Agreement (SLA)**: Request from the manufacturer a service-level agreement for identifying, notifications, and patching critical vulnerabilities once they publically disclosed. This will help you ensure IIoT devices selected for your organization will stand up to regulatory scrutiny and patch compliance initiatives. Also, make sure these questions are asked during an RFP or procurement process to ensure the vendor has the proper maturity for managing risks. In most critical infrastructure environments, these devices are expected to last for decades.
- **Remediate Security Flaws**: Document a process and ensure all IIoT devices can be patched promptly if a flaw is found and without extensive disruption to the business. Some devices are very difficult to remediate and may have hidden labor costs to manage one at a time. This includes making sure that you maintain the latest firmware on all IIoT devices to mitigate any emerging threats that could be leveraged against the devices.
- **Role Based Access Control (RBAC)**: Authentication into IIOT devices should be flexible enough to allow integration into an Active Directory, a Radius Server, or other third party Identity Services. As a long term goal, all credentialed access to IIOT devices should be centrally managed and properly organized within an existing identity governance solution. If the devices do not support identity store based integration, it may be a tempting target for threat actors to store rogue accounts. Finally, if the IIOT device does support RBAC, ensure all accounts follow the model of least privilege and the administrative account is secured with a privileged access management solution.

IIoT devices are just another piece of technology that businesses have enabled for critical infrastructure to better monitor and manage production environments. Truthfully, they are not as robust or mature compared to the server and desktop counterparts, and everything from default credentials to vulnerabilities represent a real risk to an environment. IIOT devices need restrictions, governance, and should always be monitored.

A true cyber war could be nearly as devastating as conventional bombs and weapons in terms of human life and destruction to infrastructure and the economy. The war itself would use technology as weapons and software as munitions. History has shown us how these attack could occur and best practices from information technology have taught us how to build a resilient strategy for protecting resources. The targets will be critical infrastructure to cause the most damage to the general population and cause social upheaval. Our mission now is to learn from history, understand the recommendations for mitigation, and apply them to models and live infrastructure that supports our daily lives.

References

1. Smith. (2017). https://www.beyondtrust.com/blog/entry/7-steps-cyber-attack-can-protect-windows-privileged-accounts.
2. Lyngaas (2021). https://www.cyberscoop.com/oldsmar-water-plant-botnet-dragos/.
3. Turton (2021). https://www.bloomberg.com/news/articles/2021-06-04/hackers-breached-colonial-pipeline-using-compromised-password.
4. Collier (2021). https://www.nbcnews.com/tech/security/meat-supplier-jbs-paid-ransomware-hackers-11-million-n1270271.
5. Seals (2019). https://threatpost.com/triton-ics-malware-second-victim/143658/.
6. Edwards (2019). https://www.businessinsider.com/gnss-hacking-spoofing-jamming-russians-screwing-with-gps-2019-4.
7. Khandelwal (2019). https://thehackernews.com/2019/03/ghidra-reverse-engineering-tool.html.
8. Stoneff (2018). https://www.beyondtrust.com/blog/entry/so-youre-interested-in-a-cyber-security-career.
9. BeyondTrust (2019). https://www.beyondtrust.com/blog/entry/file-integrity-monitoring-explained.
10. Khandelwal (2019). https://thehackernews.com/2019/03/hacking-implantable-defibrillators.html.
11. Wikipedia https://en.wikipedia.org/wiki/Barnaby_Jack.
12. Haber (2016). https://www.beyondtrust.com/blog/entry/making-cyber-security-priority-healthcare-organizations.
13. Miller (2019). https://www.beyondtrust.com/blog/entry/hardcoded-and-embedded-credentials-are-an-it-security-hazard-heres-what-you-need-to-know.
14. Seals (2019). https://threatpost.com/north-koreas-hidden-cobra-strikes-u-s-targets-with-hoplight/143740/.
15. Cyber Security and Infrastructure Security Agency (2019) https://www.cisa.gov/uscert/ncas/analysis-reports/AR19-100A.
16. BeyondTrust https://www.beyondtrust.com/resources/glossary/ransomware.
17. Atherton (2018) https://www.vox.com/world/2018/3/28/17170612/russia-hacking-us-power-grid-nuclear-plants.
18. White Hours (2021) https://www.whitehouse.gov/briefing-room/presidential-actions/2021/05/12/executive-order-on-improving-the-nations-cybersecurity/.
19. Freguson (2021) https://www.govinfosecurity.com/nist-releases-critical-software-definition-for-us-agencies-a-16952.
20. Office of Budget and Management (2007) https://obamawhitehouse.archives.gov/sites/default/files/omb/assets/omb/memoranda/fy2007/m07-16.pdf.
21. Zetter and Greenberg (2015) https://www.wired.com/2015/06/opm-breach-security-privacy-debacle/.
22. Section 508 https://www.section508.gov.
23. Miller (2021) https://www.beyondtrust.com/blog/entry/what-is-least-privilege.
24. Haber (2020) https://www.beyondtrust.com/blog/entry/cybersecurity-strategies-to-stop-lateral-movement-attacks-leave-your-adversaries-marooned.
25. VMware https://www.vmware.com/resources/security/the-ominous-rise-of-island-hopping-counter-incident-response-continues.html.
26. BeyondTrust https://www.beyondtrust.com/resources/glossary/systems-hardening.
27. Roccia (2020) https://www.mcafee.com/blogs/other-blogs/mcafee-labs/cybercriminals-actively-exploiting-rdp-to-target-remote-organizations/.
28. Skulkin (2021) https://www.group-ib.com/resources/threat-research/ransomware-2021.html.
29. Health IT Security https://healthitsecurity.com/news/fbi-cisa-warn-apt-hackers-chaining-vulnerabilities-in-cyberattacks.
30. BeyondTrust https://www.beyondtrust.com/resources/glossary/privileged-access-management-pam.

31. Miller (2019) https://www.beyondtrust.com/blog/entry/hardcoded-and-embedded-credentials-are-an-it-security-hazard-heres-what-you-need-to-know.
32. BeyondTrust https://www.beyondtrust.com/resources/whitepapers/microsoft-vulnerability-report.
33. National Vulnerability Database https://nvd.nist.gov/vuln/search.
34. Maude (2021) https://www.beyondtrust.com/blog/entry/will-darkside-pipeline-ransomware-attack-fuel-cybersecurity-upgrades-for-critical-infrastructure.
35. Beyond Trust https://assets.beyondtrust.com/assets/documents/JIT_PAM_WP.pdf.
36. Haber (2020) https://www.beyondtrust.com/blog/entry/least-privilege-application-management-a-lesson-learned-from-solarwinds-orion.
37. Internet Security Report (2020) https://www.watchguard.com/wgrd-resource-center/security-report-q4-2020.
38. Pitfield (2020) https://www.beyondtrust.com/blog/entry/how-trusted-application-protection-builds-on-application-control-endpoint-privilege-management.
39. Matt **(2020)**https://www.beyondtrust.com/blog/entry/how-separation-privilege-improves-security.
40. Miller (2019). https://www.beyondtrust.com/blog/entry/superuser-accounts-what-are-they-how-do-you-secure-them.
41. **Beyond Trust**https://www.beyondtrust.com/solutions/cloud-security.
42. https://ics-cert.us-cert.gov/.
43. https://ics-cert.us-cert.gov/alerts.
44. Haber (2021). https://www.beyondtrust.com/blog/entry/dangers-of-iot-privilege-management-blind-spots-exposed-in-verkada-security-camera-breach.

Morey Haber is Chief Security Officer at BeyondTrust. He has more than 25 years of IT industry experience and has authored four Apress books: Cloud Attack Vectors, Privileged Attack Vectors, Asset Attack Vectors, and Identity Attack Vectors. Mr. Haber originally joined BeyondTrust in 2012 as a part of the eEye Digital Security acquisition. Morey currently oversees BeyondTrust security for privileged access management, cloud security, and remote access solutions.

The Cyber-Grand Strategy Gap

Jacob Shively

Abstract This chapter finds that America's central cyber strategic challenge is a massively skewed risk/reward calculation that favors peer competitors. Leading proposals to address this imbalance are technical and operational, but such solutions are inadequate for an unbalanced strategic environment. This study applies a grand strategy framework to discussions of national security and cyber articulated by senior officials, military commanders, and other experts at the April 2021 Great Power Competition conference. For attackers, the consequences of being seen as a cyber threat by the United States are relatively minor compared to the rewards of hacking US systems, stealing intelligence and intellectual property, conducting information warfare, and developing capabilities to quickly devastate critical infrastructure. In short, the benefits of violating US cyber systems are specific and valuable. The costs are diffuse and, attackers seem to agree, manageable. These are classic conditions favoring offensive behavior. In response, professionals at the conference recommended operational and technical solutions. These focused on versions of deterrence, namely *collective defense* and *defending forward*. Here, collective defense refers to cooperation and collaboration among private US actors and the US government. It is designed to convince attackers to redirect their efforts because US systems are resilient and difficult to penetrate. Defending forward refers to US agencies launching persistent, offensive attacks that keep adversary resources distracted and adversary operators worried about US reprisals. Viewed in the context of grand strategy, these are necessary but not sufficient policy solutions. US planners need to bridge the gap between a current strategic context that favors challengers and a future in which that balance favors US defenders. As in prior eras of technological change, policy makers likely need to pair technical and operational prowess with statecraft and other diplomatic tools.

Keywords Grand strategy · Great powers · Cyber warfare · Cyber security

---

J. Shively (✉)
University of West Florida, Pensacola, FL 32514, USA
e-mail: jshively@uwf.edu

A. Farhadi et al. (eds.), *The Great Power Competition Volume 3*,
https://doi.org/10.1007/978-3-031-04586-8_2

Grand Strategy and Cyber

Grand Strategy typically refers to a government's overarching plan to achieve its goals—particularly national security—in the international system.[1] Cyber poses classic concerns addressed by grand strategy. A digitally networked world facilitates global interaction, poses new and emerging threats to state infrastructure, and defines a growing slice of the global economy. Indeed, cyber now sets many of the stakes that define contemporary great power competition. Studies of long-term geopolitics consistently show that great powers have risen and declined based partly upon the opportunities and threats posed by emerging technologies. [5, 6] Grasping and dominating a given era's technical innovations lend significant, even definitive, advantage to the states wielding them. In a July 2020 interview, for example, Gen. Keith Alexander (former director of the National Security Agency and Commander, U.S. Cyber Command) argued that emerging networked technologies like 5G and artificial intelligence would be "critical for our future economic success," and they raise "core national security issues" [7]. This type of technology directly shapes the international system's "elements of power." Any state falling behind on these innovations "will have a tough time competing in future rounds."

In the case of cyber, no single strategic concept covers threats from traditional geopolitical adversaries, organized crime, and domestic infrastructure. Often, cyber threats cannot be disaggregated from traditional diplomacy, political economy, and strategic priorities. For instance, spanning the Obama, Trump and Biden administrations, Russian government actors and their proxies effectively seized the Crimean peninsula from Ukraine and supported an ongoing nationalist revolt in that country's eastern provinces; staked out support against US partners in the Syrian civil war; committed a series of successful or attempted assassinations of Russians in Western countries; effectively hacked US political party officials and spread disinformation during the 2016 US presidential campaign; and sanctioned or conducted cyber attacks ranging from crime to espionage. In response, Congress and the Executive Branch moved to impose sanctions, expel diplomats, and work up a "defend forward" policy of actively seeking out known cyber threats rather than responding to specific attacks.

These challenges exemplify a geopolitical environment dominated by a grand strategic overlap between military, diplomatic, economic, and cyber capabilities. It is an asymmetrical situation. Attackers and challengers enjoy strategic advantages in this environment. For instance, US and partner government responses have not appeared to shift the apparent cost–benefit calculations for hacking and disinformation. Indeed, traditional state responses to all kinds of threats incentivize their adversaries to reach for nontraditional responses. The US foreign policy and national security establishments spent the years after 9/11 recalibrating for asymmetrical and insurgent challenges, and they are now moving to reorient back to "peer competitors" and "strategic competition," but through all this, policy planners and leaders have tended to treat cyber as an adjunct, technical issue in existing domains such

[1] Defining grand strategy remains an ongoing debate, but for recent examples, see [1–4].

as espionage, transnational crime, and operational security. In short, grand strategy remains ripe for changes that integrate cyber.

Speakers at the 2021 Great Power Competition Conference primarily emphasized external strategic cyber threats emanating from the United States' leading peer competitors, China and Russia. This reflected the event's priorities, but it also captured broader conventional wisdom. For example, the Biden administration's March 2021 Interim National Security Strategic Guidance specifically highlighted "both Beijing and Moscow" for "invest[ing] heavily in efforts meant to check U.S. strengths and prevent us from defending our interests and allies around the world" [8, 9]. In general, speakers portrayed a threat landscape defined by traditional nation-state actors who, in cyberspace, push all the advantages of an asymmetric challenger. They enjoy potential access to US systems through a growing number of entry points. They also hold strong incentives, as challengers, to probe and attack US systems without fear of major or uncontrolled escalation. Still, asymmetrical threats only exist when one side holds clear advantages. One recent study found that the United States remained the only government on Earth dominant in every category of cyber power, ranging from strategy and doctrine to offensive cyber capability [10]. In short, as the world's leading power online and offline, the United States faces all the normal advantages and challenges of a great power alongside the emerging and uniquely domestic or internal threats posed by cybersecurity.

Observing Grand Strategy and Cyber

Grand strategic issues implicitly defined many of the GPC speakers' analyses. The following section reviews their observations and treats them as a sample of the contemporary "state of the art" among senior officials and other policy leaders. This is an inductive exercise. It is aimed at identifying patterns and trends as they emerged among the conference experts. These observations fell into a range of overlapping threats and challenges: traditional geopolitics; "if you build it, they will come;" network penetration, norms and information warfare; domestic impediments and needs; and network resilience. The subsections below detail each of these.

Afterwards, this analysis applies the author's own tripartite division for the grand strategy concept. Grand strategy, it argues, can be comprised of three overlapping elements: CONTEXT—MEANS—ENDS. Typically, scholars and practitioners define any strategy, at its core, as a means-ends calculation. The strategist defines a goal, or set of goals. Then, in a world of limited or fixed resources, assesses the capabilities, tools, and so forth that are available for building a plan to achieve that goal. For grand strategy, this typically means that government decision makers can assess their state's overall capabilities in order to pursue international goals. Though sensible, the means-ends framework leaves out a common but often implied aspect of strategy: the context or environment in which the means are identified and the ends defined. For instance, a business strategy for a multinational corporation might carefully evaluate its logistics, standards, capital, and customers; however, it would be ill-served if it failed also to recognize that operating across different sovereignties and cultures—the global context—may alter the strategy's basic conditions.

The following sections summarize the conference speakers' comments as they relate to grand strategy. It identifies common themes and, where relevant, disagreements or contradictions. In turn, it asks the same questions of each panel and speaker: what is the strategic context and ends? What are the means? This allows the observer to gather systematic observations and draw answers. These findings then set up possible answers for how these individuals define the strategic ends of cybersecurity as well as the proposed strategies or other mitigating actions and policies for national cybersecurity. With all that in hand, the study can turn back to whether and to what degree these leaders fit cybersecurity into a larger grand strategy.

Traditional Geopolitics

Two prominent senators, Marco Rubio (R-FL) and Mark Warner (D-VA), kicked off the conference. China and Russia, they agreed in separate prerecorded statements, posed specific and persistent cyber threats for the United States. Rubio [11] invoked a stark menace from China. "The Chinese Communist Party," (CCP) he stated, "Has no moral or legal restraints on its pursuit of dominance." Drawing upon Cold War rhetoric, he warned that "a genocidal, authoritarian regime is working tirelessly to rewrite the rules of the international system." In this context, Beijing's cyber threat ranges from the type of information suppression tactics the CCP uses at home to stealing intellectual property (IP) from US organizations to potentially infiltrating software by penetrating supply chain networks. Senator Warner [12] presented a less bombastic but no less dire assessment. The core challenge, he explained, is that US adversaries enjoy an asymmetric advantage, and unlike the Soviet Union, China "is actually trying to compete with us in the economic domain and technology domains." For instance, whereas US innovators may spend years developing new capabilities, adversaries have many possible penetration points to steal that intellectual property (IP). Moscow, meanwhile, plays a spoiler role by spreading mis- and disinformation through social media. Like Beijing, it also poses a pervasive espionage threat, as the SolarWinds hack illustrated.

The senators reflected a broad consensus. Along with many conference speakers, they assumed that the United States and the rest of the world are sliding into a period of inevitable power, technology, and political transition. Separately, from his vantage at US Southern Command, Adm. Faller already had argued that traditional great power competition for influence and strategic chokepoints was persistent and increasing even in the United States' global backyard. [13] A March 2021 report by the Congressional Research Service [14], found that both the Obama and Trump administrations based their long-term strategic planning on assumptions of renewed great power competition. All this would transition the United States away from two decades of prioritizing asymmetrical challenges. After the interventions of the 1990s and the response to 9/11, strategists and decision makers assumed that nonstate actors and weak states defined the most pressing threats to US security and interests. Now, the government would need to turn its resources back to an older strategic practice: preventing or deterring a Eurasian regional hegemon. The CRS report observed that US force structure has long reflected two assumptions: (1) Eurasia represents the world's core population and resource center, so developments there can threaten vital US interests, and (2) no one actor in Eurasia will reliably staunch the emergence of a hegemon. Such framing is politically safe in Washington D.C. because it is familiar and, for that reason, allows greater bipartisan agreement. By contrast, during his panel discussion, Wilson [15], President of the Joint Special Operations University, expressed skepticism about this received wisdom. During an earlier recorded talk, he said the "so-called return to great power competition isn't." [16] In his reading, traditional powers like China and Russia may pose serious threats, but the postwar liberal international order—along with its norms and rules—is already succumbing to a wide range of insurgent, political, and technological challenges.

Whatever the status of international systemic change, conference speakers agreed that cyber threats pose a strategic challenge. "Full spectrum" cyber competition will only exaggerate a putative great power transition. "The locus of struggle for power," Gen. Paul Nakasone (Commander, U.S. Cyber Command, and Director, National Security Agency/Chief, Central Security Service) explained in an earlier article, "Has shifted toward cyberspace, and from open conflict to competitions below the level of armed attack." [17] Jan Tighe (VADM, USN, Ret.) observed that in the grand strategic context, cybersecurity was likely "a(n) existential threat to our nation" [18]. "Today peer- and near-peer competitors operate continuously against us in cyberspace," [17] she had explained during a previous interview, "These activities are not isolated hacks or incidents, but strategic campaigns."

'If You Build It, They Will Come'

Conference speakers seemed to presume an "if you build it, they will come" logic: if cyber access and other network vulnerabilities exist, then competing states will seek to exploit them. During his plenary comments, Adm. Faller [19] unfolded a massive paper schematic of computer networks related to his command and stated that, not unlike a large number of forward operating bases in a combat theater, US cyber infrastructure suffered "too much surface area." Later, in his keynote address, Gen. Nakasone [20] characterized cyber as "a new strategic environment where states can increase their power, degrade the power of others, and gain a strategic advantage without resorting to armed conflict." Similarly, Gen. Alexander elsewhere argued that cyber attacks in recent years had "fundamentally shifted our understanding of the threat environment." [21] NotPetya and WannaCry, in particular, showed that actors with limited resources relative to the United States could cause a huge amount of collateral damage, cripple public and private sector networks, and prosecute extensive intellectual property theft. All of which presents "an existential economic threat to the public and private sectors" [22]. By exploiting social media to drive wedge issues and provoke distrust, adversary states and organizations "are using our own value system against us." In addition, Gen. Alexader explained in prior remarks, China leads other actors in extracting data from public and private actors in order to train machine learning algorithms. [21] According to Gen. Nakasone [20], China in particular is increasingly assertive economically, diplomatically, militarily and technologically and is "our number one pacing challenge." Russia has become a committed disruptor "seeking to undermine the integrity and legitimacy of our political system." Iran, meanwhile, displays "capability and intent to strike the US," and North Korea has shown an ability to threaten key nodes in the global financial system. Adm. Faller [19] summarized his interpretation of the environment. In places like Venezuela, "we can quickly be overwhelmed by the generation of false narratives in cyberspace" promulgated by China, Russia, and Iran. Along with blatant IP theft, what was happening in cyber was not merely competition. It was conflict [23].

Perhaps the most obvious example of this perceived threat was speakers' treatment of network penetration as a strategic, not merely an operational, challenge. Cybersecurity involves the threat of unauthorized or malicious actors accessing data systems and information networks. From a certain perspective, this is a narrow and technical problem: it is about these systems and how humans interact with them. Speakers at the conference, however, seemed to agree that cybersecurity long ago had moved from a frustrating technical challenge to a high-level strategic vulnerability. Gen. Nakasone [20] observed that lone hackers, extremist groups, organized criminals, nation states and others benefit from "the inherent functionality built into operating systems" and the fixed reality of human error. Together, these actors pose a persistent threat to their targets' social and national infrastructure. In an earlier interview, he outlined the historical path to this point. Over the 2010s, "nation-states progressed from exploitation, to disruption and finally to destructive attacks against us in cyberspace." [17] Initially, the threat centered on traditional espionage goals, usually accessing classified systems and information. Through 2012–2013, Iran's penetrations of New York financial networks revealed a disruptive cyber capacity. A year later, Tehran's data deletion attack against a US casino and North Korea's cyber attack against Sony Pictures ramped up to a destructive level. In the years since the 2016 presidential election, corrosive attacks had been added to the repertoires of adversary governments, which could and did "weaponize information in order to conduct influence campaigns, steal intellectual property, or leverage someone's personally identifiable information." US cyber networking will only grow, [20] so "no matter how good we are, these challenges will endure and will only increase in both scale and scope."

Public–Private Networks

Working across public and private actors, coordinating between agencies, identifying shared practices, and so forth were also framed as strategic challenges in themselves. In particular, high levels of integrated public–private infrastructure have facilitated innovation, but they also present a massive number of penetration points for adversaries as well as pernicious coordination challenges. In themselves, these are not direct, external strategic threats, but they create the conditions in which great power threats emerge and network penetration is a major threat. Further, public–private coordination is not simple for open societies with constitutionally-constrained governments. "This is," Gen. Nakasone explained, "Primarily a private industry-driven domain" in which "90 percent of the networks—the critical infrastructure—resides in the private sector, not in the public." [17] Private sector software, financial, logistics, manufacturing and other organizations directly and indirectly related to national security are under constant pressure, yet few of those actors feel free or incentivized to inform the government about embarrassing systems breaches. In prior eras, US and allied businesses were physically removed from the espionage and other threats posed by foreign governments. Intellectual property theft was limited to what could be physically stolen or photographed. Sabotage of critical infrastructure would have been, effectively, an act of terrorism or war. This policy model has persisted. "Organizations," [24] Gen. Alexander explained in an earlier interview, "Frequently must defend themselves, often with limited resources, against well-funded nation state groups who plan, code and attack in teams over long periods of time." Private actors remain a tempting target for adversarial governments because hurting them can "cause us pain" without a fear of immediate reactions from the US government. At the same time, without direction, government agencies are likely to interpret reality in their own interests rather than build bridges with the private sector. [25] As another conference speaker described the situation, if an adversary fires a missile at a US city, everyone knows who is responsible for stopping or responding to that threat. By contrast, if an adversary compromises the networks of thousands of US businesses, neither government nor business is certain who should respond and in what way.

Network Penetration, Norms and Information Warfare

In this environment, several traditional challenges took new forms. Warfighting capabilities, of course, remained a primary concern. O'Hanlon [26], a director of research at the Brookings Institution, commented that cyber incentivizes adversaries to strategize preemptive, decisive cyber attacks. Kaplan [27] Overconfidence is a common precursor of war, and cyber access encourages overconfidence. Chris Inglis (former Deputy Director of the National Security Agency and later President Biden's lead in the Office of the National Cyber Director) observed that adversaries might avoid a traditional conflict escalation by, for example, targeting private infrastructure rather than traditional military capabilities. [28] Still, several speakers agreed that the United States could stumble into a permanent disadvantage in a shooting war because it is digitally connected and might be vulnerable to early, crippling cyber attacks. [7, 29] Even more dramatically, Tighe warned, "future defensive decisions will have to be made at machine speed." [30] This 'machine speed' menace concerned many speakers. The relevant AI technology requires years to develop, they warned, so any state already behind on day one of a conflict will face compounding failures with an interconnected force structure. [31] In a prior interview, Tighe (2017a) also argued that "the point of advanced analytics and AI compression of the decision cycle is specifically to buy decision makers time." Ironically, attackers with the same capacity can iterate attacks faster than defenders can respond.

Many speakers also identified relatively weak or nonexistent cybersecurity norms as a strategy-level concern. As a naval commander, for instance, Tighe had already observed that "we've had an awful lot of examples of what, 10 years ago, we assumed would be construed as an act of war, and, in a lot of cases, there has not been a response, either a military response or a diplomatic response." [32] In other words, the norms of war and coercion surrounding cyber are unclear and—currently—unenforced. Similarly, others commented, vague or nonexistent norms on topics like cyberespionage and disinformation encouraged adversaries to exploit those gaps. [12, 33, 34], Senator Rubio [11] stated that deterrence required "maturing the doctrine of promulgating cyber norms," which help eliminate confusion and clarify threats. As Wingfield observed elsewhere, "Our strategy hinges on collaborating with our allies and partners … to ensure cyberspace remains a safe, secure, and open engine of innovation and prosperity." Fixed norms, however, may be difficult to achieve. During his panel comments, he explained that defenders want flexibility and fixed, covering norms, but such public constraints may be counterproductive at an operational level. Cybersecurity is often a "science of single instances," thus, specifying cyber capabilities may reveal trade secrets. [35] Quoting former Secretary of Defense James Mattis, Wingfield argued that cyber defenders seek to be 'strategically predictable but operationally unpredictable.' In the cyber realm, that may rule out a rich roster of fixed norms even if it allow actors like China and Russia to claim that "cyber is a law-free zone."

Finally, for many speakers, information warfare represented a growing strategic concern. After the 2016 election season, US intelligence agencies agreed that Russian-affiliated actors had actively worked through social media to spread mis- and disinformation. They were, Chris Inglis [19] had previously observed, "trying to hack us," the people, not just the technical networks. Traditionally, Americans enjoyed a form of immunity from foreign information warfare. Gaining access to influential media across the United States' large territory and varied population had been prohibitively difficult until social media companies left users with effectively zero barriers to entry. Now, both Senators Rubio and Warner warned that peer competitors like Russia and China could easily apply information warfare methods that they honed on their own populations and regions. Their goal, as described by Lt. Gen. Michael Groen (Director, DoD Joint Artificial Intelligence Center), is to prevent democracies from exercising core political functions. In other words, adversaries are "hacking our democracy" [31]. Such influence operations, Gen. Nakasone [20] concluded, "are not going to dissipate." Lt. Gen. Crall (Director for Command, Control, Communications, and Computers Cyber and Chief Information Officer, Joint Staff) offered the most sustained presentation of this problem. Whereas the US government, journalists and other actors need time to vet and promulgate accurate information, actors spreading disinformation can rapidly inject, hone, and repeat false, inflammatory, and misleading information. [36] "Speed is more important than perfection," he explained, and is particularly useful when the goal is simply to undermine trust in the political system. Gen. Richard D. Clarke (Commander, U.S. SOCOM) characterized this threat as a psychological battlefield. [37] "Social media now underpins social, civic, political, and commercial life globally," he explained, and in this environment, an "outcome isn't determined by mastery of the facts but rather the back-and-forth battle of the psychological, political and increasingly algorithmic manipulation."

In Sum

Overall, Lt. Gen. Crall [36, 33] summarized, the United States' cyber strategic challenge is a massively skewed risk/reward calculation for peer competitors. For them, the consequences of the United States identifying them as a cyber threat are relatively minor compared to the rewards of breaking into American systems, stealing intelligence and intellectual property, conducting information warfare, and developing capabilities to quickly devastate critical infrastructure. In short, the benefits of violating US cyber systems are specific and valuable. The costs are diffuse and, attackers seem to agree, manageable. These are classic conditions favoring offensive behavior. Speakers appeared to adopt the premise that, in cyber, "if you build it, they will come." In other words, whatever the status of great power relationships, competitive and adversarial governments will always seek to exploit cyber vulnerabilities. Indeed, US cyber systems are actively attacked and penetrated every day, and the forms and severity of attack have proliferated. As former Assistant Secretary of the Treasury Leslie [38] observed, interconnectivity means that there are no isolated threats. Other threats included quickly falling behind during the early phases of a militarized conflict and the parlous state of international cyber norms. Many speakers also worried about the pernicious political and psychological challenge of information warfare.

The Strategic State of the Art from GPC

The following section reviews the GPC speakers' proposals for solutions to the strategic challenges posed above. These proved to be wide-ranging, from a well-developed "defending forward" policy to a broad "deterrence" concept to an almost unwieldy but widely-agreed need for organizational and institutional adaptation. In each case, speakers seemed to agree that, on the one hand, US national security professionals and policy makers need to increase technical capabilities and experiment with new public–private networks even as, on the other hand, they try to gauge the motivations and behavior of global peer competitors. At the highest levels, grand strategy can provide a framework guiding both technical innovation and investment as well as geopolitical agendas. In some cases, these approaches can be integrated, in others they run contrary to one another.

Defending Forward

The United States launched its most assertive cyber strategy to date under the Trump administration. "Defending forward" emerged in 2018 during the lead-up to the US midterm election. Fearing even greater interference from foreign actors—primarily Russia—US national security officials sought and the administration allowed a more proactive, even preemptive, treatment of known malicious cyber actors. [39] Rather than defending US systems, cybersecurity professionals would seek out and actively inflict damage or take other actions against likely threats. Gen. Nakasone had spent several years promulgating this approach. Strategists should assume, he explained, that the United States is in "constant contact with our adversaries." ([17], 4) In cyberspace, barriers to entry are low and any newly developed superiority is rapidly diffused. For all these reasons, "advantage favors those who have initiative" and "is gained by those who maintain a continual state of action." Unlike with nuclear weapons, Nakasone argued, it is "the *use* of cyber capabilities that is strategically consequential." Interaction with adversaries determines subsequent moves.

Nakasone, writing with Michael Sulmeyer [40], made the case for a fully developed 'defending forward' strategy in a 2020 *Foreign Affairs* article. The approach is defined by a "persistent engagement" doctrine, designed to create a deterrent effect. With most of the negative impacts of cyber attacks "occur[ing] below the threshold of traditional armed conflict," they argued, government agencies need to compete actively rather than simply prepare for and deter future challenges. Previously, Nakasone [41] explained that persistent engagement seeks to "degrade the infrastructure and other resources that enable our adversaries to fight in cyberspace," and that "over time, a persistence force, operating at scale with U.S. and foreign partners, should raise the costs that our adversaries incur from hacking the United States." By contrast, a reactive policy of responding to attacks and building network resilience "committed the fundamental flaw in military operations of holding one's forces in reserve past the point of decision." At Cyber Command, this meant actively monitoring networks themselves rather than just "the walls around them" and adopting a "zero trust" approach to "every host, server, and connection." Leading up to the 2018 and 2020 US elections, "persistent engagement" also involved "hunt forward" operations across multiple governments. Defeating adversaries this way is impossible, but it can force them to squander time and resources playing defense and shoring up their own vulnerabilities. Other conference speakers supported this approach, though they often treated it as one among other solutions. Gen. Crall [33], for instance, said that defend forward and hunt teams are "absolutely needed," in part to map adversaries' networks and threat profiles. McConnell [25] made a similar argument that the US should "own" adversaries' networks, though he saw this as part of a larger cooperative, collective security approach.

Deterrence

Most GPC speakers reached for deterrence as a strategic agenda, but it remained a relatively broad concept. Typically, deterrence involves high front-end costs as a government invests in new capabilities, but it offers long-term stability as other actors decide that direct challenges and attacks are too costly relative to whatever benefit they might gain. In short, deterrence aims to shift an adversary's cost–benefit calculation away from offensive attacks. A major challenge for cybersecurity among great powers is shifting adversaries' obvious incentives for low-risk, offensive cyber attacks to incentives for self-regulation based upon fear of reprisals, political consequences, and so forth. One year prior to the conference, a high-profile study made this case. Chartered in the 2019 National Defense Authorization Act, the U.S. Cyberspace Solarium Commission determined that "deterrence [in cyberspace] is possible." Such deterrence "relies on a resilient economy" and "require[s] private-sector entities to step up and strengthen their security posture" [42]. The report set out a layered approach. The United States should push responsible behavior in cyberspace by working with allies and partners. It should "deny benefits" to adversaries by collaborating with the private sector to secure critical networks and create great national resilience. It should, finally, impose costs on cyber adversaries by "maintaining the capability, capacity, and credibility needed to retaliate against actors who target America in and through cyberspace." The defending forward approach, for instance, seeks to achieve this latter goal with direct and constant engagement.

Writer and commentator Fred Kaplan [27] set out the challenges for cyber deterrence. Observers reach for comparisons with nuclear weapons, which were another dramatic and new technology, yet cyber threats defy nearly every aspect of nuclear deterrence. Any given cyber strike is low stakes, and they are as much or more about espionage and other types of information competition as they are about destruction. Planners, Mike Rogers [43] observed, need to identify what specific action or behavior they seek to deter. A major attack like shutting down a power grid may prompt a kinetic response, but everything below that—as well as escalation to that point—remains ill-defined as a threat. Stopping all attacks is impossible, and claiming to halt attacks but failing to achieve it will undermine credibility. Recent administrations have grappled with strategic cybersecurity nominally at a White House level, but agencies largely have been left to define their own doctrinal and tactical agendas. Kaplan argued that a high level government initiative to define first principles and agendas of deterrence is required.

Short of a concerted, presidential-level intervention, conference speakers set out their own interpretation of current and recommended forms of cyber deterrence: denial and punishment. [44] According to Chertoff, advertising capabilities and responding to attacks with espionage, releasing sensitive information about adversary governments, targeted sanctions, and other proportionate reactions cannot keep adversaries out of US systems, but they may convince those actors to seek out other targets or carefully calibrate their attacks. He likened the approach to visible anti-theft tools in automobiles, which cannot stop an expert thief but may convince the actor to decide the object is not worth the effort. Former Department of Homeland Security Director McKenzie [34], meanwhile, said that the deterrence target is not foreign agencies or malicious programs per se but the "cognitive space between the ears of the potential opponent." For this reason, he argued, the United States should spend more energy using rather than hiding its capabilities. If allies and other partners agree to coordinated responses to threats and attacks, such collective security can shift the cost–benefit calculation for mutual adversaries.

Collective Defense

Conference speakers almost unanimously agreed that operational and organizational reforms were necessary to address cyber threats. Their proposals included passing new legislative mandates, building resilience, training personnel, and developing institutional responses to foreign disinformation campaigns. Options ranged from public education to the U.S. Special Operations Command's Joint MISO WebOps Center [19, 33, 36, 37]. "Collective defense," in particular, emerged as a solution category for public–private coordination. Forging coordination and collaboration between private actors and the government should, speakers reasoned, establish far more strategic resiliency and capacity. Shared information and collective responses might even increase deterrence. Unfortunately, no obvious model exists for tight public–private integration. In theory, the Cybersecurity and Infrastructure Security Agency (CISA) might serve this role, but as an organization, it lacks strong enforcement capabilities. Within the military, these cooperation functions are more feasible. The Joint Force J6 Command, Control, Communications, & Computers/Cyber, for instance, operates under the Joint Chiefs of Staff and facilitates integrated planning, and possibly training, across commands [34, 36]. Such examples, however, prove the rule: though sprawling, the US military is an organizational hierarchy with some basic interoperability and shared authority under the Executive Branch. Extending such a model to achieve a public–private collective defense will be challenging.

Still, speakers emphasized practical reforms for organizations and procedures to smooth public–private coordination. Wingfield [35] argued that clearly identifying lines of responsibility among governments and other actors will facilitate cooperation and allow various actors—whether governmental or private—to play to their strengths. As Tighe [29] pointed out, whereas information sharing on cyber threats and attacks already is a common practice—such as with CISA—such practices are not collective defense and are impaired because they are not happening at speed. Other proposed models included an "air traffic control for cyber," an NTSB (National Transportation Safety Board) for cyber that monitors supply chains, or even, as proposed by Sen. Warner [12], an "infinite level reporting system" that can "pulse the entire system" regarding threats and attacks [24, 19]. The military itself has been experimenting with public–private outreach. DreamPort, for instance, was designed to facilitate civilian-military collaboration and was established as a joint venture between the State of Maryland and USCYBERCOM. [20] Notably, cooperation with foreign governments remained a relatively underdeveloped option. Despite extensive collaboration with NATO allies and coordination with partner governments through regional military commands, speakers tended to focus on home-bound US national capabilities and collective defense [34].

Findings

Cyber poses potentially dire and persistently specific threats to US national security interests. Addressing those threats remains a tangle of options and practices. By using the CONTEXT—MEANS—ENDS framework described above to evaluate the commentary on GPC panels and speakers, this study seeks to organize these options. The following section distills the speakers' comments on these topics. In many cases, strategies are implied by the speakers, so the author extrapolates these based upon the challenges and solutions presented. Table 1 summarizes these findings. It also reveals that the strategic lynchpin—the overarching "Means"—remains underdeveloped. This is essential for formulating a cyber strategy that might fit within a broader grand strategy.

Table 1 Identifying the context, means, and ends of cyber in grand strategy as described by the April 2021 GPC speakers

Context	Means	Ends
Traditional geopolitics	Deterrence, Dominance	Favorable cost–benefit calculation, stability
"If you build it, they will come"	Defending forward, Collective defense	Favorable cost–benefit calculation, Slow/stymie true peers
Public–private networks	Collective defense (policy, organizational reforms)	Reliability, transparency/communication
Network penetration	Defending forward, Collective defense	Favorable cost–benefit calculation
Nonexistent, weak and/or unenforced norms	Defending forward, Collective defense	Clear expectations and enforcement
Information warfare	Defending forward, Collective defense, institutional & SOP reforms, education/training	Confidence in political system/savvy public
Cost–benefit environment strongly favors challengers	????	**Balanced or favorable cost–benefit calculation for defenders**

Context

Speakers tended to agree on *context*, or the major aspects of the United States' cyber threat and geopolitical environment. Cyber threats were portrayed as a strategic-level challenge because they foster an unfavorable cost–benefit environment in a world defined by fears of traditional "great power competition." When earlier eras saw weapons, communications, economic, governance, and political innovations, those innovations defined the era's geopolitics. Cyber, today, unites all these categories and is defined by low barriers to entry. Following postwar US strategy, most speakers assumed that the United States would continue to maintain a military, political, and economic order that protects US prosperity and global dominance. This order is designed to prevent any one state from emerging as a true peer competitor able to exert hegemonic influence in Eurasia, build a competing political system, match global US military capabilities, and so forth. Skeptics worry that this is a form of "neo-primacy" that will overextend the United States, but whatever its merits, it enjoys the weight of inertia [45]. The challenge in a cyber-networked world is that traditional power advantages are necessary but not sufficient to achieve dominance or even certain types of deterrence.

In this context of power dominance, cyberspace and other networking undergird many aspects of modern power, and this technology favors challengers and offensive behavior. Most conference speakers assumed, for instance, an "if you build it, they will come" environment. Irrespective of geopolitical factors, governments and other actors will seek to penetrate cyber systems because the targets are easy to achieve relative to the likely costs. For the United States, this vulnerability is compounded by the cyber environment's public–private nature. Most US cyber infrastructure is private, and even US military software and other supply chains extend deeply into the civilian sector. Certainly the US economy and other infrastructure are often managed through cyber and other information networks. Despite this, coordination and responsibility between government agencies, state and local governments, and private actors remains unclear. Some bodies do exist and some coordination does occur, but the US lacks an overarching, national strategic approach. Recent years of domestic political tumult also led experts to worry about information warfare. Never before have US adversaries been able to conduct mis- and disinformation campaigns at marginal cost, sitting in their own capitals, and reaching unknown numbers of eyes. Speakers also worried that if an armed conflict erupted with a peer competitor, early network penetration would hobble the US response. Finally, at home and abroad, some experts argued that the world's relatively thin, or even nonexistent, cyber norms allow "bad actors" to act with impunity while businesses, partner governments, and others grope for standards and expectations around which they can coordinate.

Ends

Broadly, speakers tended to agree on the major strategic *ends* of national cybersecurity: a stable cyber environment characterized by clear and widely-held norms, by political systems and infrastructures that are resilient against digital penetration or manipulation, and by—perhaps above all—a cost–benefit situation that dissuades rather than invites putative attackers. At minimum, as seen above, these goals involve convincing peer competitors and other actors that the costs of directly challenging the United States outstrip the benefits. Unfortunately for US national security professionals, as a growing "element of national power," cyber still favors challengers [9]. In the areas of network vulnerability and penetration, for instance, the challenge facing cybersecurity defenders is widely-known: with networks impossible to fully harden against all possible attacks, defenders need a means to shift the thinking and, in turn, the behavior of actors attacking those networks.

Stated differently, purely technical solutions to network penetration are impossible. The strategic goal, therefore, is to convince likely attackers to avoid, narrow, or reduce their attacks. Similarly, with public–private coordination as well as coordination across local, state and federal government, many speakers called for more reliable and pervasive collective defense through more comprehensive coordination systems. Speakers also hoped to create broad confidence in domestic political systems and international cyber norms. Ultimately, they reasoned, state security is only as viable as the political systems underpinning it.

Means

Finally, *means* represent the applied core of strategy. Overall, several common approaches appeared among the speakers. At a general level, most GPC speakers seemed to agree that the United States should work to preserve its dominant power position. In theory, this allows maximal security and leverage, and it is a key component of geopolitical deterrence. Unfortunately, asymmetrical cyber threats offer to undermine the United States' traditional power advantages, so many speakers proposed some version of "defending forward." By actively pressuring and even attacking cyber adversaries, this aggressive approach aims to push adversaries into a defensive posture and lead to revised cost–benefit calculations. The effect, runs this logic, will be similar to traditional deterrence. Defending forward also fits an historical progression from a more reactive to more proactive posture. Over the quarter century preceding the conference, increasingly ambitious cyber deterrence strategies had failed to deliver a clear and reliable security [46]. "Defending forward" for many planners was a natural next step.

Many speakers also called for collective defense, or tight coordination among US government and private actors. Speakers cited well-established, existing practices, but no general system or policy exists. In theory, a more united front would serve as a deterrent by increasing the capabilities of every participating party and, in turn, increasing costs for attackers. Alongside collective defense, speakers proposed a raft of institutional reforms and updated operating procedures. These addressed both coordination as well as more abstract challenges, ranging from educating and training anyone who interfaces with the cyber environment to a wider social awareness of information warfare. To a lesser extent, speakers also referred to coordination and cooperation with foreign partner governments. Presumably, this would be a major component of any sustained strategic solution, but it was not well developed during the conference. Often, these suggestions were a wish list rather than a coherent strategic approach. Still, taken together, they represent a guiding set of approaches related to defending forward and collective defense, all in an effort to shift adversaries' cost–benefit calculations.

Discussion

Cyber shifts cost–benefit calculations. It is an asymmetrical tool that deliberately elides traditional power advantages. In response, GPC speakers called for a more integrated collective defense and a more aggressive "defending forward" to shift the cost–benefit calculations of adversarial governments. They recognized that cyber, along with other computing and networking capabilities, now shapes every aspect of national power. On the one hand, the United States enjoys a high level of technical innovation and networking. The US private sector is robust and delivers a vast amount of critical infrastructure and even direct services for the US government. Americans themselves are deeply networked. On the other hand, that very connectivity increases US threat exposure. Peer competitors and other governments are able to push into US systems, hack US intellectual property, and threaten critical infrastructure with relatively little fear of escalation or retaliation.

In this environment, speakers articulated, first, an overriding traditional strategy goal and, second, an overriding cybersecurity strategy goal: (1) maintain traditional US geopolitical dominance, and (2) shift the cost–benefit situation against asymmetrical challengers. On the second goal, speakers varied on their operational, institutional, normative, and technical recommendations, but most agreed that, at a minimum, the US government needs to develop a robust, integrated collective defense capability. Many also endorsed "defending forward," which—theoretically—would keep adversaries on the defensive as they try to fend off aggressive US threats and penetrations. Implementing these approaches in the context of a dominance- or deterrence-oriented grand strategy is a natural fit. It requires building technical and personnel capacity along with constant engagement. Indeed, these approaches represent an extension or reapplication of existing strategic thought. Starting in the early postwar period, the United States has built and maintained institutional capacity and strategies of deterrence that rest partly upon leading technical prowess.

Unfortunately, these strategic ambitions for cyber rest on a major assumption: that efforts such as collective defense and defending forward are sufficient to shift the cost–benefit calculus of peer competitors. The logic undergirding these initiatives appears sound because those efforts are likely to raise the costs of penetrating US digital systems and networks. In reality, evidence for these strategies remains thin and can only be extrapolated from prior technological waves. The threats are new, so planners and policy makers are reaching for established strategic theories and logics. Deterrence is particularly appealing because the experience of the twentieth century suggests that unquestioned military and technical dominance coupled with credible threats, as during the Cold War, is likely to deter peer competitors. By mid-century, leading powers seemed to have reached a consensus regarding "rational" self-interest and the dangers of military escalation. Leaders everywhere seemed to have had learned the same lessons about deterrence. By contrast, conciliation, withdrawal, and/or norm agreements engaged without adequate supporting power is often associated with the calamitous interwar period. Brinksmanship and offensive posturing, meanwhile, are often associated with the outbreak of World War I.

Like today, those periods are associated with growing mass politics, communications innovation, and military capabilities; nevertheless, the cyber environment may still be qualitatively different. It involves far lower barriers to entry than technologies of earlier eras, allows remote access, and is difficult to trace. It also rewards success with crippling leverage over the adversary or with almost unthinkably large troves of data and intelligence. To date, cyber threats have almost always increased and proliferated. Collective defense may help private companies, defending forward may slow network penetration, building norms may reduce some threat categories. Combined, though, these strategic means do not systematically shift a global cost–benefit environment defined by low barriers to entry, expanding penetration points, and low fear of escalation or reprisal. The parts do not necessarily combine into a viable whole. This is a grand strategic challenge. To date, however, most of the recommended solutions are targeted at lower levels. This is understandable: commanders and civilian managers need to reach for accessible solutions. Still, as parts, capacity and mid-level strategies do not necessarily sum to a strategic solution that addresses systemic threats.

Conclusion

This chapter separated speakers' approaches to cyber and strategy into three categories: *context*, *means*, *ends*. It identified several common themes, and it uncovered a foundational gap in US cyber strategic thinking. In general, speakers agreed that a core strategic challenge in a world of growing peer competition is that the cost–benefit calculation among adversaries easily favors more rather than less cyber penetration, intelligence hacking, and other attacks. Even as the United States continues to focus on its core, traditional geopolitical advantages, there is an "if you build it, they will come" quality about cyber infrastructure: if it exists, someone will try to penetrate it. If it exists in abundance, potential penetrations are numerous. Public–private networks, weak norms, information warfare are all challenges that flow from these basic realities. Speakers recommended a wide range of solutions. These focused on collective defense and defending forward, which are variations of deterrence: compelling attackers to redirect their efforts because US systems are difficult to penetrate and resilient, because of a fear of reprisal, or because offensive US attacks keep adversary resources distracted.

Though sensible, I argue that these efforts are necessary but not sufficient. Governments may need to reach outside cyberspace for solutions to cyber threats. From the vantage of grand strategy, this study finds that shifting peer states' cost–benefit calculations is a systemic challenge. Professionals at the conference understandably reached for operational and technical solutions. These solutions advance that cause, but they are unlikely to change overall state behavior. Cyber deterrence, for instance, may be a necessary equilibrium for long-term global stability, but the concept is typically drawn as a metaphor from conventional power capabilities rather than reimagined on new terms for cyber. "Defending forward" would require an asymptotically increasing amount of energy to directly challenge the major threat actors, to say nothing of the wider universe of cyber threats. At the level of tactics and technical challenges, the threat from peer competitors, organized crime, and political actors is the same. At the grand strategy level, however, a large portion of threats could be minimized or expectations set with other governments. For instance, the Geneva Conventions cover traditional state behavior even if they are ignored by insurgents and terrorists. Such agreements are still valuable. In a similar fashion, the US should continue to push its partners, most likely starting with NATO, to formulate an international cybersecurity regime even as Washington, separately, begins to enforce its preferred norms and standards. Ideally, presidential administrations will work out these standards and lead their application. To be clear, this is an argument for enforcement and action with norms and standards as the guiding framework. In reality, senior military leaders, such as the Joint Chiefs, could develop and implement these standards with private sector partners and foreign militaries as a "proof of concept." In addition, ongoing diplomacy with peer competitors and, at home, ongoing experimentation will help accelerate the process of evolving viable strategy. As Sen. Rubio commented, this "doesn't necessarily mean that we need to fight cyber with cyber." Multiple instruments of national power are likely required to shift cost–benefit calculations among sovereign states. These parting recommendations are loose, but they serve as an illustration. More work is required to further disentangle technical and operational cybersecurity from grand strategy. In principle, deterrence and its adjunct approaches are a viable solution to many cyber threats. In practice, leaders and policy makers would do well to recognize that cyberstrategy defies prior lessons.

References

1. Brands, H. (2014). *What good is grand strategy? Power and purpose in american statecraft from Harry S. Truman to George W. Bush.* Cornell University Press: Ithaca, NY.
2. Balzacq, T., Dombrowski, P., & Reich, S. (2019). *Comparative grand strategy: A framework and cases.* Oxford University Press.
3. Posen, B. R. (2014). *Restraint: A New Foundation for U.S. Grand Strategy.* Ithaca, NY: Cornell University Press
4. Silove, N. (2018). Beyond the buzzword: The three meanings of 'grand strategy.' *Security Studies, 27*(1), 27–57.
5. Modelski, G., & Thompson, W. R. (1988). *Seapower in global politics, 1494–1993.* Palgrave.
6. Levy, J. S., & Thompson, W. R. (2014) *The arc of war: Origins, escalation, and transformation.* Chicago: University of Chicago Press.
7. Alexander, K., & Tighe, J. (2020). 5G technology: An interview with two top cyber warriors. *Proceedings U.S. Naval Institute, 146*(7), 1409. www.usni.org.
8. Biden, J. R. Jr. (2021). *Interim national security strategic guidance.* Washington, DC: The White House. https://www.whitehouse.gov/wp-content/uploads/2021/03/NSC-1v2.pdf.
9. Biden Faces Major Cybersecurity Challenges, Says Former NSA Director. (2021). Interview. *Bloomberg Technology.* Retrieved January 21, from https://www.bloomberg.com/news/videos/2021-01-22/biden-faces-major-cybersecurity-challenges-says-former-nsa-director-video.
10. International Institute for Strategic Studies. (2021). *Cyber capabilities and national power: A net assessment.* International Institute for Strategic Studies. Retrieved March 21, from file:///C:/Users/jshively/Downloads/Cyber/20capabilities/20and/20national/20power/20-/20a/20net/20assessment.pdf.
11. Rubio, M. (2021). Opening remarks. In *Great Power Competition Conference.* Cyber Florida. Retrieved April 15, from https://www.youtube.com/watch?v=g-OWBsES1xY&list=PLHwh9SRof1I__oLOWSnOZZ8IrW_WbPRx4.
12. Warner, M. (2021). Opening remarks. In *Great Power Competition Conference.* Cyber Florida. Retrieved April 15, from https://www.youtube.com/watch?v=TEkGvUU8gqg&list=PLHwh9SRof1I__oLOWSnOZZ8IrW_WbPRx4&index=2.
13. Interview. (2020). Defense writers group. Retrieved December 2, from https://cpb-us-e1.wpmucdn.com/blogs.gwu.edu/dist/2/672/files/2020/12/DWG-with-Admiral-Craig-S.-Faller.pdf.
14. Congressional Research Service. (2021). Renewed great power competition: implications for defense—issues for congress. Retrieved March 4, https://assets.documentcloud.org/documents/20508479/renewed-great-power-competition-implications-for-defense-issues-for-congress-march-4-2021.pdf.
15. Wilson, I. (2021a). Implications of CSD for SOF with Dr. Isaiah "Ike" Wilson III (Part 4 of 4). Retrieved March 19, from https://www.youtube.com/watch?v=6sAk2CQEDk4.
16. Wilson, I. (2021b). Panel: Our response options. In *Great Power Competition Conference.* Cyber Florida.
17. An Interview with Paul M. Nakasone. (2019). *Joint Force Quarterly.* 1st Quarter. https://ndupress.ndu.edu/Portals/68/Documents/jfq/jfq-92/jfq-92_4-9_Nakasone-Interview.pdf.
18. Tighe, J. (2017). Cyber warfare in the maritime domain. In *Center for strategic and international studies, Maritime security dialogue.* Washington, DC: CSIS Headquarters. Retrieved September 14, from https://csis-website-prod.s3.amazonaws.com/s3fs-public/event/170915_Cyber_Warfare_in_the_Maritime_Domain.pdf.
19. Inglis, J. C. (2019). *Security in the age of cyber and great power competition.* Address to Baltimore Council on Foreign Affairs. Retrieved September 24, 2019, from https://www.youtube.com/watch?v=VRYOBNpQ2pI.
20. Nakasone, P. (2021). Opening keynote: Cybersecurity and the great powers. In *Great Power Competition Conference.* Cyber Florida. Retrieved April 15, https://www.usf.edu/arts-sciences/great-power-competition/.

21. Alexander, K. B., & Jaffer, J. (2020). Navigating the security maze in a new Era of cyberthreats. DarkReading. Retrieved December 9, 2020, from https://www.darkreading.com/vulnerabilities---threats/navigating-the-security-maze-in-a-new-era-of-cyberthreats-/a/d-id/1339628.
22. Al-Mansoury, L. (2021). Top international cyber experts call for national cyber security centers. *Kuwait Times* Retrieved March 23, https://news.kuwaittimes.net/website/top-international-cyber-experts-call-for-national-cyber-security-centers/.
23. Faller, C. S. (2021). Plenary speech. Cybersecurity and the great powers. In *Great Power Competition Conference*. Cyber Florida.
24. Alexander, K. (2021). Panel discussion: The future of cyber conflict. Cybersecurity and the great powers. In *Great Power Competition Conference*. Cyber Florida.
25. McConnell, J. M. (2021). Panel discussion: Cybersecurity & strategic deterrence. Cybersecurity and the great powers. In *Great Power Competition Conference*. Cyber Florida.
26. O'Hanlon, M. E. (2021). Panel discussion: The cyber threat. Cybersecurity and the great powers. In *Great Power Competition Conference*. Cyber Florida.
27. Kaplan, F. (2021). Panel discussion: The cyber threat. Cybersecurity and the great powers. In *Great Power Competition Conference*. Cyber Florida.
28. Inglis, C. (2021). Panel discussion: The future of cyber conflict. Cybersecurity and the great powers. In *Great Power Competition Conference*. Cyber Florida.
29. Tighe, J. (2021). Panel discussion: The cyber threat. Cybersecurity and the great powers. In *Great Power Competition Conference*. Cyber Florida.
30. Interview. (2017b). On the precipice: The future of cyber & intelligence with vice admiral Jan Tighe. *Over The Horizon Journal* Retrieved February 27, from https://othjournal.com/2017/02/27/inteview-vice-admiral-jan-tighe/.
31. Groen, M. (2021). Panel discussion: The future of cyber conflict. Cybersecurity and the great powers. In *Great Power Competition Conference*. Cyber Florida.
32. Gehrke, J. (2017). *Admiral: US tolerated cyber 'acts of war' over last decade*. Washington Examiner. Retrieved September 14, from https://www.washingtonexaminer.com/admiral-us-tolerated-cyber-acts-of-war-over-last-decade.
33. Crall, D. (2021b). Panel discussion: Cybersecurity & strategic deterrence. In *Great Power Competition Conference*. Cyber Florida.
34. McKenzie, K. F. Jr. (2021). Closing keynote: Cybersecurity and the great powers revisited. In *Great Power Competition Conference*. Cyber Florida.
35. Wingfield, T. (2021). Panel: Our response options. In *Great Power Competition Conference*. Cyber Florida.
36. Crall, Dennis (2021a). Plenary speaker. Cybersecurity and the Great Powers. Great Power Competition Conference, Cyber Florida. 15 Apr.
37. Clarke, R. D. (2021). Plenary speaker. Cybersecurity and the great powers. In *Great Power Competition Conference*. Cyber Florida.
38. Ireland, L. (2021). Panel discussion: The cyber threat. Cybersecurity and the great powers. In *Great Power Competition Conference*. Cyber Florida.
39. Graff, G. M. (2020). The man who speaks softly—and commands a big cyber army. *Wired* Retrieved October 13, from https://www.wired.com/story/general-paul-nakasone-cyber-command-nsa/.
40. Nakasone, P. M., & Sulmeyer, M. (2020). How to compete in cyberspace. *Foreign Affairs*, *99*(4).
41. Naksone, P. M. (2019). A cyber force for persistent operations. *Joint Force Quarterly*. 1st Quarter. https://ndupress.ndu.edu/Portals/68/Documents/jfq/jfq-92/jfq-92.pdf.
42. King, A., & Gallagher, M. (2020). United States of America cyberspace solarium commission. https://drive.google.com/file/d/1ryMCIL_dZ30QyjFqFkkf10MxIXJGT4yv/view.
43. Rogers, M. (2021). Panel: The Cyber Threat. In *Great Power Competition Conference*. Cyber Florida.
44. Chertoff, M. (2021). Panel discussion: Cybersecurity & strategic deterrence. In *Great Power Competition Conference*. Cyber Florida.

45. Itzkowitz Shifrinson, J. R. (2020). Neo-primacy and the pitfalls of US strategy toward China. *The Washington Quarterly.*, *43*(4), 79–104.
46. Jasper, S. (2017). *Strategic cyber deterrence: The active cyber defense option*. Rowman & Littlefield.
47. U.S. Cyber Command news release. (2020). Hunt Forward Estonia: Estonia, US strengthen partnership in cyber domain with joint operation. Retrieved December 3, from https://www.cybercom.mil/Media/News/Article/2433245/hunt-forward-estonia-estonia-us-strengthen-partnership-in-cyber-domain-with-joi/.

Jacob Shively is an associate professor of University of West Florida where he teaches international relations. Dr. Shively's recent book project, "Make America First Again," develops a new analytical framework called grand strategy analysis (GSA) and applies it to the first two years of the Trump administration. Dr. Shively has also presented at international and national conferences; published scholarly reviews, articles and encyclopedic chapters; conducted public talks; and delivered briefings to national security professionals.

Cybersecurity and Strategic Deterrence: Changing Adversary's Risk Versus Reward Calculations

Hon J. Mike McConnell, VADM, USN, RET, and Mark Grzegorzewski

Abstract SolarWinds, Microsoft Exchange Server, JBS, Colonial Pipeline, Kaseya…cyber-attacks, cyber intrusions and exploitation breaches keep mounting, not only from cybercriminals but from nation-states, most notably Russia and China. While cyberattacks against U.S. companies and U.S. Government organizations are not new, the scale and frequency are increasing at an alarming rate (SonicWall in 2021 SonicWall cyber threat report. SonicWall, 2021, [1]; Internet Crime Complaint Center (IC3) in 2020 Internet crime report, 2020, [2]). With the U.S. becoming increasing "digitally dependent," this problem has reached strategic proportions. Why is this the case? How can the most military capable country in the world be so exposed to cyberattacks and not forcefully respond? We proffer the answer can be found in General (GEN) Paul Nakasone's 2018 Senate confirmation as Commander of USCYBERCOMMAND. Responding to Sen. Dan Sullivan's line of inquiry as to why the United States is the "cyber punching bag of the world," GEN Nakasone responded, "I would say right now they do not think that much will happen to them." He added, "they don't fear us" and "the longer that we have inactivity, the longer that our adversaries are able to establish their own norms" (C-SPAN in NSa nominee testifies senate armed services committee, March 1, 2018, [3]) In essence, the U.S. is not deterring its adversaries in cyberspace since they do not fear a consequence. Part of their risk-reward calculation surely assesses that although the U.S. may have some of the most exquisite cyber capabilities, it also has many more unpatched cybersecurity vulnerabilities and consequently lacks cyberspace resiliency. Thus, adversarial states assess that the reward of carrying out an action through cyberspace far exceeds any potential cost (Cyber Florida in 2021 great power competition conference: strategic deterrence panel, April 24, 2021, [4]).

Keywords Cybersecurity · Cyber adversary · Risk versus Reward calculation · Cyber space

H. J. M. McConnell (✉) · VADM, USN, RET
University of South Florida, Tampa, FL 33620, USA
e-mail: MikeMcConnell@usf.edu

M. Grzegorzewski
St. Petersburg College, St. Petersburg, Tampa, FL, USA
e-mail: Grzegorzewski.Mark@spcollege.edu

A. Farhadi et al. (eds.), *The Great Power Competition Volume 3*,
https://doi.org/10.1007/978-3-031-04586-8_3

Introduction

While there are certainly factors constraining how the U.S. can respond, it should not mean the U.S. government (USG) could not respond beyond its current efforts. The cliché that the U.S. lives in the largest glass house and is therefore reluctant to throw stones can only hold for so long. It is well past time the USG fortify its glass house, through cyber resiliency, so it can make its adversaries fear consequences for their actions. In this chapter, we propose a way in which the United States can respond, namely through cyber resiliency (i.e., deterrence by denial), to build towards establishing a more effective overall deterrence posture.

In what follows, we extend the important insights shared during the "Cybersecurity and Strategic Deterrence" panel. First, we address the risk inherent in the U.S.' large attack surface. This overview provides context to our argument that increased cyber resiliency could bolster overall U.S. national security. Second, we present our thesis to the reader. Our central premise is that the U.S. needs to strengthen its cyber resiliency to deter more effectively, and we demonstrate this claim by illuminating the scope of the problem and providing specific recommendations for changing the risk versus reward calculations of U.S. adversaries. We also address challenges to our recommendations. We then layout what is meant by deterrence and cyber deterrence, so that the reader can have a deeper appreciation of its complexity and understand how cyber resiliency is just one part of a larger calculation. Finally, we critique the nuclear deterrence-cyber deterrence analogy before moving to a summation of our chapter.

The Attack Surface

In a networked world where every device is connected to another device, every device is a point of vulnerability and the human that makes and then operates those devices is yet another point of vulnerability.[1] This explosion in risk mainly is the result of U.S. technology companies focusing on getting their product out the door as rapidly as possible to quickly return a profit, and smaller technology companies having a checkbox mentality when it comes to cybersecurity. This means the interests of technology companies and the security needs of technology users are not aligned. The situation leaves many individual points of vulnerability for adversaries to exploit at little cost and relatively little effort. Consequently, why would a state focus on developing a large cyber arsenal, akin to stockpiling nuclear weapons, when in fact they can exploit the many holes in an adversary's cybersecurity? Many technology companies around the world use American made products that have security holes in their design. Why develop your own weaponized code to match the complexity of the U.S. cyber arsenal when instead you can exploit the ubiquitous security flaws in American made technology products?

[1] Jardine [5].

To strengthen the USG's position in its ongoing strategic competition and enable its offensive cyberspace capabilities, the U.S. must first address its vulnerabilities. The USG, by most accounts, has the strongest cyber arsenal in the world.[2] Yet, every year, the U.S. suffers the most significant cyberattacks in the world.[3] Something must change. The U.S. must become more resilient in cyberspace so that a single point of failure will not cause cascading effects. We propose that by increasing American cyberspace resiliency, the U.S. can in fact raise its collective security. Building this resiliency is not only the responsibility of the USG and corporate America. It is a responsibility of every American. The security of the U.S. is only as strong as the least resilient citizen.

The Case for Cyber Resiliency

In a thoughtful article addressing cyberspace deterrence, Dr. Joseph Nye of Harvard disentangles cyberspace defense from cyberspace resiliency.[4] For the former, he regards it as building security architecture to defend against attacks, like building a wall around a castle for protection. For the latter, he portrays it as withstanding an attack and having the ability to recover. In cyberspace this could be analogous to backing up your systems if data is stolen or deleted. Another point of reference is the National Institute of Standards and Technology (NIST). The NIST defines cyber defense as "actions normally created within DoD (Department of Defense) cyberspace for securing, operating, and defending the DoD information networks. Specific actions include protect, detect, characterize, counter, and mitigate."[5] By comparison, the NIST defines cyber resiliency as "the ability to anticipate, withstand, recover from, and adapt to adverse conditions, stresses, attacks, or compromises on systems that use or are enabled by cyber resources."[6] Clearly, both in the views of Nye and the NIST, there is a nuanced distinction between cyber defense and cyber resiliency. The distinction is about a change of mindset from preventing most cyber intrusions to realizing that despite an organization's best efforts it will inevitably be compromised, and having a plan in place to recover one's systems.[7]

We find common ground with Nye in his assertion that the U.S. needs to increase its cyber resiliency to act as a more effective deterrent. If the U.S. increases its cyber resiliency, adversaries will be forced to expend more resources when operating against U.S. information systems. As the cost of cyber intrusions increases, adversaries will be less inclined to attack U.S digital infrastructure. Most importantly, once U.S. hard and soft infrastructure becomes more resilient, the USG can

[2] Voo et al. [6].

[3] Ang [7].

[4] Nye [8].

[5] Computer Security Resource Center [9].

[6] Computer Security Resource Center [10].

[7] Crosston [11].

increase the scale and scope of its own cyber operations against adversaries and more effectively deter them through punishment. Strengthening cyber deterrence through cyber resiliency would therefore open more options in cyberspace for the USG.

To build the cyber resiliency we advocate, organizations must model the 3 Ps: Predict, Prioritize and Practice.[8] This involves predicting where a breach could occur, prioritizing which vulnerabilities are the most damaging, and practicing the procedures in place to respond to a breach. While a 3 Ps model is an inherently attractive model, it is not foolproof. In fact, all 3 Ps are influenced by the main cyber threat vector: the human. In that vein, we put forward three specific recommendations on how the U.S. can increase its cyber resiliency by focusing not just on making digital infrastructure more resilient but also the wetware (humans) and policy/law.

The Human Dimension

As we pool individual sovereignty to form a nation-state that provides security for all, we cannot allow some unsecure individuals to jeopardize the overall security of all. It takes a collective to form a nation-state, and individuals have responsibility to each other to make sure they are not the weakest link in the chain. One weak link, due to the interconnectivity of society, has the power to undermine U.S. national security. Just as in the Cold War,[9] the U.S. needs to realize that to lead in a world of strategic competition, it must convince its citizens to engage in a whole-of-society approach to deter against adversaries.

Speaking to the fragility of American's collective digital unpreparedness, a Pew Research poll recently found survey participants scored lowest regarding digital security and privacy.[10] The results are not shocking. To anyone listening, U.S citizens constantly throw around buzzwords, like the cloud, 5G, and the blockchain, without really having a clear idea of what they are articulating. If U.S. citizens are not using the same terms of reference, how can we expect them to properly respond to calls for cyber resiliency? What is badly needed is a common vernacular and understanding of cyberspace so that all U.S. citizens can do their part in strategic competition.

As such, American citizens must constantly bolster their cybersecurity skills to make overall U.S. defenses stronger and more resilient against computer network attack and exploitation. According to an 18-month study conducted by the cybersecurity firm Proofpoint, 99% of cyberattacks require human intervention to execute.[11] Human intervention can range from social engineering to plugging in an unauthorized USB drive. This finding implies that when considering what cyberspace infrastructure needs protection, those considerations must also include humans.[12] While

[8] Kumar [12].

[9] Brands [13].

[10] Vogels, and Anderson [14].

[11] Proofpoint [15].

[12] Truitte [16].

U.S. cyber systems certainly have many holes in need of patching, it still requires a human to gain access to a vulnerability. Therefore, humans are on the front line of cyber defense, and the U.S. will struggle with building resiliency without producing cyber-informed citizens.

Recommendation #1

Educate U.S. citizens starting at the grade school level about cyberspace and cybersecurity. Despite the complexity of the technology, it is all about the human. If the human does not know how to operate the technology, they put all others at risk. This recommendation is analogous to high schools offering driver's education courses.[13] Society would not tolerate putting a 16-year-old teenager on the road operating a nearly 4000-pound vehicle without knowing how to drive a car or knowledge of the rules of the road. Although an imperfect comparison to operating a 10-pound desktop computer, the implications of making a mistake in cyberspace rival those of sending out an inexperienced driver on the road.

While our focus is primarily on creating safer experiences for everyone online, the education we propose could even extend to identifying mis- or disinformation online. In fact, efforts are underway to educate U.S. citizens on disinformation to "understand digital media's influence on public dialogue and opinion and develop solutions to disinformation."[14] In short, it is about more than just making hardware and software more resilient. It is also about making the human mind more resilient.

Education need not be punitive but rather informative and integrative. Framed as an educational opportunity to bolster U.S. security through a "whole-of-society" approach, this endeavor could both serve to inform Americans on basic cybersecurity practices and provide buy-in to the American national project. That is to say, the experience could be framed as all Americans doing their small part to secure their individual systems to contribute to collective defense of the country. This framing could be contrasted against the scourge of China, Russia, Iran, and North Korea in cyberspace not doing their part to develop a more overall secure Internet but rather using cyberspace to drag the rest of the world down to their substandard level and diminish cybersecurity worldwide. Education could tie domestic collective security measures to America's narrative as a beacon of hope in the world. This framing could include the world viewing America as leading the world towards a more secure Internet, as well as applying internally a positive form of civic nationalism where fellow citizens look to their compatriots to be responsible Americans and secure their networks. Ultimately, cyberspace education would both target of soft and hard systems. It would promote cyber awareness in general, so individuals do not inadvertently allow cyber intrusions into their systems. It would also produce a more civic-minded individual who can protect against the influence of other malign

---

[13] Cyber Florida [4].

[14] Mitchell [17].

actors online and a more critical thinker who will challenge what they read on the Internet to avoid to proliferation of disinformation.

Collective Security at Network Speed

For several reasons, ranging from U.S. political culture to the construction of the U.S. economic system to policies and legalities, there is a dividing line between the public and private sector. This divide is based upon a series of choices made by humans over time. While we do not argue that the divide between the public and private sector is a bad thing, we do think it needs to be reconceptualized to meet the security needs of the present. Choices Americans made in the past regarding security between public and private can be re-constructed, just as the concepts of public and private were previously constructed.

Choices about how to separate the public from the private sphere started around the Late-Medieval Period (sixteenth century) to the Enlightenment (seventeenth century)[15] and with the rise of Liberalism (eighteenth century).[16] Breaking the old European feudal system required a new compact amongst citizens. As collectives created their own states, they began making a distinction between what was for everyone, "public," and what had not been transferred to their newly constructed governments, which they labeled "private." As states were called upon to raise militaries to defend themselves, they also had to raise taxes to fund their collective security. In fact, security was one of the primary reasons individuals came together to create a state. In determining what was taxable property, the state looked to an individual's private wealth they had amassed under the security of the state (public) and taxed them a percentage of their wealth. To the state, the citizens that were taxed based upon this wealth were simply making a private gift to the state in return for the security provided by the state.[17]

Without recounting the entire history of Liberal, capitalist democracies and the private–public divide, this example demonstrates how these seemingly natural concepts of public and private were developed over time by humans making choices based upon the needs of their day. When a state, or its citizens', needs are not met, it should reevaluate its choices and make modifications to improve its position. This can include re-conceptualizing what it holds as public and what it deems private. When alterations are not made to meet the needs of the day, an actor will be overcome by a wiser, more pragmatic adversary.

How does this relate to the U.S. today? We are not arguing for abolition of private rights or even erasing the line between public and private. We are advocating increasing collaboration on cybersecurity between the public and private sectors. This will not be an easy task. U.S. private organizations focus on making profit while the

[15] Horwitz [18].

[16] Starr [19].

[17] Starr [19].

USG focuses on effectively providing security services. If the USG wants private organizations to take security more seriously, that shift will impact their profits. This may mean government support will be needed as an enticement for businesses to better secure their networks.

It also may mean rethinking the solid red line separating the two regarding classified intelligence sharing and thinking how the public and private can collaborate to further the interest of both sectors. As noted during the panel, today we live in a world in which nation-states frequently attack private sector entities.[18] These U.S. businesses should not be forced to fend for themselves against the capabilities of nation-states. We hold the USG has an obligation to report this information to companies regardless of security classification. Cyberspace intelligence classification is notoriously overclassified.[19] It is time the USG update its classification guidance so that the private sector can improve its security. In doing so, the U.S. will raise its collective security by sharing critical information with its private sector partners. Again, for this to occur, the U.S. must think differently about its social constructed public and private spheres.

Recommendation #2

The U.S. needs to reconsider the nature of its partnerships with the private sector. Far too often the public and private sectors operate in stovepipes.[20] This needs to change. In the U.S., the public and private must understand they have overlapping interests. If there is a continuing lack of dialogue between the public and private spheres, or if they continue operating as competitors rather than collaborators, foreign adversaries will continue to exploit gaps in U.S. defenses.[21] Adversaries understand the U.S. military will most likely not respond to attacks against the U.S. private sector. This calculation is partly due to the U.S. having specific legal authorities for its military to operate within (Title 10) and the intelligence agencies to operate within (Title 50), as well as other laws such as the Posse Comitatus Act which limits the involvement of the military in domestic affairs. These laws are not meeting contemporary security needs. It is time to discuss what needs to change. Our adversaries see the seams between the public and private and are actively exploiting them.

After reconceptualizing the responsibilities of the public and private spheres, the USG needs to invest more in government and private sector collaboration. More employees from the corporate world must take tours within government and more government employees must take a tour within corporate America. Both spheres are facing similar challenges but addressing those challenges differently due to their own capabilities, constraints, and limited perspectives. Only by further pooling

[18] Cyber Florida [4].

[19] Healey and Jervis [20].

[20] Doherty et al. [21].

[21] Williams [22].

resources[22] and working with private industry to share publicly derived sensitive (classified) information can the U.S. make its infrastructure more resilient. Currently, the USG knows more than the private sector about an expected executed attack. How can the private sector be more resilient if it cannot make accurate predictions about its vulnerabilities? By sharing sensitive warning data with the private sector, it will allow them to be more resilient to adversarial attacks, and in the process raise the collective security of everyone. This sharing of information need not be a one-way street. The USG can also leverage the trust many U.S. citizens have in private sector companies. In practice, the U.S. could allow the public sector organizations to take the lead in attributing a cyberspace operation or allow the private sector to confirm the USG's attribution. It is a win–win outcome that allows the public to increase its trust with the USG and the USG building network resiliency through privately identifying those vulnerabilities that are likely to be attacked.

It's Time to Modernize Digital Laws

As recounted by VADM McConnell during the panel, if the United States were under physical attack by an adversarial state, the USG would send in the Army, Navy, and Air Force to protect itself.[23] When compared to a private business under virtual attack by an adversarial nation, the DOD would likely claim the act to be outside of its legal authority to react.[24] What is different between these two hypotheticals? For one, the attack would take place in a physical domain rather than a man-made domain. Secondly, the target of the attack would be the private sector rather than a center of industry or a military installation. If these reasons for a lack of response come across as ludicrous, they are. The U.S. cannot continue to treat adversarial actions in one domain as legitimate and attacks in cyberspace that cause immense hardship as undeserving of a response. The old ways of thinking, about war and peace, crime and conflict, and private and public, continue to keep the USG from properly responding, and deterring, adversarial actions.

Exploiting these constructed divides, nation-states are choosing to launch cyber-attacks using U.S. servers located inside the U.S. to attack companies and government agencies, thereby demonstrating that U.S. adversaries understand current U.S. law restricting U.S. Intelligence Agencies from engaging in domestic activities.[25] Congress should debate and consider changing current restrictions on U.S. Intelligence Agencies, the nation's most capable cyber defenders, to allow for searches of malware intrusions in U.S. critical infrastructure. Any legislative changes would have

[22] Atkins and Lawson [23].

[23] Cyber Florida [4].

[24] Cyber Florida [4].

[25] Schneier [24].

to be mindful of protecting U.S.' citizen's privacy and includes strong legal restrictions and intrusive oversight by Congress. Such changes could serve the nation's strategic needs in the evolving digital age.

Recommendation #3

With the increasing digital dependency of the nation and a shrinking timeframe to respond to cyber intrusions, the U.S. must rethink the constructed security rules that separate and divide "domestic" from "foreign." Congress needs to craft legislation that will allow some intelligence agencies to look at certain sectors metadata, like critical infrastructure. In addition, Congress needs to write laws that will allow malware metadata to be shared between the government and the private sector. As more USG organizations opt to move to the cloud, the line between the government and private industry is already changing. As such, the U.S. is at a point of shared digital dependency and increasing vulnerability. If the USG has the best cyberspace capabilities, why not look for indications and warnings in these shared systems and then harness the best in the USG and private industry to respond to the attack? The argument from the Intelligence Community will be that it must protect sources and methods. We ask, what good are sources and methods if they consistently handcuff the U.S. from taking meaningful action?

U.S. citizens, and the U.S. government, must realize it is literally in battle for its way of life. The lines between the authoritarian systems and liberal-capitalist democracies could not be starker. In this battle for the future, why is the U.S. fighting with outdated rules that meaningfully constrain its actions?[26] The U.S. government appears stuck in the way of doing the same thing, the same way because that is the way things have "always" been done. If the U.S. sticks with its current posture, it will fall behind in strategic competition. U.S. citizens must have a very public debate about the tradeoffs between security and freedom. If U.S. citizens decide they want the capabilities of the USG to defend private networks, the public must demand Congress enact laws with strong oversight and transparency. Certainly, there will be some apprehension over concerns of data privacy. In response, there must be laws passed in conjunction that focus on safeguarding individual privacy. Conversely, the U.S. will not "win" in competition if it forsakes its national identity and the rule of law in favor of total security. Rather, a new balance must be struck. The U.S. cannot continue to fight with one hand tied behind its back. It must reconsider the balance between freedom and security in its digital age.

Once the public decides, the Executive branch and Congress must find the right balance by crafting legislation that does not restrict its most capable cyber resources, while also protecting individual liberties. Ultimately, if not addressed, the current

[26] Octavian Report [25].

situation will remain untenable, and adversaries will continue to probe U.S. critical infrastructures and resources.[27] Updating rules and legislation to cyber-enabled conflict will allow the U.S. to get ahead of threats rather than treating cyber intrusions as isolated incidents.

Challenges

Critics may take issue with the serious measures we advocate in this chapter. Detractors may see existing U.S. laws, institutions, and cybersecurity frameworks as sufficient to address threats emanating from cyberspace,[28,29] We lay forth three specific recommendations, that while not perfect, are a step in the right direction. The U.S. current equities process does not adequately weigh the risk of doing nothing, which drives policy inaction. The USG must re-think the risk of doing and accept that there is no perfect answer that will address all cyberspace vulnerabilities and concerns. Every correction act will entail risk and lead to unforeseen consequence. If the USG continues to remain largely static in this dynamic environment, the many incremental cyberspace actions against it will have strategic level impacts (i.e., quantity will become quality). The USG must not let a perfect solution be the enemy of the good solution. Even small actions can have large impacts. Therefore, the USG should embrace a test and learn culture to cyberspace and pilot solutions within the national, state, city, and municipal governments. Some proposals may work and others will not. Yet, if any of these recommendations succeed, they could have big impacts on U.S. cybersecurity and overall deterrence. The risks of the U.S. acting will change our adversary's deterrence calculations, whereas not publicly acting against their cyberspace activities will signal that their behavior is acceptable.

The time is now to start small with pilot programs in individual states to educate U.S. citizens starting at the grade school level about cyberspace and cybersecurity. The USG should try different cyberspace-related curricula to see what works and what does not. The same can be done in collaborating with the public and private sectors. They can try out different partnerships and working arrangements to see what is successful and what is ineffective. Meanwhile, politicians and civil servants must engage in serious political debate about modernizing our digital laws. Even if the U.S. decides not to change existing laws, it must at least calculate what doing nothing will entail. Whatever it takes, the U.S. must stop being paralyzed by inaction due to uncertainty and lack of operational clarity. The U.S. cannot afford not to act. Time is short, and U.S. adversaries are controlling the operational tempo. It is time to retake control and strengthen the U.S. deterrence posture through cyber resiliency.

---

[27] Greenberg [26].

[28] Valeriano and Maness [27].

[29] Gartzke [28].

A Complex Concept: Cyber Deterrence

Increasing cyber resiliency will not single-handedly improve overall deterrence. Deterrence is complex concept with many known unknowns. Hence, it is imperative to understand the broader conceptualization of deterrence. The academic literature on cyber deterrence is vast. In only taking a sampling of the literature, it breaks out into four dominant themes: conceptualizing cyber deterrence; cyber deterrence challenges; operationalizing cyber deterrence; and norms in cyber deterrence. In conceptualizing cyberspace, many scholars take issue with the exclusive focus on cyberspace deterrence.[30,31,32,33] Cyber deterrence scholars see deterrence as occurring across all domains and not just an exclusive focus on the cyberspace domain. In fact, Jacquelyn Schneider takes the argument a step further and asks are we speaking of "cross-domain deterrence of cyberspace operations" or "cyberspace operations as a tool of cross-domain deterrence"?[34] These are critical, foundational questions when conceptualizing cyberspace deterrence. Starting with the former question as opposed to the latter will lead to very different answers. While the academic debate about the proper starting point continues, Pentagon leaders have reportedly settled on the second question as the proper way to conceptualize deterrence.[35] Consequently, in this chapter we too conceptualize of cyberspace operations as a tool of cross-domain deterrence.

The Department of Defense defines deterrence as "the prevention of action by the existence of a credible threat of unacceptable counteraction and/or belief that the cost of action outweighs the perceived benefits."[36] Specifically concerning cyber deterrence, the DoD provides a much murkier overview by calling it a comprehensive strategy to "deter aggression, including cyberattacks that constitute a use of force, against the United States, our allies, and our partners."[37] This unclear position provides an opening to U.S. adversaries as it leaves large swaths of corporate America labeled as unprotected by the USG since cyber espionage against the private sector may not rise to a use of force. U.S. adversaries are likely uncertain about the gap between what the USG claims as its cyber deterrence policy and what actions the USG deems unacceptable. Further, an adversarial state may initiate an action that is not regarding as deserving a response from the U.S. military or intelligence community (e.g., cyber economic espionage). The problem with attribution, threshold, and whether it was a legitimate target may mean the largest and most sophisticated cyber arsenal is not deployed in responding to actions taken below the

30 Jardine [5].

31 Denning [29].

32 Nye [8].

33 Andres [30].

34 Gartzke and Lindsay [31].

35 Pomerleau [32].

36 U.S. Department of Defense [33].

37 U.S. Department of Defense [34].

level of armed conflict. Therefore, confusion persists about exactly what is meant by cyber deterrence and how it would be operationalized.

As articulated by the "Cybersecurity and Strategic Deterrence" panel, there are several ways in which to conceptualize and operationalize deterrence. An actor may deter by active defense; cybersecurity (denial); interception and warning; sanctions and punishment; criminal prosecution; leaking information (e.g. doxing); or attacking digital infrastructure (punishment).[38] For each of these deterrence options to be effective, the threatening party must be willing to respond to demonstrate their capability and credibility.[39] The party that is demonstrating their capability also has to consider what the opposing party may do in response and properly scale their response.[40] In such a case, does an nation-state adversary consider the lack of cyber resiliency in the private sector? If so, does it hold that the private sector should be more accountable for its cybersecurity and therefore a legitimate target? Or does the adversary calculate that the private sector is not accountable since the state did not mandate cyber resiliency and therefore perhaps not a legitimate target? What if the private sector employs infrastructure that the USG utilizes? Is that infrastructure a legitimate target? How responsible is the adversarial actor if the action taken against the state on the shared infrastructure has cascading effects against the private sector (e.g., a power plant that powers both a military base and a hospital)?

Taken further, how does a state conduct a digital battle assessment and respond with a proportional response? Does the attacked state keep its response in cyberspace so as not to escalate further? Does a state respond to cyber espionage against the public sector? If so, how does a state know it was cyber espionage as opposed to (digital) preparation of the battlefield? As noted by McConnell, the only thing that separates computer network exploitation from an offensive cyberspace operation is intent.[41] Therefore, each of these questions needs to be thoroughly thought through in calibrating a response to an adversarial cyberspace action. A mis-calibrated or misfired signal adds little to effective deterrence.[42]

For deterrence to be effective, the threatening party must be willing to respond.[43] If a party is making a legitimate threat, it must have the capability and resolve to carry out its threat.[44] Short of both, deterrence does not exist, and an adversary will cross red lines. Timing also matters. If a state does not respond to an act once it has laid down a red line, it demonstrates that the activity was worth the risk, and not acting sets a precedent that the same act can occur again without consequence.[45]

The offensive focus (punishment) in cyberspace has had at least two negative outcomes. First, the emphasis on offensive capability has caused a global cyber arms

---

[38] Cyber Florida [4].

[39] Cyber Florida [4].

[40] Libicki [35].

[41] Cyber Florida [4].

[42] Kolton [36].

[43] Cyber Florida [4].

[44] Hennessey [37].

[45] Cyber Florida [4].

race, which in turn has made developing malware and other vulnerabilities a very lucrative trade.[46] It is much more profitable to work as a black hat hacker selling exploits to malign actors than a white or gray hat hacker working to secure systems. Second, the focus on stockpiling cyber exploits may have caused the USG not to alert private companies about vulnerabilities since alerting them could compromise stockpiled U.S. cyber capabilities. This mindset has changed slightly recently as the USG has begun publicly issuing warnings about cyber threats.[47,48]

Perceptions of the Target State

U.S. adversaries, namely China, Russia, Iran, and North Korea view cyberspace very differently than the USG. Focusing on just China and Russia, they view cyberspace more holistically than does the United States which separates warfare into separate domains.[49] For instance, the Chinese and Russian governments' view cyberspace in terms of information control. Both the Xi and Putin-led governments view the use of cyberspace actions, to include espionage, information warfare, and offensive cyberspace operations as part of their governments applying information as an instrument of war against the U.S.[50] These states view themselves as currently engaged in an information war with the U.S., and that their actions are not raising to a level in which the U.S. military would respond. These adversaries surely believe their current behavior is worth the risk.

The United States and its adversaries also differ in cyberspace regarding whom they consider valid actors in that space. To the USG, it endeavors to keep the Internet a commercial space wherein the only entities engaging in offensive cyberspace actions are the Department of Defense and some U.S. intelligence agencies. According to the so-called Gerasimov Doctrine, Russia sees the United States as working through non-governmental actors,[51] such as the International Republican Institute, and using U.S. technology companies to spread U.S. influence. Hence, to the Russians, it is fair game to employ their own non-governmental actors, to include "patriotic hackers" and cyber criminals, against the United States as part of their overall defense strategy or using disinformation platforms like Russia Today (RT) to spread their preferred narrative. To Russia, they are mirroring the U.S., and using all elements of national power to win the information battle before a kinetic battle can occur.

Perceptions of the actor also are driven by the motivations of the state. What motivates the state? Insecurity? Prestige? Material gain? Ideational factors? How the state views itself will color how it views other actors in the system. Returning

---

[46] Perlroth [38].

[47] U.S. Department of Justice [39].

[48] Vavra [40].

[49] Grzegorzewski and Marsh [41].

[50] Grzegorzewski and Marsh [41].

[51] Denison [42].

to the Russia example, it views itself as under threat due its current weak position in the international system. The Russian state excuses its weakness by blaming its adversaries for its difficulties rather than focusing on its own structural problems. Moreover, in a world in which Russia went from superpower to a middle power overnight, it sees conspiracy all around it to explain its shocking fall in the international system. If the International Republican Institute (IPR) comes to Ukraine to oversee free and fair elections,[52] the Russian government interprets it to mean the IPR are there to subvert the election in favor of the Western candidate (as opposed to the people turning against a corrupt regime propped up by Russia). In affirming it matters as to what motivates a state, this observation does not mean that the USG should change its actions to placate Russia's misperceptions. It does mean that the USG should factor in how an adversary views actions in the system. If the U.S. does not understand how Russia views the world, it risks sending a signal that does generate the desired outcome. A nation-state cannot effectively influence an actor if it does not truly understand what motivates that actor.

Signaling

As noted, misperceptions fuel antagonisms between the U.S. and Russia. The Russian government, as outlined in the Gerasimov Doctrine, wildly misperceives how the U.S. organizes itself in defense of the nation. Given this wildly inaccurate misperception of the USG, there will surely be other signals missed in employing a deterrence strategy. As noted in January 2021, the U.S. may in fact have signaled to America's adversaries about its intentions in cyberspace.[53] These signals of U.S. resolve and capability, which were very public, committed the U.S. to a course of action if its adversaries did not cease their cyberspace operations and activities. It is debatable whether the Russians understood these signals.

Conversely, signals can also not be received; may not be strong enough to deter an adversary; or an adversary may choose to go after other targets. For instance, while in 2016 the USG did not find "persistent Russian cyber efforts to gain access to election infrastructure," the Russian government did work to "undermine public confidence in the electoral process, and exacerbat[e] sociopolitical divisions in the U.S." These outcomes may demonstrate that deterrence was present in some form against election (hard) infrastructure, while at the same time Russia perceived the U.S. information space (soft) was a legitimate target of influence.

For an effective signal to be received by adversaries, the U.S. must understand their perceptions and vice-versa. Per Nye, states can achieve this understanding through what he labels entanglement.[54] Entanglement entails increased interactions with adversaries, and over time these iterative exchanges make signaling in and

[52] Druckman [43].

[53] Grzegorzewski [44].

[54] Nye [8].

through cyberspace more predictable, and stable. While we wholeheartedly support hard power in international affairs, leading the digital future will require much more than offensive cyberspace capabilities. The U.S. must also convince others to follow. The U.S. must advocate via soft power the idea of a free and prosperous Internet.[55] In that vein, the entanglement proposed by Nye would both allow for a better understanding of cyberspace signaling and could shape what is deemed as acceptable deterrence measures in cyberspace.

The Nuclear Weapons Analogy

Finally, any serious discussion on cyber deterrence must address the nuclear deterrence analogy. VADM Mike McConnell in a 2010 *Washington Post* op-ed compared cyber deterrence to the nuclear deterrence exercised during the Cold War in "terms of the potential economic and psychological effects."[56] The comparison to Cold War nuclear deterrence was critiqued for "import[ing] the results of the Cold War era process of strategy development into contemporary discussions of cyber war" and for "see[ing] global cyber conflict in bi–polar terms."[57] The critique itself misses the point, taking the analogy of nuclear deterrence far too literal. The op-ed was meant to alert the U.S. to the dangers of not considering how cyberspace operations impact U.S. national security. The reference to nuclear deterrence was a call for action and meant to stress the urgency of the situation. While that op-ed was written in 2010, the USG still has lagged in incorporating cybersecurity and cyber operations into overall deterrence.

The reasoning we proffer, in cyberspace the USG has relied too heavily upon the analogy of nuclear deterrence.[58] While nuclear deterrence during the Cold War was incredibly successful (in which luck played no small part), applying that deterrence model to cyberspace is not a perfect fit.[59,60] For instance, one of the main lessons senior leaders appear to have learned from Cold War nuclear deterrence is that successful deterrence is demonstrated by stockpiling offensive cyber capabilities.[61] This rationale dates back to the Cold War where having more nuclear weapons than an adversary provided a state with second strike capability thereby making it irrational for the first state to launch an attack and not survive the return fire. In cyberspace, this is a mistaken parallel. Having the most and best cyber capabilities in an arsenal will not necessarily deter an adversary from acting. In cyberspace, cyber resiliency is a much effective deterrent since defense can offset a cyber capability

[55] Cilluffo et al. [45].

[56] McConnell [46].

[57] Lawson [47].

[58] Iasiello [48].

[59] Andres [30].

[60] Goodman [49].

[61] Perlroth [38].

whereas an additional offensive capability cannot offset an adversary's offensive cyber capability, and in fact may drive an escalation cycle.

Also, in cyberspace, there remain troubles with attribution. During the Cold War, if a nuclear weapon was fired, the world could determine where it came from. The nature of cyberspace, with its ubiquitous connections, makes attribution difficult (but not impossible). As adversaries take over friendly networks, use increasingly complex encryption, use of proxies, and even conduct false flag attacks,[62] a cyberattack may remain a mystery, as well as the offending actor. The digital stealth of cyber-capable nation-states usually means that, at least initially, only one party knows that an action occurred.[63] As noted, when this action is not responded to in a timely matter or made public, the time lapse in responding may indicate to the offending party that a target was not worth defending.

While the nuclear weapon analogy is not perfectly transferrable, it still does have some limited usefulness. The analogy still holds that to respond to action you need to have near perfect attribution. A state would not launch a retaliatory nuclear strike unless it was 100% certain of where the attack originated from. That is to say, the USG would not retaliate with D4M (deny, degrade, disrupt, destroy, manipulate) cyberspace operations unless it was certain a specific adversary launched an attack. The nuclear analogy also holds in that the USG must re-engineer its systems to harden against attack. The USG must re-consider the implications of a weapon that, unlike nuclear weapons, are available to anyone with an Internet connection and device. As such, it is about changing the frame of reference by appealing to the stunningly destructive way in which digitized weapons can impact the world in milliseconds just as nuclear weapons could change the world within 30 min.

Conclusion

Establishing cyberspace operations as a tool of cross-domain deterrence remains a challenge for policymakers and decision-makers. For over the past 30 years, the world has been living in a digital age in which space and time is compressed. In this digital age, it has presented a great many opportunities and challenges to both peoples and states. These opportunities have allowed for the dispersion of power in the system away from a hierarchically ordered system to a flatter international system. Consequently, the U.S. can no longer maintain dominance through stockpiling the most and the best military capabilities. Cyberspace has opened new avenues for actors across the world to compete with and challenge the U.S. The USG must rethink both how it conducts itself in cyberspace and how to deter both new and old actors from taking actions that would harm U.S. interests. In this chapter we proposed three ways to increase cyber deterrence: education, public–private partnerships, and modernization of laws. Each of these suggestions would increase U.S. digital resiliency and,

62 Greenberg [50].

63 Chen [51].

in the process, increase the chances that the USG could deny, and deter, an actor in cyberspace. As it stands, the U.S. cannot effectively deter through cyberspace out of concern that our losses would be disproportionate to the damage it inflicts. Therefore, the U.S. must begin today to change the risk versus reward calculations of U.S. adversaries, not by focusing on offensive capabilities but rather by reinforcing its own wetware, relationships, and infrastructure.

References

1. SonicWall. (2021). *2021 SonicWall cyber threat report*. SonicWall. https://www.sonicwall.com/resources/white-papers/2021-sonicwall-cyber-threat-report/.
2. Internet Crime Complaint Center (IC3). (2020). *2020 Internet crime report*. https://www.ic3.gov/Media/PDF/AnnualReport/2020_IC3Report.pdf.
3. C-SPAN. (2018, March 1). *NSa nominee testifies senate armed services committee*. http://www.c-span.org%2Fvideo%2F%3F441917-1%2Fnsa-nominee-testifies-senate-armed-services-committee&t=Pending+Nominations%20(C=SPAN%20Video%20Library).
4. Cyber Florida. (2021, April 24). *2021 great power competition conference: strategic deterrence panel*. https://www.youtube.com/watch?v=pAJqsDV5lMs.
5. Jardine, E. (2020). *Optimizing cyber deterrence*.
6. Voo, J., Hemani, I., Jones, S., DeSombre, W., Cassidy, D., & Schwarzenbach, A. (2020). *National cyber power index 2020*. Belfer Center for Science and International Affairs, Harvard Kennedy School.
7. Ang, C. (2021, May 10). *The most significant cyber attacks from 2006–2020, by Country*. Visual Capitalist. https://www.visualcapitalist.com/cyber-attacks-worldwide-2006-2020/.
8. Nye, J. S., Jr. (2016). Deterrence and dissuasion in cyberspace. *International Security, 41*(3), 44–71.
9. Computer Security Resource Center. (2022). *Glossary*. Information Technology Laboratory. https://csrc.nist.gov/glossary/term/cyberspace_defense.
10. Computer Security Resource Center. (2022). *Glossary*. Information Technology Laboratory. https://csrc.nist.gov/glossary/term/cyber_resiliency.
11. Crosston, M. D. (2011). World gone cyber MAD: How "mutually assured debilitation" is the best hope for cyber deterrence. *Strategic Studies Quarterly, 5*(1), 100–116.
12. Kumar, S.U. (2020, May 26) Cyber resilience: A new way of looking at cybersecurity. *Security Magazine*. https://www.securitymagazine.com/articles/92456-cyber-resilience-a-new-way-of-looking-at-cybersecurity.
13. Brands, H. (2021, August 9). *To counter China, U.S. needs to mobilize academia*. Bloomberg Opinion. https://www.bloomberg.com/opinion/articles/2021-08-09/to-counter-china-u-s-needs-to-mobilize-academia.
14. Vogels, E. A., & Anderson, M. (2019, October 9). *Americans and digital knowledge*. Pew Research Center. https://www.pewresearch.org/internet/2019/10/09/americans-and-digital-knowledge/.
15. Proofpoint. (2021). *Threat report: The human factor 2021*. Proofpoint. https://www.proofpoint.com/us/resources/threat-reports/human-factor
16. Truitte, K. (2019, August 23) An unpatchable exploit: The human vulnerability in cybersecurity. *Georgetown Security Studies Review*. https://georgetownsecuritystudiesreview.org/2019/08/23/an-unpatchable-exploit-the-human-vulnerability-in-cybersecurity/.
17. Mitchell, K. (2021, July 8). GW students mobilize to counter COVID-19 misinformation on social media. *GW Today*. https://gwtoday.gwu.edu/gw-students-mobilize-counter-covid-19-misinformation-social-media.

18. Horwitz, M. J. (1981). History of the public/private distinction. *University of Pennsylvania Law Review, 130*, 1423.
19. Starr, P. (2007). *Freedom's power: The true force of liberalism*. New York: Basic Books.
20. Healey, J., & Jervis, R. (2021, March 22). *Overclassification and its impact on cyber conflict and democracy*. Modern War Institute. https://mwi.usma.edu/overclassification-and-its-impact-on-cyber-conflict-and-democracy/.
21. Doherty, P., Pflaging, J., Ware, B., & Weatherford, M. (2021, May 12). *Cybersecurity opportunities for the public and private sectors: 100 days in cybersecurity perspectives on the biden administration*. Cynergy Partners (white paper). https://www.coalfire.com/documents/whitepapers/cynergy-partners-white-paper.
22. Williams, B.D. (2021, March 25). *Nakasone warns adversaries hack unseen in US*. Breaking Defense. https://breakingdefense.com/2021/03/nakasone-warns-adversaries-hack-unseen-in-us/.
23. Atkins, S., & Lawson, C. (2021). Cooperation amidst competition: Cybersecurity partnership in the US financial services sector. *Journal of Cybersecurity*, 7(1), tyab024.
24. Schneier, B. (2021, May, 24). Russia's hacking success shows how vulnerable the cloud is. *Foreign Policy*. https://foreignpolicy.com/2021/05/24/cybersecurity-cyberattack-russia-hackers-cloud-sunburst-microsoft-office-365-data-leak/.
25. Octavian Report. (n.d.). *Fight of our lives: Michael Hayden on intelligence security and transparency*. Octavian Report. https://octavianreport.com/article/michael-hayden-on-intelligence-security-and-transparency/.
26. Greenberg, A. (2021). *Hackers tied to Russia's GRU targeted the US grid for years, researchers warn*. Wired. https://www.wired.com/story/russia-gru-hackers-us-grid/.
27. Valeriano, B., & Maness, R. C. (2015) *Cyber war versus cyber realities: Cyber conflict in the international system*. Oxford University Press.
28. Gartzke, E. (2013). The myth of cyberwar: Bringing war on the Internet back down to earth. *International Security, 38*(2) (Fall 2013), pp. 41–73.
29. Denning, D. E. (2015). *Rethinking the cyber domain and deterrence*.
30. Andres, R. (2012). The emerging structure of strategic cyber offense, cyber defense, and cyber deterrence. *Cyberspace and National Security: Threats, Opportunities, and Power in a Virtual World*, 89–104.
31. Gartzke, E., & Lindsay, J. R. (Eds.). (2019). *Cross-domain deterrence: Strategy in an era of complexity*. Oxford University Press.
32. Pomerleau, M. (2019, April 30). *Is there such a concept as 'cyber deterrence?'* Fifth Domain. https://www.fifthdomain.com/dod/2019/04/30/is-there-such-a-concept-as-cyber-deterrence/.
33. U.S. Department of Defense. (2019). *DOD dictionary of military and associated terms*. U.S. Department of Defense. https://www.jcs.mil/Portals/36/Documents/Doctrine/pubs/dictionary.pdf?ver=2019-08-06-101717-717.
34. U.S. Department of Defense. (2018). *Summary: Department of defense cyber strategy*. U.S. Department of Defense. https://media.defense.gov/2018/Sep/18/2002041658/-1/-1/1/CYBER_STRATEGY_SUMMARY_FINAL.PDF
35. Libicki, M. C. (2021). *Cyberspace in peace and war*. Naval Institute Press.
36. Kolton, M. (2017). Interpreting China's pursuit of cyber sovereignty and its views on cyber deterrence. *The Cyber Defense Review, 2*(1), 119–154.
37. Hennessey, S. (2017). Deterring cyberattacks: How to reduce vulnerability. *Foreign Affairs, 96*, 39.
38. Perlroth, N. (2021). *This is how they tell me the world ends: The cyberweapons arms race*. Bloomsbury Publishing USA.
39. U.S. Department of Justice. (2018, May 23). Justice department announces actions to disrupt advanced persistent threat 28 botnet of infected routers and network storage devices. *Justice News*. https://www.justice.gov/opa/pr/justice-department-announces-actions-disrupt-advanced-persistent-threat-28-botnet-infected.
40. Vavra, S. (2019, July 10). Why cyber command's latest warning is a win for the government's information sharing efforts. *Cyberscoop*. https://www.cyberscoop.com/cyber-command-information-sharing-virustotal-iran-russia/.

41. Grzegorzewski, M., & Marsh, C. (2021, March 15). *Incorporating the cyberspace domain: How Russia and China exploit asymmetric advantages in great power competition.* Modern War Institute. https://mwi.usma.edu/incorporating-the-cyberspace-domain-how-russia-and-china-exploit-asymmetric-advantages-in-great-power-competition/.
42. Denison, B. (2020, July 29). *Where US sees democracy promotion, Russia sees regime change.* Russia Matters. https://www.russiamatters.org/analysis/where-us-sees-democracy-promotion-russia-sees-regime-change.
43. Druckman, M. (2020, November 3) *Ukrainian local election results reflect country's decentralized* democracy. Ukraine Matters. https://www.atlanticcouncil.org/blogs/ukrainealert/decentralized-democracy-ukrainian-local-elections/.
44. Grzegorzewski, M. (2020, January 9). *Why US officials are revealing more about Cyber Ops.* Defense One. https://www.defenseone.com/ideas/2020/01/why-us-officials-are-revealing-more-about-cyber-ops/162341/.
45. Cilluffo, F. J., Cardash, S. L., & Salmoiraghi, G. C. (2012). A blueprint for cyber deterrence: Building stability through strength. *Military and Strategic Affairs, 4*(3), 3–23.
46. McConnell, M. (2010, February 28). Mike McConnell on how to win the cyber-war we're losing. *Washington Post.* https://www.washingtonpost.com/wp-dyn/content/article/2010/02/25/AR2010022502493_pf.html.
47. Lawson, S. (2012). Putting the "war" in cyberwar: Metaphor, analogy, and cybersecurity discourse in the United States. *First Monday.*
48. Iasiello, E. (2014). Is cyber deterrence an illusory course of action? *Journal of Strategic Security, 7*(1), 54–67.
49. Goodman, W. (2010). Cyber deterrence: Tougher in theory than in practice? *Strategic Studies Quarterly, 4*(3), 102–135.
50. Greenberg, A. (2019, October 17). The untold story of the 2018 Olympics cyberattack, the most deceptive hack in history. *Wired.* https://www.wired.com/story/untold-story-2018-olympics-destroyer-cyberattack/.
51. Chen, J. (2017). Cyber deterrence by engagement and surprise. *PRism, 7*(2), 100–107.

Hon J. Michael Mike McConnell is executive director of Cyber Florida. He is re-tired U.S. Navy vice admiral and former director of the National Security Agency (NSA). McConnell's distinguished career spans over 50 years, focused on international and foreign intelligence issues, including 29 years as a U.S. Navy intelligence officer. In 2007, he became the second U.S. Director of National Intelligence, serving under Presidents George W. Bush and Barack Obama. From 1992 to 1996, he served as NSA director. McConnell has received the nation's highest award for service in the U.S. intelligence community from both Presidents Bill Clinton and George W. Bush. He's also been vice chairman of Booz Allen Hamilton, a management and information technology consulting firm.

Dr. Mark Grzegorzewski serves as full-time faculty with Joint Special Operations University, and holds a Ph.D., M.A., and B.A. in Political Science from the University of South Florida (USF), along with a graduate certificate in Globalization Studies. Before coming to SPC, Dr. Grzegorzewski worked for the US government as an intelligence analyst. During that time, he also taught several Political Science classes at USF.

Examining Systemic Risk in the Cyber Landscape

Anthony Masys

Abstract Cyber attacks and incursions have certainly emerged as a national security issue. Globally we are seeing the effects of such attacks not only on the financial domain but also in healthcare, government and critical infrastructure (Masys in Networks and network analysis for defence and security. Springer Publishing, 2014, [19]; Masys in Sensemaking in security. Springer Publishing, 2021a, [24]). Understanding the extent of the impact of cyber incursions and attacks requires understanding the systemic cyber risks '…of risks spreading across interdependent systems' (Welburn et al., in Risk Analysis: An International Journal, 2021, [39]). The World Economic Forum (2016) [36] defines Systemic cyber risk as '…the risk that a cyber event (attack(s) or other adverse event(s)) at an individual component of a critical infrastructure ecosystem will cause significant delay, denial, breakdown, disruption or loss, such that services are impacted not only in the originating component but consequences also cascade into related (logically and/or geographically) ecosystem components, resulting in significant adverse effects to public health or safety, economic security or national security'. Lucas et al. (Advanced Theory and Simulations, vol 1, 2018, [15]) argues that '…systemic risk refers to a potential collapse of a system of potentially global importance and criticality to services that humans urgently need. This dimension of a large potential threat within a complex web of interacting elements distinguishes systemic from other types of risk'. As such traditional risk management approaches are not sufficient for dealing with them IRGC (Guidelines for the governance of systemic risks. International Risk Governance Center, Lausanne, p 5, 2018, [9]). This chapter explores systemic risk across the cyber landscape through the non-traditional security lens (Masys in Handbook of security science. Springer, 2021b, [25]) and presents applications of systems thinking, scenario planning and High Reliability Security Organizations to support systemic risk awareness and management.

Keywords Systemic risk · Cyber crime · Scenario planning · Vulnerability analysis

---

A. Masys (✉)
University of South Florida, Tampa, FL 33620, USA
e-mail: anthony.masys@gmail.com; tmasys@usf.edu

A. Farhadi et al. (eds.), *The Great Power Competition Volume 3*,
https://doi.org/10.1007/978-3-031-04586-8_4

Introduction

Events such as the WannaCry, Solar Winds cyber-attack, global terrorism, serious and organized crime, disease vectors, and natural disasters create challenges that affect both global and national security interests. Today's cyber threats both from organized crime or state actors pose threats to national security. As described in Masys [25], such events are shaping the security calculus across dimensions such as health security, economic security, food security and energy security emerging as interrelated concepts that characterize the security landscape as complex. The increased transnational flow of people, goods, money and information as products of 'globalization' has also changed the security landscape in terms of the 'globalization' of risks.

This transnational/transborder nature of security challenges our traditional views of national security characterized by state-based with military dimensions. The non-traditional security calculus thereby emerges as part of the security landscape that can often have significant national security impacts through the implications associated with systemic risks. The systemic cyber risks represent a clear and present threat to societal functions and national security. The cyber threat actors involved in such attacks can be categorized across the following: Nation-States; Proxy Organizations; Cyber crime; Hacktivist; Insider [10]. This chapter examines the systemic risks and the links with cyber security through a non-traditional security risk centric lens and explores the notion of systemic risk as part of the security calculus. As described in Masys [24], 'there is an urgent need to re-examine our exposure to cyber security risks and vulnerabilities in order to transform our understanding and perception of security. This essentially questions how we make sense of global security risks, vulnerabilities and impacts. Key is understanding the security landscape in terms of inherent systemic risks'.

The realization of systemic risks is shaping the cyber discourse as we see the impacts crossing domains such as health security, energy security, water security, food security, economic security, environmental security and human security. According to Helbing [6], we are increasingly living in a world which creates 'hyper-risks' because of numerous networks and interdependencies. In this 'hyper-connected world' with interconnected social/technical/political/economic domains, shocks to regional, national and global systems can have significant security implications. Societal functions and systems are interdependent thereby creating inherent vulnerabilities to systemic risk and cascading impacts. As cited in Masys [20: 143], the 'hyper-connectivity' characterized by the global pervasiveness of internet and cyber usage has provided a conduit for threats to national security as described in the United Nations Office on Drugs and Crime [34].

As described in Masys [19: 272], cyber attacks on or affecting critical infrastructure interdependencies supporting societal functions such as power generation grids, telecommunication, and transportation can have significant social disruption. What emerges is the realization that recent cyber attacks have revealed the inherent vulnerabilities hiding in the many interdependencies across different infrastructures. The World Economic Forum [36] defines it as follows: "Systemic cyber risk is the risk that a cyber event (attack(s) or other adverse event(s)) at an individual component of a critical infrastructure ecosystem will cause significant delay, denial, breakdown, disruption or loss, such that services are impacted not only in the originating component but consequences also cascade into related (logically and/or geographically) ecosystem components, resulting in significant adverse effects to public health or safety, economic security or national security'. Lucas et al. [15] argues that '…systemic risk refers to a potential collapse of a system of potentially global importance and criticality to services that humans urgently need. This dimension of a large potential threat within a complex web of interacting elements distinguishes systemic from other types of risk'. Traditional risk management practices are not sufficient for dealing with them [9: 5]. Systemic Risk perception and communication play an important role in the management of cyber threats and risks. With this in mind, systems thinking and complexity science inform the analytical and perceptual framework and provide a foundational perspective in which to bring multidisciplinary discourse to understanding systemic risks.

Systemic Risks

The COVID-19 pandemic has certainly stress tested our societal systems and had reignited interest and research into systemic risks and certainly illustrates the transnational nature of today's security landscape [24, 25]. As described in Saha et al. [29: 112].

'COVID-19 has firmly established itself as the single largest security disrupter of this century in the non-traditional sense. It has necessitated a recalibration of securitisation framework…'. Systemic risk emerges from the interconnectedness of the networks underlying a system and result from the interactions of individual risks resulting in cascades of secondary and tertiary failures. Renn [28] describes the characteristics of systemic risks comprised of: high complexity, trans-boundary effects, stochastic relationships, and nonlinear cause-effect patterns with tipping points and often associated with less public attention than they deserve. Systemic risks are defined in Centeno et al. [3: 68]: '… as the threat that individual failures, accidents, or disruptions present to a system through the process of contagion. This is the risk that unexpected and unlikely interactions may lead to unpredicted threats to system survival'. With this in mind, the IRGC [9] states that 'external shocks to interconnected systems, or unsustainable stresses, may cause uncontrolled feedback and cascading effects, extreme events, and unwanted side effects, implying that the potential for cascading disruption is a growing and critical concern for many facets of daily life'. The impacts of such shocks can be realized across the spectrum of non-traditional security domains: health security, economic security, food security; water security; energy security; environmental security; human security. The interconnectedness associated with societal systems makes predicting the probability, cost and magnitude of risk problematic. The COVID-19 pandemic certainly resonates with this problem space of systemic risks [24, 25].

Cyber Systemic Risk

Whether theft-related, disruption related or damage related, cyber attacks can be impactful across societal systems. As detailed in Montasari et al. [26: 71], 'the speed, diversity and frequency of attacks are generating cyber security challenges that have never been witnessed before'. WEF Global Risks Report [37] ranks cyber risks as among the top global risks. In one estimate, the global cost of cyber crime by 2025 is said to be upwards of $10 Trillion.[1] Recent cybersecurity events have highlighted and elevated the profile of cyber risk and systemic cyber risk. WannaCry, Solar Winds and the ransom attack on Colonial Gas come to mind. The WannaCry ransomware attack of May 2017 affected computer systems in more than 150 countries. The WannaCry cyber incident, which originated outside the financial system, caused significant disruption across several countries and industrial sectors, with an impact on critical infrastructure and public services and are discussed in detail [4].

In late 2020, Solar Winds, the American cyber-security community, discovered a widespread breach of private-sector and government networks including the Department of Homeland Security and Treasury Department. In 2021, the cyber attack on Colonial Pipeline took down the largest fuel pipeline in the U.S. and led to shortages across the East Coast. These examples are illustrative of the systemic vulnerability that exists across interdependent critical infrastructure and societal functions.

Systemic Risks and Nontraditional security Landscape

Threats and risks to national security have become increasingly non-military in nature. The recent ransomware cyber attack on May 7, 2021 targeting the US Colonial Pipeline that impacted computerized equipment managing the pipeline highlights the fragility and vulnerability of critical infrastructure revealing national security implications. Such events challenge traditional notions of security and highlight the requirement to better understand systemic risks and the inherent societal vulnerabilities, intervention strategies and resilience by design. Helbing [6: 51] argues that:

> Today's strongly connected, global networks have produced highly interdependent systems that we do not understand and cannot control well. These systems are vulnerable to failure at all scales, posing serious threats to society, even when external shocks are absent. As the complexity and interaction strengths in our networked world increase, man-made systems can become unstable, creating uncontrollable situations even when decision-makers are well-skilled, have all data and technology at their disposal, and do their best.

Critical infrastructure is nationally defined (see Table 1a, b for CI sectors: Canada and US). Within the context of COVID-19 we have seen how not only the virus

---

[1] https://securityboulevard.com/2021/03/cybercrime-to-cost-over-10-trillion-by-2025/.

Table 1 a, b: Critical infrastructure sectors for US and Canada

United States critical infrastructure sectors[2]
Chemical sector
Commercial facilities sector
Communications sector
Critical manufacturing sector
Dams sector
Defense industrial base sector
Emergency services sector
Energy sector
Financial services sector
Food and agriculture sector
Government facilities sector
Healthcare and public health sector
Information technology sector
Water and wastewater systems sector
Nuclear reactors, materials and waster sector
Transportation system sector
Canada[3] critical infrastructure sectors
Energy and utilities
Finance
Food
Transportation
Government
Information and communication technology
Health
Water
Safety
Manufacturing
Energy and utilities
Finance
Food
Transportation

(exogenous shock) but also management of the pandemic (endogenous shock) has created cascading impacts across societal interdependent functions [27]: Public Health, Medical Care; Water and Sanitation; Shelter and Clothing; Food and Nutrition; Energy Supplies; Public Works and Engineering; Social Structure; Logistics

[2] https://www.cisa.gov/critical-infrastructure-sectors.

[3] https://www.publicsafety.gc.ca/cnt/rsrcs/pblctns/srtg-crtcl-nfrstrctr/index-en.aspx.

And Transportation; Security; Communications; Economy; and Education, which are dependent on functioning critical infrastructure.

From the view of societal functions and critical infrastructure, a fundamental property of interdependent networks is that failures within one network may cascade across nodes in other networks [19]. Investigating systemic risk or vulnerabilities goes beyond the usual agent-consequence analysis and focuses on interdependencies and spill-over risk clusters [11: 37]. This is certainly illustrated with the COVID-19 pandemic as discussed in detail in Masys [24, 25]. Weick and Sutcliffe [38: 2] highlight how such an event can be '...considered as an abrupt and brutal audit: at a moment's notice, everything that was left unprepared becomes a complex problem, and every weakness comes rushing to the forefront'. The WannaCry, Solar Winds and Colonial Pipelines cyber attacks certainly reflect this.

One of the primary features underlying non-traditional security is the inherent global interdependency and interconnectivity. Foreshadowing the current impact of COVID-19, Helbing [6: 51] argues that: '...When analysing today's environmental, health and financial systems or our supply chains and information and communication systems, one finds that these systems have become vulnerable on a planetary scale. They are challenged by the disruptive influences of global warming, disease outbreaks, food (distribution) shortages, financial crashes, heavy solar storms, organized (cyber-) crime, or cyberwar'. This points to a requirement for a paradigm shift in how we view the world and in our thinking. As discussed, traditional risk management approaches are ill suited for systemic risks. Network thinking, as described in Barabasi [1] opens novel perspectives to understanding complex systems. This network thinking mindset leverages a topological analysis which is based upon classical graph theory through which interesting properties of the structure of a network system can be revealed.

Building upon the network mindset, the application of Actor Network Theory (ANT), as described in Masys [19], can be instrumental in assessing vulnerabilities (structure and dynamics) associated with complex societal systems. Such analysis leverages the application of scenario planning as described in Masys [16] and network analysis Masys [17, 18] to address such questions as:

How do we define the boundaries of the system
What are the relevant threats and hazards associated with the system (an examination of the space of possibilities)
How is resilience realized within the system
How do the interdependencies shape the structure and dynamics of the system
How do we manage the uncertainties

Such an approach to vulnerability analysis examines the robustness and resilience of the network against random failures and targeted attacks. This is of particular importance considering the impact that critical infrastructure failures can have on national security, economy, health, safety and social well-being. Hence risk management within the cyber systemic risk landscape requires a systems thinking lens to support the development of a High Reliability Security Organizations (HRSO) [23].

Discussion

As described in Xu and Masys [40: 178] 'today there exists a growing dependence on the continuous flow of essential goods and services provided by infrastructure systems such as energy (electricity, oil and gas supply), information and telecommunication, transportation (by road, rail, air and sea) and water supply including waste water treatment'. Kroger and Zio [11: 1] regard these critical infrastructure '...so vital to any country that their incapacity or destruction would have a debilitating impact on the health, safety, security, economics, and social well-being'.

A network mindset that encompasses a systemic analysis and vulnerability analysis approaches and methods have been covered in detail in Disaster Forensics [21] and Xu and Masys [40] and Masys [22] leveraging Actor Network Theory. As noted in Goldin and Mariathasan [5: 66] and captured in the disaster forensics approach '...systemic analysis must examine nodes, pathways, and the relationships between them, because catastrophic changes in the overall state of the system can ultimately derive from how it is organized- from feedback mechanisms within it and from linkages that are latent and often unrecognized'. Systems thinking is rooted in this approach.

Systems Thinking

Systems thinking emerges as a key lens in which to better understand systemic risks. Such an approach supporting vulnerability analysis focuses '...in seeing interrelationships rather than linear cause-effect chains and seeing processes of change rather than snapshots' [30]. In this sense, systems thinking reveals the complexities and interdependencies in systems. Cyber attacks half way around the world can cause disruptions globally through supply chain disruptions. Like the COVID-19 pandemic, this poses a national security threat given the globalized world we live in. With such interconnectivity 'a cyber attack anywhere is a cyber attack everywhere'. Systems thinking is thereby about complexity framing [21, 22, 24]. The notion of hyper-risks [6, 19] captures well the interconnectivity and complexity of the security threats. Opening each security node reveals an imbedded interconnected network of societal infrastructure and functions that connects across the various non-traditional security domains. We have seen how the COVID-19 crisis stress-tested national and global societal systems revealing the challenges associated with understanding systemic risks and disaster risk reduction. Complex interdependencies across societal functional systems precipitated cascading primary, secondary and tertiary effects across health, education and economic sectors [24].

Hynes et al. [8: 68] argue that:

> Systems thinking provides a methodology to achieve a better understanding of the behavior of complex systems and to improve the assessment of the consequences of policy interventions. Serious disease outbreaks such as Covid-19 are the result of systemic properties, and in this

> case, emergence in particular. Emergence describes a process whereby a situation arises through the interaction of a number of actors and influences, without any intention to create that situation. Helbing [6] and others have noted that the consequences of failing to appreciate and manage the characteristics of complex global systems and problems can be immense'.

The paradigm of systems thinking permits a view of the world as a complex system in which as noted by Sterman [32: 10] we come to the understanding that 'you can't do just one thing' and that 'everything is connected to everything else'. This is supported by Senge [26: 73] who is of the opinion that the discipline of the systems approach lies in a shift of mind: in seeing interrelationships rather than linear cause-effect chains and seeing processes of change rather than snapshots. Systemic risks are opaque to a linear mindset. Levine et al. [13: 7] argue that '...a system perspective can often reveal how behaviour that is competent from the standpoint of each individual actor does not contribute to achieving the overall goals which collectively all the actors in the 'system' say they are working towards, in different ways.

The systems thinking mindset thereby brings a lens to reveal what has previously been opaque in terms of systemic risks. Leveraging a network/systems thinking mindset, vulnerability analysis can explore the interdependencies and interconnectivity that provide pathways to cascading disruptive events. Scenario planning emerges as a key approach to better understand the complex dynamics of systemic cyber risks.

Scenario Planning

Scenario planning represents a set of tools and methodology to facilitate cyber systemic risk and vulnerability analysis with the purpose not to predict, but to facilitate a change in the mindset. In risk, crisis and disaster management, the use of scenarios stretch and refocus thinking and mental models thereby challenging the assumptions concerning linearity to recognize the fundamental indeterminacy associated with the emergence of black swans and events that stress test societal systems (WanneCry and Solar Winds). A scenario is essentially a story, describing potential future conditions and their emergence to facilitate sense making [24] and to inform decision making. The thought process involved in scenario planning supports "thinking the unthinkable" exploring uncertainty and challenging mental models and assumptions in order to recognize alternate futures in a space of possibilities. As described in Masys [16], risks to critical societal functions may derive from upstream critical infrastructure interdependencies.

Considering the lessons learned from recent cyber attacks, the comments of Kunreuther and Useem [12: 20] certainly resonate:

> ...we need to look further in the future, after the end of the immediate response and into the longer process of recovery. And we need to look further into the past—into things that can be done to avert events (by preventing them altogether), to shape them so that their

consequences are not so severe if they do occur (through efforts of mitigation), or to permit more rapid recovery from disaster consequences that still occur in spite of these efforts.

Hence the application of scenario planning becomes an enabler for risk and vulnerability analysis. In support of scenario planning, red teaming gives voice to alternative perspectives of the problem space. Further details regarding scenario planning and red teaming approaches are covered in Masys [16].

IRGC [9] described seven interlinked steps to support resilience and governance of systemic risks. Of note is the application of scenario planning.

Explore the system, define its boundaries and dynamics
Develop scenarios considering possible ongoing and future transitions
Determine goals and the level of tolerability for risk and uncertainty
Co-develop management strategies dealing with each scenario
Address unanticipated barriers and sudden critical shifts
Decide, test and implement strategies
Monitor, learn from, review and adapt [9: 18].

In crafting scenarios for cyber systemic risks, what we have learned from cyber attacks is that 'Cyber risks cannot be managed in silos, fragmented among specific individuals or departments (e.g., IT department, finance team, legal, etc.) responsible for a piece of an organization's risks with little or no in-between interaction. By leveraging data from entities within and outside their circle, organizations can fully realize the possible extent of their vulnerabilities (if exploited), such as to other sectors or industries; identify clusters of common vulnerabilities and drivers of risk; and evaluate investments in cyber controls to holistically and collectively manage these risks.'[4] With the scenarios we wish to explore not only what the cyber threat is but also what is does across societal systems.

Such complex security challenges within a complex risk network structure and behavior can stress test societal systems and reveal 'latent risk' such as the lack of preparation, insufficient vulnerability analysis and response. This was realized in the COVID-19 pandemic both nationally and globally. Toft and Reynolds [33] use the term 'isomorphic learning' to argue that lessons derived from one setting can be relevant to settings that are similar (i.e. isomorphic) in some respect. Hence, COVID-19 pandemic provides a valuable lens in which to understand cyber threats and their impact on societal systems. Here the national and global experience of COVID-19 on societal systems maps well to understanding inherent systemic risks and vulnerabilities in critical infrastructure. Understanding local, national and global vulnerabilities and fragility in our interdependent societal systems is clearly a learning point from our experience of COVID-19. Burwell et al. [2] argues that:

[4] https://www.cisa.gov/systemic-cyber-risk-reduction.

COVID-19 has confirmed the global vulnerabilities that were repeatedly identified in high-level reports, commissions, and intelligence assessments on pandemic threats for nearly two decades prior to this pandemic. COVID-19 has underscored several truths about pandemics and revealed important shortcomings in current global and national capacities to prepare for, detect, and respond to them.

Fig. 1 Cyber systemic risk management: resilience by design; Resilience by intervention

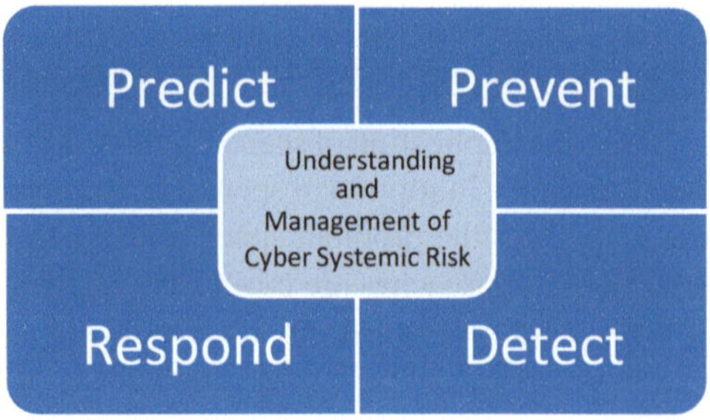

Leveraging lessons learned from the impact of COVID-19 on national and global societal systems, Fig. 1 shows the key elements/phases that comprise systemic risk management of cyber threats. Central to the model is the requirement for understanding the nature of systemic risk. This feeds into and leverages insights from the 4 phases of Predict, Prevent, Detect, Respond. Essentially we approach the cyber systemic risk management with a focus on resilience: Resilience by Design and Resilience by Intervention.

As described in Linkov et al. [14].

The underlying principle of resilience-by-design is a system must be designed to recover its critical functions from disruption on its own or else the system will fail.

The underlying principle of resilience-by-intervention is an external resource must be envisioned and be available to enable a system to recover from cascading and systemic disruptions or else the system will fail.

With this in mind, an overarching organizational architecture and posture that supports the management of cyber systemic risks is the development of High Reliability Security Organizations (HRSO) described in Masys [23]. The concept of an HRSO is derived from the notion of High Reliability Organizations (HROs). Despite the existence of accidents stemming from poor safety cultures, there are organizations that avoid accidents although they are operating in conditions characterized by volatility, uncertainty, complexity and ambiguity. Such organizations are called High Reliability Organizations (HRO). As detailed in Masys [23], 'the roots of the HRO paradigm stem from the examination of aircraft carrier operations, air traffic control and nuclear power operations. Such organizations take a proactive posture in managing complexity and risk. As noted in Spann [31], such organizations '…not only avoid dire consequences by catching problems early, but they have actually proven over time to have disproportionately fewer problems. These organizations do this by consistently noticing the unexpected, reporting it in an honest way, responding quickly and appropriately, learning from the things they did, and improving the process for the next time a challenge arises'.

The HRO construct is divided into five concepts that help organizations manage the threat of failure; absorb damage and surprises [38]. These concepts entail the management of crisis through "Anticipating the Unexpected" and "Containing the Unexpected." These principles of the HRO have been adapted for security operations and conceptualized as High Reliability Security Organizations (HRSO) (see Masys

Table 2 Derived from Masys [23] presents the key principles of an HRSO

High Reliability Security Organization (HRSO)
HRSO principle 1: Preoccupation with failure HRSO explore the threat and intelligence domain to determine the space of possibilities. They encourage the reporting of errors and miscalculations with regards to security assessments, they elaborate experiences of a near miss for what can be learned, and they are wary of the potential liabilities of success, including complacency, the temptation to reduce margins of security
HRSO principle 2: Reluctance to simplify HRSOs take deliberate steps to create more complete and nuanced pictures of what they face and who they are as they face it. Knowing that the world they face is complex, stable, unknowable, and unpredictable. HRSOs position themselves to see as much as possible. They welcome diverse experience, skepticism toward received wisdom
HRSO principle 3: Sensitivity to operations They are attentive to the front line, where the real work gets done. …Anomalies are noticed while they are still tractable and can still be isolated. All of this is made possible because HRSOs are aware of the close ties between sensitivity to operations and sensitivity to relationships
HRSO principle 4: Commitment to resilience The essence of resilience is therefore the intrinsic ability of an organization (system) to maintain or regain a dynamically stable state, which allows it to continue operations after a major mishap and/or in the presence of a continuous stress. The hallmark of HRSO is not that it is incident-free but that incidents don't disable it. Resilience demands a deep knowledge of the threat and space of possibilities (both in time and space)
HRSO principle 5: Deference to expertise HRSOs cultivate diversity, not just because it helps them notice more in complex security environments, but also because it helps them do more with the complexities they do spot. Rigid hierarchies have their own special vulnerability to error and misjudgment
HRSO principle 6: Adaptability HRSOs embrace a Safety II [7] approach that recognizes the need for dynamic and adaptive behavior that is derived from the first 5 principles. HRSOs understand that work as imagined often deviates from work as done. The volatility, uncertainty, complexity and ambiguity that characterize the counter terrorism environment require a certain degree of adaptation to support the very fluid environment. Mindfulness [38] and reflective practices [19] figure prominently

[23]). Here we view the term 'security' in the HRSO context as applying to the non-traditional security paradigm discussed earlier. This thereby reveals an analysis and understanding of systemic cyber risks and the impact on societal systems leading to strategies to support systemic resilience by design and by intervention (Table 2).

Conclusion

The COVID-19 pandemic has clearly shown that:

> The linear agent-consequence analysis is no longer valid in this highly interconnected world where risks migrate and evolve [16, 19]. As such our understanding of their complex aetiology and management must be considered from a 'networked mental model' that recognizes the interdependency-induced vulnerability embedded in the complex structures, processes and

dynamics of modern social-technical systems [35]. However, currently there is a dearth of understanding regarding security, vulnerabilities and the nature of system interdependences in critical infrastructures as it pertains to hyper/hybrid risks and their implications on the security agenda [19]. (cited in Xu and Masys [40])

The COVID-19 pandemic has illustrated the inherent complexity, interconnectivity, vulnerability and fragility that reside within our societal systems whereby systemic risks permeated and defined the security landscape. Applied to the cyber security landscape, understanding the extent of the impact of cyber incursions and attacks requires understanding the systemic cyber risks '…of risks spreading across interdependent systems' [39]. The World Economic Forum [36] defines Systemic cyber risk as '…the risk that a cyber event (attack(s) or other adverse event(s)) at an individual component of a critical infrastructure ecosystem will cause significant delay, denial, breakdown, disruption or loss, such that services are impacted not only in the originating component but consequences also cascade into related (logically and/or geographically) ecosystem components, resulting in significant adverse effects to public health or safety, economic security or national security'. Understanding the inherent vulnerabilities and fragility in our interdependent and interconnected societal systems is key to 'future proofing' them against cyber incursions. Key to that is the application of systems thinking, scenario planning and the establishment of the key principles of High Reliability Security Organizations.

References

1. Barabasi, A. L. (2003). Linked. Plume, New York: Penguin Group.
2. Burwell, S. M., Townsend, F. F., Bollyky, J. T., Patrick, S. M. (2020). Improving pandemic preparedness lessons from COVID-19. Independent Task Force Report No. 78. https://www.cfr.org/report/pandemic-preparedness-lessons-COVID-19/pdf/TFR_Pandemic_Preparedness.pdf.
3. Centeno, M. A., Nag, M., Patterson, T. S., Shaver, A., & Windawi, A. J. (2015). The emergence of global systemic risk. *Annual Reviewing Sociology, 41*, 65–85.
4. ESRB. (2020, February). *Systemic cyber risk*. https://www.esrb.europa.eu/pub/pdf/reports/esrb.report200219_systemiccyberrisk~101a09685e.en.pdf.
5. Goldin, I., & Mariathasan, M. (2014). *The butterfly defect: How globalization creates systemic risks, and what to do about it*. Princeton University Press.
6. Helbing, D. (2013). Globally networked risks and how to respond. *Nature, 497*, 51–59.
7. Hollnagel, E. (2012). *A tale of two Safeties*. http://www.resilienthealthcare.net/A_tale_of_two_safeties.pdf.
8. Hynes, W., Trump, B. D., Kirman, A., Latini, C. & Linkov , I. (2021). Complexity, Interconnectedness and resilience: Why a paradigm shift in economics is needed to deal with Covid 19 and future shocks. In: I. Linkov, J. M. Keenan, B. D. Trump (Eds.), COVID-19: Systemic risk and resilience. Springer Publishing.
9. IRGC. (2018). *Guidelines for the governance of systemic risks*. Lausanne: International Risk Governance Center (IRGC).
10. Kaffenberger, L., & Kopp, E. (2019, September). Cyber risk scenarios, the financial system, and systemic risk assessment. Cyber Policy Initiative Working Paper Series|"Cybersecurity and the Financial System" #4.
11. Kroger, W., & Zio, E. (2011). *Vulnerable systems*. Dordrecht: Springer Publishing.

12. Kunreuther, H. & Useem, M. (Eds.) (2009). Learning from catastrophes: Strategies for reaction and response. Upper Saddle River, NJ: Wharton School Publishing.
13. Levine, S., Crosskey, A., & Abdinoor, M. (2011) System failure? Revisiting the problems of timely response to crises in the Horn of Africa. Humanitarian practice network, Overseas Development Institute.
14. Linkov, I., Trump, B. D., Golan, M., & Keisler, J. M. (2021). Enhancing resilience in post-COVID societies: By design or by intervention? *Environmental Science & Technology, 55*(8), 4202–4204.
15. Lucas, K., Renn, O., & Jaeger, C. (2018). Systemic risks: Theory and mathematical modeling. *Advanced Theory and Simulations, 1*. https://doi.org/10.1002/adts.201800051.
16. Masys, A. J. (2012). Black swans to grey swans-revealing the uncertainty. *International Journal of Disaster Prevention and Management, 21*(3), 320–335.
17. Masys, A. J. (2010, June). *Fratricide in air operations: Opening the black box- revealing the social*. Ph.D., Dissertation, University of Leicester, UK.
18. Masys, A. J. (2011). The emergent nature of risk as a product of heterogeneous engineering. In: S. Bennett (Ed.), *Innovative thinking in risk, crisis and disaster management*. London: Gower Publishing.
19. Masys, A. J. (2014). Critical infrastructure and vulnerability: A relational analysis through actor network theory. In: A. J. Masys (Ed.), *Networks and network analysis for defence and security*. Springer Publishing.
20. Masys, A. J. (2015) The cyber-ecosystem: Enabling resilience through the comprehensive approach. In A. J. Masys (Ed.). *Disaster management: Enabling resilience*. Springer Publishing.
21. Masys, A. J. (Ed.). (2016a). *Disaster forensics: Understanding root cause and complex causality*. Springer.
22. Masys, A. J. (Ed). (2016b). *Exploring the security landscape: Non-traditional security challenges*. Springer.
23. Masys, A. J. (2018). Designing high-reliability security organizations for the homeland security enterprise. In A. Siedschlag, & A. Jerkovic (Eds.), *Homeland security cultures: Enhancing values while fostering resilience*. London, New York: Rowan & Littlefield International.
24. Masys, A. J. (2021a). The security landscape—Systemic risks shaping non-traditional security. In A. J. Masys (Ed.), *Sensemaking in security*. Springer Publishing.
25. Masys, A. J. (2021b). Non-traditional security: A risk centric view. In A. J. Masys (Ed.), *Handbook of security science*. Springer.
26. Montasari, R., Hosseinian-Far, A., & Hill, R. (2018). Policies, innovative self-adaptive techniques and understanding psychology of cybersecurity to counter adversarial attacks in network and cyber environments. In H. Jahankhani (Ed.), *Cyber criminology*. Springer Publishing.
27. NA. (2014). 4. The transectional structure of society: The basic societal functions. *Scandinavian Journal of Public Health, 42*, 36–47.
28. Renn, O. (2021). New challenges for risk analysis: Systemic risks. *Journal of Risk Research, 24*(1), 127–133.
29. Saha, S., & Chakrabarti, S. (2021). (2021) The non-traditional security threat of COVID-19 in South Asia: An analysis of the Indian and Chinese leverage in health diplomacy. *South Asian Survey, 28*(1), 111–132.
30. Senge, P. M. (1990). *The fifth discipline: The art and practice of the learning organization*. London, UK: Century Business.
31. Spann, D. (2006). Coping with the unexpected: Integrating HRO and Agile. *Cutter IT Journal, 19*(2006), 32–37.
32. Sterman, J. D. (2000). *Business dynamics: Systems thinking and modeling for a complex world*. Boston: McGraw-Hill Publishing.
33. Toft, B., & Reynolds, S. (2005). *Learning from disasters: A management approach*, 3rd edn. Houndmills, UK: Palgrave Macmillan.
34. UNODC. (2013, February). *Comprehensive study on cybercrime draft*. http://www.unodc.org/documents/organized-crime/UNODC_CCPCJ_EG.4_2013/CYBERCRIME_STUDY_210213.pdf.

35. Vespignani, A. (2009). Predicting the behavior of techno-social systems. *Science, 325*, 425–428.
36. WEF. (2016). *Understanding systemic cyber risk*. https://www3.weforum.org/docs/White_Paper_GAC_Cyber_Resilience_VERSION_2.pdf.
37. WEF. (2021). *The global risks report*. https://www3.weforum.org/docs/WEF_The_Global_Risks_Report_2021.pdf.
38. Weick, K. E., & Sutcliffe, K. M. (2007). *Managing the unexpected: Resilient performance in an age of uncertainty* (2nd ed.). California, USA: Wiley.
39. Welburn, J. W., & Strong, A. M. (2021). Systemic cyber risk and aggregate impacts. *Risk Analysis: An International Journal*. https://doi.org/10.1111/risa.13715
40. Xu, T., & Masys, A. J. (2016). Critical Infrastructure vulnerabilities: embracing a network mindset. In A. J. Masys (Ed.), *Exploring the security landscape: Non-traditional security challenges*. Springer Publishing.

Dr. Anthony Masys is an Affiliate Associate Professor and former Director of Global Disaster Management, Humanitarian Assistance and Homeland Security. A former senior Air Force Officer, He is Editor in Chief for Springer Publishing book series: Advanced Sciences and Technologies for Security Applications and holds various advisory board positions with academic journals and books series.

Countering Violent Extremism in Central Asia and South Asia: Islamophobia and Cyber-Radicalization in the Digital Era

Adib Farhadi

Abstract Widespread political and economic uncertainty following the COVID-19 pandemic, paired with increased access to digital messaging and online social media platforms, has rendered vulnerable populations in Central Asia and South Asia (CASA) even more susceptible to misinformation, radical propaganda, and population targeting by violent extremists. More, studies show that violent extremism is inextricably linked with Islamophobia. Violent extremist recruiters frequently capitalize on publicized Islamophobic events to spread digital misinformation and lure disenfranchised recruits, particularly among youth populations. A debilitated Afghanistan only compounds these issues in CASA. The growing humanitarian crisis in the wake of U.S. military withdrawal, leaves this impoverished nation ripe for the proliferation of violent extremist activity that will reach far beyond its borders. Weaponized cyber-misinformation is a moving target that threatens even rural populations. Effective deterrence calls for novel multilateral efforts between great and local powers, both on and offline, to dispel skewed narratives and reinforce positive counter-narratives. While expanding access to digital communications in CASA presents obvious challenges for countering violent extremism, it likewise affords vital new opportunities for cooperation between global and regional powers to reach previously unreachable, vulnerable populations.

Keywords Great power competition · Islamophobia · Countering violent extremism · Radicalization · Central Asia and South Asia · Afghanistan · Counter terrorism · Cyber messaging

---

A. Farhadi (✉)
University of South Florida, Tampa, FL 33620, USA
e-mail: Farhadi@usf.edu

A. Farhadi et al. (eds.), *The Great Power Competition Volume 3*,
https://doi.org/10.1007/978-3-031-04586-8_5

Introduction

Widespread political and economic uncertainty following the COVID-19 pandemic, paired with increased access to social media and digital messaging platforms in rural areas, has rendered already vulnerable populations in Central Asia and South Asia (CASA) more susceptible to misinformation and population targeting by violent extremists. In Bangladesh, for instance, Islamic radicalization has "surged in the COVID-19 period as more terrorist organizations take to the internet to propagate their cause," according to Shafi Md Mostofa, Assistant Professor of Religion and Culture at the University of Dhaka. He explains that the destabilizing effects of COVID-19 alongside "persistent social deterioration will aggravate young people even more and create the perfect conditions for terrorist organizations to recruit them" (2020).

Exacerbating underlying regional vulnerabilities is the destructive stigma, pervasive in popular thought and amplified in global digital media, that Islam is largely synonymous with violent extremism. This demoralizing stigma further marginalizes already at-risk youth populations by reinforcing suspicion of "outsiders" and expanding target audiences in the region. These vulnerable youths are often subjected to misinformation and the radical counter-narrative that "Westerners" seek to stifle or even eradicate the Islamic religious way of life.

Compounding these factors since the withdrawal of U.S. troops from Afghanistan, is the popular claim that the U.S. was "defeated" by the Islamic Emirate of Afghanistan (popularly known as the "Taliban"). Frequently touted online by Islamist propagandists, this claim has amplified Islamist appeal in cyberspace, heightened recruitment efforts and enrollment, and spurred a dangerous new era of violent radicalization. Violent extremists capitalize on these divisive narratives in CASA and beyond to convince target audiences that the religion of Islam itself is under attack and must be protected at all costs. For example, the Islamic State-Khorasan (ISIS K) targets youth populations in Afghanistan and the wider region. Muggah has explained that ISIS K's "attacks against U.S., Pakistani, Russian, and other countries' security forces—skillfully shared on social media—may solidify its base and attract new recruits sympathetic to its harsh brand of political Islam." Moreover, "the group's active digital outreach could likewise inspire organized or lone-wolf attacks not just in Afghanistan, but also in neighboring countries" [1].

The defensive stance of violent extremists has been inadvertently legitimized in the "West" by mainstream media and radicalization theories that support the notion that radical religiosity, specifically within the Islamic religion, is a primary catalyst for acts of violent extremism [10]. Amnesty International recognizes that Muslims are often targets of hate crimes arising from anti-terror campaigns, reporting that "discrimination against Muslims in European counter-terrorism efforts has helped to create an environment in which Muslims are more likely to be the subject of hate speech and attacks" [2]. Additionally, anti-Muslim hate crimes, often prompted by misinformation, elicit retaliation by violent Islamic extremists who then use these hate crimes to justify further acts of violence at home and abroad—and the vicious cycle of retribution is continued.

While deterring the spread of violent radicalization globally, particularly in CASA, has been a vital national interest for decades, increased regional access to online social media and messaging platforms has escalated this issue to a near-crisis level. Additionally, violent extremist recruiters have become more adept at understanding and utilizing specific population dynamics and human trauma variables to target online audiences through sophisticated digital messaging. According to Dr. Aleksandra Nesic of the J.F.K. Special Warfare Center, "ISIS and Al-Qaeda […] understand the need that is missing [in their target audience] and then tailor their messaging to fulfill the need" (2020). Even in remote areas of CASA, many communities are swiftly gaining access to the internet and online social networking platforms that violent extremists commonly utilize to spread radical propaganda and misinformation. This is also true internationally in diaspora populations that maintain communication, influence, and interests in their native countries [12].

Before any lasting and meaningful progress against violent radicalism can be made in CASA, false narratives surrounding Islam, violence, and the "West" must be expeditiously addressed and countered with regional local leaders and populations. The negative portrayal and false narrative that the religion of Islam is synonymous with violent extremism (popularized in the "West"), inhibits building key alliances with local communities and tribal leaders who are essential to countering violent extremism, especially in rural areas. In their comprehensive RAND report, *The Muslim World After 9/11*, Rabasa, et al. suggest that "tribal culture and affiliation can play a substantial role in the extent to which radical, violent forms of Islamism prosper in some regions of the Middle East and South Asia" (2004).

Additionally, there is the ongoing problem of mitigating "skewed perspectives" that arise from the prolific output of digital misinformation on social media and other cyber-messaging platforms regarding current events, religious tenets, the scope of insurgencies, etc. Misinformation campaigns are not limited to events in CASA or issues of violent Islamic extremism. Rather, they remain "moving targets" in global cyberspace, ranging from international and local news to politics, and most recently, events relating to the COVID-19 pandemic and the U.S. withdrawal from Afghanistan. False narratives that conflate Islam with violent extremism are only one example of how weaponized digital misinformation can become dangerous. Speaking at the third biannual Great Power Competition Conference (GPCC), General Richard Clarke, Commander, U.S. SOCOM, stated that, "social media has reshaped political warfare, allowing the delivery of information operations effects at unimaginable speed and scale, empowered by tech in cyberspace" [7].

Currently, Afghanistan and its neighbors in CASA face the greatest imminent threat of accelerated violent extremism since the withdrawal of U.S. military troops in late 2021 and the subsequent reinstatement of the Islamic Emirate of Afghanistan (IEA). The "military victory" of the IEA is currently being exploited on social media to embolden support in the region and beyond. The BBC reports that tweets from Taliban members, "boasted about the group's recent victories—sometimes prematurely—and pushed several hashtags, including #kabulregimecrimes (attached to tweets accusing the Afghan government of war crimes); #westandwithTaliban (an attempt to drive grassroots support); and #نَصْرٌ_مِّنَ_اللهِ_وَفَتْحٌ_قَرِيبٌ (help from God and victory is near). The first hashtags at least trended in Afghanistan" (2021).

Afghanistan's present volatile and vulnerable state, now on the verge of becoming a failed state, renders it ripe for a significant resurgence in violent extremism. In recent history, Afghanistan has found itself in the crosshairs of violent extremists who capitalized on its fractured ethnic identity and weak governance by taking refuge in the country to further their violent agenda against the "west". The most notorious and tragic manifestation of violent extremism planned and carried out from Afghanistan resulted in 9/11. The attacks were orchestrated by Osama bin Laden and Al-Qaeda, resulting in a U.S. military response in Afghanistan and the longest war in U.S. history.

Outcomes of violent extremism, exacerbated by an Afghanistan left fractured and vulnerable in the past, should serve as a cautionary tale for future cooperation between great powers and local leaders in CASA. To avoid such devastating outcomes in the future, the U.S. must remain strategically engaged in Afghanistan. Continued U.S. engagement in Afghanistan benefits all stakeholders in the region, including powerful influencers like China, Russia, and Pakistan. Now is the time for expedient, multilateral cyber-initiatives that focus on mitigating misinformation to affirm U.S. intentions for the continued support of the Afghan people, and to dispel the notion that the "West" is the enemy (and a "weakened" one at that) of Islam or Afghanistan.

While rapidly evolving technology offers violent extremists unmatched opportunities for messaging and population-targeting in CASA and globally, it likewise affords unprecedented opportunities for great power cooperation to mitigate violent radicalization more effectively across vulnerable youth populations. Shared cyber platforms can be employed to positively engage local leaders and vulnerable populations in the region and to encourage unity against violent extremism. Afghanistan must also be re-engaged through economic development such as the New Silk Road Initiative (NSRI) that can stave off an imminent state failure. While Afghanistan remains critically debilitated, heightened threats from violent extremist organizations (VEO) in cyberspace and on the ground remain potent. Policymakers, reticent to engage in initiatives that may strengthen the Islamic Emirate of Afghanistan (IEA) position, can balance future risks with contingent development strategies, as well as short-term humanitarian aid, that increase strategic leverage and oversight.

This chapter broadly surveys how Islamophobia and weaponized misinformation interact with social media and digital messaging in cyber space to catalyze violent Islamic radicalization. The chapter further suggests that effective mitigation of violent extremism in cyberspace necessitates a cooperative, inter-state approach to countering violent extremism initiatives by regional actors and great powers. Such initiatives may include (1) multilateral efforts aimed at dispelling Islamophobic rhetoric linking the religion of Islam with violent extremism; (2) building trust and alliances with local populations and tribal leaders to mitigate violent extremist rhetoric both online and offline; and (3) sustained cooperation between great powers and regional actors in areas of cyber intelligence sharing, cyber mitigation, and cyber counter-messaging in CASA.

Stigmatized Islam and Radicalized Cyberspace

The Rise of Modern Islamophobia

Following the tragic events of 9/11 in 2001, discourse and theories on radicalization and violent extremism have increasingly centered on the role of religion (Islam) in the development of violent extremism. Behavioral models ranging from Moghaddam's staircase to Hafez's puzzle model (Hafez and Mullins 2015 as referenced in Farhadi 2020) have largely focused on social theory aspects of radicalization, particularly the social process of religious indoctrination. Though most modern radicalization theories agree that violent Islamic extremism arises from a skewed understanding of the Islamic faith, the direct correlation between Islam and violent extremism emerging from such studies persists in mainstream Western thought today.

Social theories of radicalization that support the notion that religion (Islam) is the essential impetus for acts of violent extremism are often used to support radical Islamophobic ideologies [10]. The conflation of Islam with violent extremism often leads to the false narrative that Islamic extremism is the most prevalent form of violent extremism today. Modern theories of radicalization, developed to address and mitigate threats from violent extremists, may themselves, unwittingly include Islamophobic language that further legitimizes the false perception that Islamic extremism is more radical, more dangerous, and more prevalent than other forms of violent extremism. Additionally, "Americans have consumed media headlines about the Patriot Act, the Iraq and Afghanistan wars, violent extremist organizations, and the Muslim ban, all perpetuating an association between Muslims and terrorism" [18].

While theories that emphasize the prominent role of religion (Islam) in "violent radicalization" may have negatively influenced public perceptions of religion and Islam in general, the emergence and proliferation of online social media platforms such as Facebook and Twitter, and the resulting rapid spread of misinformation, have facilitated increasingly more virulent expressions of Islamophobia worldwide.

Recent statistics show that there are "over 3.30 billion [users on] Facebook, Instagram, WhatsApp or Messenger each month and according to Facebook…over 2.80 billion monthly active users and growing as of January 27, 2021" [24]. Such statistics demonstrate that social media is far from waning in popularity or its ability to influence global society. Social media usage is expanding by approximately 12% each year on average globally. Exact social media and internet usage statistics for predominantly rural states such as Afghanistan are less clear. However, recent trends confirm expanding access to internet services, primarily through increased availability and use of cell phones in rural areas. A recent report shows that Afghanistan had 7,337,489 regular internet users, and in 2018, "over 4.7 million of the population of Afghanistan had access to the internet" (Internet World Stats 2021). Additionally, Facebook now has "3,848,400 users in Afghanistan," according to one online source [21].

Digital social media platforms offer users a significant amount of anonymity, allowing for a freer sharing of radical views and statements that might otherwise remain hidden during face-to-face group gatherings. The continuous exchange of radical and often violent anti-Islamic rhetoric with other "like-minded" users often creates a dangerous false sense of individual and group primacy that can lead to "mob rule" victimization, both online and offline. Countering the continuous output of Islamophobic hate speech and misinformation on digital platforms lies at the core of mitigating violent Islamic radicalization and its dangerous outcomes globally. Ironically, it is often the emotional and psychological phenomena directly resulting from Islamophobic victimization and targeting, including feelings of alienation and helplessness, that can and do render victims of hate speech more susceptible to the counter-narratives (equally isolating) of violent extremists and possible radicalization [23].

Islamophobia in Cyberculture

With the constant stream of digital news today, Islamic hate speech is on the rise among social media outlets, with serious consequences for victims in the "real" world. A recent study by Castano-Pulgarín et al. found that "the most attacked religion in the world is Islam and it seems to be motivated by an Islamophobic sentiment, favored by the cultural processes of globalization and digital media circulation (Horsti 2017)" [5, p. 2]. Regarding outcomes from news events, the authors concluded that, "Overall, these expressions found Muslims being demonized online through negative attitudes, discrimination, stereotypes, physical threats and online harassment, which all had the potential to incite violence or prejudicial actions because it disparages and intimidates a protected individual or group" [5, p. 4].

According to Irene Zempi and Imran Awanin, in their book entitled *Islamophobia: Lived Experiences of Online and Offline Victimization*, "…given that they [the victims] were targeted because of the 'visibility' of their Muslim identity —which was easily identifiable because of their Muslim name and/or Muslim appearance either online or offline—participants were unable to take comfort in the belief that what happened to them was simply random and 'could have happened to anyone'… [and] at the same time, online Islamophobic hate crimes made participants particularly fearful due to the anonymity that the internet provides its users" [[23], p. 2].

One striking illustration of the widespread online victimization of Muslims was the proliferation of the hashtag "#StopIslam", popularized in 2016 on Instagram. A recent study on the "demonization" of Islam, and the #StopIslam hashtag campaign specifically, revealed that "prejudices and stereotypes about the Arab world continue to expand and spread in the digital environment so that in recent years, the social alarm has increased over the ability of these media to increase and propagate hate speech. The current media ecosystem offers social intercommunication, which has affected how societies are built, social relations, and the concept of "us" and "them". Additionally, researchers suggested that hate speech has become so normative within social media communities that platforms such as Instagram "accept it in their behavior policies" [6, p. 15].

While some theorists claim that "discrimination" (here, Islamophobia) is not a significant factor in the cycle of violent radicalization, recent research demonstrates a distinct correlation between anti-Muslim and pro-Islamic extremism in online search queries, suggesting that Islamophobic "content" may catalyze interest in Islamic radicalism. "There are higher reports of news regarding people joining ISIS in communities where people express a strong anti-Muslim sentiment, finding that 'anti-Muslim searches are strongly associated with pro-ISIS searches, particularly in communities with high levels of poverty and ethnic homogeneity'" (Fox 2018 as referenced in [10]).

Regarding the central role of Islamophobia in the cycle of Islamic radicalization, Clinton Watts, a violent extremism expert with the Foreign Policy Research Institute, states, "this [Islamophobic rhetoric] plays [right] into the hands of [violent radical jihadists] because it doubles down on Al Qaeda's justification for attacking the United States, to begin with" (Almedia 2017 as referenced in [10]). Violent extremist organizations such as the Islamic State of Iraq and the Levant (ISIS) intentionally target marginalized Islamic youth populations with ultra-slick, glamorous online campaigns to attract potential Jihadis [3]. Violent extremists seeking to recruit potential radicals through online internet platforms rely heavily on Islamophobic digital "content" to support their own violent rhetoric. They frequently use news events such as Muslim travel bans in the U.S. to legitimize the counter-narrative that all Westerners see Muslims as terrorists and ultimately seek to eradicate Islam.

Utilizing various digital channels, sophisticated and predatory recruiters attract potential followers by targeting and exploiting distinct population dynamics. They then focus on individual traits and emotional needs, such as the desire for greater peer acceptance, to cultivate a contrived sense of intimacy and community through online correspondences. Recruiters create a "safe space" to slowly introduce their radical counter-narratives and begin the deliberate process of indoctrination into a culture of radicalized violence. Husain Ehsani from the University of Baghdad notes that ISIS employs "sophisticated digital recruitment and radicalization efforts through virtual coaches" on social media platforms with encrypted messaging [9].

To effectively counter violent extremism, policymakers must recognize the overarching role of Islamophobia in the cycle of violent radicalization, and more specifically, the dangerous impacts generated on social media platforms utilized by violent extremists. "Not only does Islamophobia contribute to alienation and discrimination, which are popularly cited grievances that leave individuals vulnerable to radicalization, but the direct correlation between a rise in hate crimes with a rise in ISIS support shows that Islamophobia may also exacerbate the radicalization process" [10].

Social Media and Radicalization

The ability to reach marginalized populations grows as regional access to the internet rapidly expands in developing nations. Today, little question remains that violent extremist groups utilize social media to attract and recruit potential radicals. It is unclear exactly how these groups determine which platforms are most effective for their specific agendas. Unlike ISIS, most armed groups strive to maintain a lower international profile on social media but have become more adept at utilizing these platforms for attracting regional support [8].

According to Laura Courchesne and Brian McQuinn, on the relationship between social media and armed groups, "[T]here are over 1,456 armed groups operating in civil wars in Mali, Libya, and Syria. Almost all use social media to target regional and local audiences but vary in their choice of platform. Social media is also providing novel and underexplored funding channels". Further, "the next global threat to exploit social media will emerge from the groups currently avoiding detection or attention by platforms" (2021).

In a worrying recent development, the private company Electronic Horizons Foundation (EHF) launched a new, more secure, cloud platform to facilitate digital interactions and online activity among ISIS supporters globally [13]. EHF's inaugural announcement stated that "[I]n light of recent developments in the media arena and the restriction of technology companies to content, we resorted to developing solutions that provide a space for propagation between the fellow supporters and the general Muslim community, so that the benefits may prevail." As explained by Bridget Johnson of Homeland Security Today, "The Electronic Horizons Foundation launched in January 2016 as an IT help desk of sorts to walk ISIS supporters through how to encrypt their communications and otherwise avoid detection online while coordinating with and recruiting jihadists" (2021). Sophisticated measures such as this, taken to ensure a meaningful online presence with targeted audiences, clearly demonstrate the critical role of social media in garnering support for groups like ISIS and other violent extremists.

A poignant example of how violent extremists aptly manipulate social media is the bold, though now defunct, ISIS Android Twitter app known as "The Dawn of Glad Tidings," released in 2014. The app allowed ISIS recruiters to randomly upload their propagandist tweets using followers' accounts (often thousands at one time) along with news and other media. According to J.M. Berger, Associate Fellow at The International Centre for Counter-Terrorism, "It's one of many tools that ISIS uses to manipulate the perception on social media that their content is bigger and more popular than it might be if you were looking at just their organic supporters" (Berger as cited by [20]).

Groups like ISIS are savvy and strategic in appealing to wider audiences using sensationalist and violent propaganda aimed mainly at younger audiences. Often, their messaging has less to do with fulfilling a religious call to duty and more to do with the "noble aim" of defending group principles using violence. A recent study that examined the content of 20,000 ISIS followers' accounts on Twitter between

2016 and 2017, determined that consistent with the nature of ISIS social media content, the users (ISIS supporters) with the most significant number of followers frequently posted jargon related to revenge, violence, and extremism. By contrast, those ISIS supporters with fewer followers were engaged in dialogue primarily related to religious concerns, such as "religious justifications for the use of violence" [22]. In other words, violent radicalizers strategically broadcast violent imagery and language online to elicit widespread attention that will yield a handful of individuals susceptible to recruitment.

Though IS factions and supporters have now migrated the bulk of digital communications to more sophisticated encrypted messaging platforms, such as Threema, they continue to post new recruitment and violent teaser videos aimed at younger target audiences on Facebook, TikTok, and Telegram, among other sites [4]. Rita Katz, director of the SITE Intelligence Group, predicts that groups like ISIS will continue to grow their presence on social media: "At this stage, with the addicted usage of Twitter by the jihadists, I'm not sure how it can be stopped…for the last three years we have been warning that social media is becoming a hydra, an uncontrollable source for jihad, due to its easy use, rapid updates with easy connection and unlimited viewers" (Katz as cited by [20]).

While far from exhaustive, these findings broadly illustrate the scope of social media employed by extremist groups like ISIS, including the sophisticated measures undertaken to engage and maintain their online audiences. Additionally, these findings suggest the efficacy of violent propaganda to attract attention (and subsequent followers) to social media sites, while highlighting the less significant role of religiosity or religion in attracting (at least initially) online followers to the cyberculture of violent extremism. Further research is necessary to determine how less-prominent violent extremist groups in CASA and beyond utilize social media to effectively motivate regional supporters and gain funding, while remaining, at least in part, "under the radar" [8].

Weaponized Misinformation and Messaging

One of the more challenging aspects of countering violent extremism is mitigating the proliferation of digital misinformation custom-tailored to specific populations and online demographics. In the case of Islamic violent extremist recruitment, aspects of the Islamic cultural narrative (such as Sharia) are often intentionally skewed and presented online to entice young male populations to enlist in a "just cause." Those who succumb to radicalization may not hold grievances related to any specific extremist agenda but are often instead attracted to the notion of defending a just cause in and of itself, a "cause" primarily supported by skewed narratives. In this case, misinformation becomes a dangerous weapon employed by violent extremists to elicit a desired emotional response.

According to Oliver Roy from the European University Institute in Italy, "Radicalization is [typically] a youth revolt against society, articulated on an

Islamic religious narrative of jihad…It is not the uprising of a Muslim community victim of poverty and racism: only young people join, including converts who did not share the 'sufferings' of Muslims in Europe. These rebels without a cause find in jihad a 'noble' and global cause and are consequently instrumentalized by a radical organization (Al Qaeda, ISIS) that has a strategic agenda" (Roy as cited by [19]).

"Weaponized" digital misinformation campaigns, employed by violent extremists and their followers to whip up emotion, are prevalent on social media platforms such as Facebook and Twitter, which have recently become more widely accessible in rural areas across CASA through the use of "smartphones." Naturally, as rural populations gain greater access to the internet and social media outlets, opportunities for violent extremists to reach larger groups of uneducated youth populations (those most susceptible to radicalization) will increase exponentially. Imran Awan of Birmingham City University emphasizes the influential role of information in social media, stating, "ISIS tactics of propaganda, recruitment, and radicalization all emerge within the online virtual space. The power of social media for groups such as ISIS is immense, as demonstrated when the Iraqi government blocked access to many social media accounts because they were being used to plan attacks" [3, p. 147].

To address the problem of mitigating online misinformation as it relates specifically to violent Islamic extremism, Roy suggests that "beyond increasing our intelligence capacity…we need to debunk the myth that radical terrorists are heroes and subvert the idea that the Islamic State is successful and impervious to our attacks. What's more, we need to foster the idea that Islam is a normal part of society, not a dangerous or oppressed minority. "Instead of 'exceptionalizing,' we should 'normalize'" (Roy as Cited by [19]). Ideally, this strategy can be extended to rural youth populations throughout CASA by employing regional entities who have established relationships with tribal and community leaders. While reaching urban youth with counter-messaging is essential, finding novel pathways to rural communities that have recently gained access to social media and online platforms is critical.

From Madrassas to Airwaves: Radicalization Among Afghan Youth Populations

Afghanistan has a long and troubled history with violent extremism. It's most notable and tragic culmination resulted in the attacks of 9/11, which were planned and executed by Osama bin Laden and Al Qaeda while taking refuge in Afghanistan. Unfortunately, at the end of the "longest war" between the U.S. and Afghanistan—a war initiated in direct response to the attacks of 9/11—the same pressing issue of preventing violent extremist groups from establishing a "home base" in a weakened Afghanistan (now facing an unprecedented humanitarian crisis) still exists today.

The many geopolitical and economic factors contributing to Afghanistan's present weakened state and ongoing vulnerability to violent extremism are well documented. The country's landlocked status and relative poverty arising from lack of sufficient market access are not least among them. Equally damaging, if less obvious, is Afghanistan's fractured cultural and ideological identity, exacerbated by decades of untenable foreign occupation and violent insurgencies. In conjunction, a continued lack of access to formal education has left rural male youths (a significant portion of the Afghan population at approximately 70%) underexposed to Islamic "thought" besides that offered by their local tribal leaders and mullahs. These local leaders may or may not be motivated at times by the financial support, protection, or ideological sway of various violent extremist groups vying for local influence. Rural male youth populations remain the most susceptible to the impact of violent extremism in Afghanistan— often falling prey to the romanticized, misleading promises of a new life of wealth, honor, and worldly adventure [11].

A special report by the United States Institute for Peace (USIP) in 2015, including interviews and data collected in the Afghan provinces of Nangarhar, Balkh, and Herat, "to elicit views on extremist groups, both violent and nonviolent and factors thought to induce youth to join such groups," found that independent local mullahs (sometimes acting as proxies for organized extremist groups) had the most significant influence on rural youth populations. Additionally, the report found that rural youth populations were more susceptible to recruitment by violent extremist groups while educated urban populations were more inclined toward participation in nonviolent Islamist groups. According to the 2015 report, "It is [often] unaffiliated extremist mullahs, rather than the Taliban or other organized groups, who are believed to be [historically] the principal sources of radicalization… interviewees also noted that the outreach of these mullahs extended beyond the mosque and the madrassa, citing computer and English-language courses run by one cleric and naming several poor neighborhoods of Mazar-e Sharif as places where violent thought was being taught along with Excel and PowerPoint" [11, p. 8].

The report further suggests that difficulty in countering violent extremism in Afghanistan "is partly the result of the absence of a clear, indigenous anti-extremist narrative" (Reza Fazli, Casey Johnson, and Peyton Cooke). The study claims that while the broader narratives of Islamic extremist groups typically treat the plight of Muslims globally, "the overarching narrative [in Afghanistan] is [usually] jihad against the 'occupying' U.S.-led coalition and the 'un-Islamic' Afghan government. The violence is more about fighting against uninvited guests than for a particular ideology" [11, p. 1]. The necessity of establishing a unified counter-narrative in Afghanistan, accessible to rural populations and designed to supplant the rhetoric of violent extremists, cannot be overemphasized here.

Though radicalization among Afghan youth populations has traditionally occurred at grassroots levels and in-person (through local madrassas, night letters, and community proxies) in rural communities, digital messaging by violent and nonviolent extremists are now on the rise, with rural areas in Afghanistan swiftly gaining access to cellular and internet services (Reza Fazli, Casey Johnson). Internet services are technically available in all thirty-four provinces in Afghanistan. Additionally, "cell phone messaging has grown in scope and appears to be especially powerful in light of the penetration of cell phones in rural areas and their increasing use among literate and illiterate alike" [11, p. 11]. The USIP further suggests that "Islamist groups [may] have a role to play as partners in developing programs to counter the ideas of violent extremist groups. Though such a partnership is ambitious, that nonviolent groups are united in their stance on not using force suggests they can be engaged to make youth (potential recruits) understand that the use of force is un-Islamic" [11, p. 14].

Conclusion

The interplay of Islamophobia, digital messaging, and social media plays a critical role in the cycle of violent radicalization. The growing problem of cyber-extremism in CASA—primarily due to expanding internet access—points to a need for multilateral cooperation to more effectively address Islamophobia and violent Islamic radicalization in the region. This need is particularly acute during the present uncertain transition period in Afghanistan.

As widespread access to digital media increases in CASA, including and especially in rural areas, so does the effective reach of violent extremist groups such as ISIS, ISIS-K, and Al Qaeda. These VE groups now rely heavily on social media and digital messaging to reach target audiences in rural areas that until recently did not have access to online sources. Further, rural male youth populations are historically most susceptible to violent radicalization in CASA, highlighting the urgency for cooperative initiatives in these areas [11]. As Afghanistan faces a growing humanitarian crisis and volatile political uncertainty, the U.S. must act swiftly to mitigate the region's inevitable proliferation of violent extremism.

Weaponized cyber-misinformation is a moving target with the ability to reach the most vulnerable rural populations in CASA and beyond. Effective deterrence of violent radicalization among youth populations in the region now calls for novel, cooperative efforts between great powers and local regional influencers online and on the ground. Both religious and civil entities, having established their unique inroads with individual local populations, must act to dispel skewed narratives and misinformation and reinforce positive counter-narratives. The United States Institute for Peace makes the following specific recommendations for local entities working on counter-initiatives in Afghanistan, "messaging materials that focus on shared values, such as peace and unity within Islam, should be created after consultation with target beneficiaries and leaders in their communities (religious leaders, tribal elders…these messages and information will need to be tailored to the target beneficiaries but should not be overtly political" [11, p. 13].

Establishing sustainable inroads with local leaders and populations in CASA will require a network of actors, including international, state, local, and regional entities, to create a chain of communication ranging from rural areas to international governing entities. Naturally, such a holistic endeavor requires committed cooperative efforts between the great powers in the region, the enlisting of local entities, and a willingness to engage in realistic dialogues with governments and local and tribal Islamic leaders who best understand the communities' grievances (Reza Fazli, Casey Johnson). In Afghanistan, for instance, each province, town, and village may have its micro-culture, replete with specific grievances, traditions, and concerns that require "insider" perspectives for developing effective counter-messaging strategies. Reflecting the need for targeted regional initiatives, the RAND corporation's report, The Muslim World After 9/11, states, "Every country in the Muslim world is riven by multiple cleavages among ethnic communities, tribes, and clans, which often constitute the principal basis of an individual's identity and the primary engine of political behavior" [17, p. 34].

While long-range outcomes in CASA remain uncertain, the present transitional period may afford a brief window of leverage to establish cooperative initiatives that effectively address disenfranchised, rural youth populations and leaders and bring strong messages of hope to a people beset by challenges. These messages must include counter-narratives that dispel the radical view that violence is the only option and affirm an Islamic cultural narrative based on non-violence. While expanding access to digital communications in rural Afghanistan and CASA presents obvious challenges for countering violent extremism, it likewise affords exciting new opportunities for cooperation between great and regional powers to reach previously unreachable, vulnerable populations.

References

1. Atiq, S. (2021). The Taliban embrace social media: we too want to change perceptions. *BBC World News Online*, 6 September, viewed 6 September 2021. https://www.bbc.com/news/world-asia-58466939.
2. Amnesty International. (2021). *Europe: How to combat counter-terror discrimination, 3 February, viewed 4 February 2021.* https://www.amnesty.org/en/latest/news/2021/02/europe-how-to-combat-counter-terror-discrimination/.
3. Awan, I. (2017). Cyber-extremism: Isis and the power of social media. *Society (New Brunswick), 54*(2), 138–149.
4. Bansal, K. (ed) (2021). Forget telegram, signal: ISIS terrorists in India, abroad using much more secure messenger. *India.com, 15 January, viewed 10 October 2021.* https://www.india.com/technology/forget-telegram-signal-isis-terrorists-in-india-abroad-using-much-more-secure-messenger-threema-4334501/.
5. Castaño-Pulgarín, S. A., Suárez-Betancur, N., Tilano, L., & Herrera López, M. (2021). Internet, social media and online hate speech systematic review. *Aggression and Violent Behavior, 58*(6). https://doi.org/10.1016/j.avb.2021.10168.
6. Civila, S., Romero-Rodriguez, L., & Civila, A. (2020). The Demonization of Islam through Social Media: A case study of# Stopislam in Instagram. *Publications (Basel), 8*(52), 1–19. https://doi.org/10.3390/publications8040052.
7. Clarke, R. (2021). Great power competition and cybersecurity. *Great Power Competition Conference Series*, University of South Florida, Tampa, FL, pp. 78–79.
8. Courchesne, L., & McQuinn, B. (2021). 'fter the Islamic state: Social media and armed groups. *War on the Rocks,* 9 April, viewed 15 May 2021. https://warontherocks.com/2021/04/after-the-islamic-state-social-media-and-armed-groups/.
9. Eshani, H. (2019). Why violent extremist groups use social media; Islamic State case study. Afghan Institute for Strategic Studies, 19 December, viewed 3 January 2020. https://aiss.af/aiss/news_details/opinions/5dfb623933fc6.

10. Farhadi, A. (2020). Post-9/11 radicalization theory and its impact on violent extremism. In: Masys, A.J. (ed.), *Handbook of Security Science,* Springer Nature, Cham, Switzerland. https://doi.org/10.1007/978-3-319-51761-2_8-1.
11. Fazli, R., Johnson, C. G., & Cooke, P. (2015). *Understanding and countering violent extremism in Afghanistan. Special Report*, 3 September, United States Institute of Peace, viewed 6 September 2021. https://www.usip.org/publications/2015/09/understanding-and-countering-violent-extremism-afghanistan.
12. Golestani, L. (2020). Contagion effect: radicalization, unrest, and competition in the Covid-19 era. *Great Power Competition Conference Series*, University of South Florida, Tampa, FL, p. 91.
13. Johnson, B. (2021). ISIS cyber group launches cloud, chat platforms to "Close Ranks". *Homeland Security Today,* 6 April, viewed 15 May 2021. https://www.hstoday.us/subject-matter-areas/cybersecurity/isis-cyber-group-launches-cloud-chat-platforms-to-close-ranks-online/.
14. Mostofa, S. (2020). Islamist cyber-radicalization in Bangladesh during COVID-19. Australian Institute of International Affairs, 20 August, viewed 22 August 2021. https://www.internationalaffairs.org.au/australianoutlook/cyber-radicalisation-bangladesh/.
15. Muggah, R. (2021). Islamic State-Khorasan's reach extends far beyond Afghanistan with a growing digital presence, the extremist group seeks to destabilize South and Central Asia. *Foreign Policy*, 9 September, viewed 11 September 2021. https://foreignpolicy.com/2021/09/09/islamic-state-khorasan-afghanistan-pakistan-central-asia-online-extremism-terrorism/.
16. Nesic, A. (2020). Contagion effect: Radicalization, unrest, and competition in the Covid-19 era. *Great Power Competition Conference Series*, University of South Florida, Tampa, FL, pp.74–75.
17. Rabasa, A., Benard, C., Chalk, P., Fair C. C., Karasik, T. W., Lal, R., Lesser, I.O., & Thaler, D. E. (2004). *The Muslim world after 9/11,* RAND Corporation, Santa Monica, CA, viewed 3 March 2020. https://www.rand.org/pubs/monographs/MG246.html.
18. Ramadan, J. (2021). Perpetuating islamophobic discrimination in the United States: Examining the relationship between news, social media, and hate crimes. *Harvard Human Rights Journal, vol. 35, viewed 5 May 2022.* https://harvardhrj.com/2021/05/perpetuating-islamophobic-discrimination-in-the-united-states-examining-the-relationship-between-news-social-media-and-hate-crimes/.
19. Swanson, A. (2015). Why young people become jihadists, according to a top expert', *The Washington Post*, 18 December, viewed 10 October 2020. https://www.washingtonpost.com/news/wonk/wp/2015/12/18/why-young-people-become-jihadists-according-to-a-top-expert/.
20. Trowbridge, A. (2014). Jihadists on the move in Iraq with weapons, hashtags. *CBS News Online*, 16 June, viewed 3 March 2020. https://www.cbsnews.com/news/isis-jihadists-on-move-in-iraq-using-weapons-and-twitter-hashtags/.
21. World Population Review. (2021). *Facebook users by country 2022*, World Population Review, viewed 3 March 2022. https://worldpopulationreview.com/country-rankings/facebook-users-by-country.
22. Tulga, A. Y., (2020). ISIS religious and extremist propaganda on social media: Dictionary-based study of Twitter. *Small Wars Journal,* 22 October, viewed 5 January 2021. https://smallwarsjournal.com/jrnl/art/isis-religious-and-extremist-propaganda-social-media-dictionary-based-study-twitter.
23. Zempi, I., & Awan, I. (2017). *Islamophobia: Lived experiences of online and offline victimisation,* Policy Press, Bristol.
24. Zephoria Digital Marketing. (2022). *The top 10 valuable Facebook statistics- Q2 2021*, 3 January, viewed 27 April 2022. https://zephoria.com/top-15-valuable-facebook-statistics/.

Dr. Adib Farhadi is an Assistant Professor, Faculty Director of Executive Education, and lead editor of the Great Power Competition book and conference series at the University of South Florida. His research is at the intersection of religion, politics, economics, and conflict, with a particular focus on the Central and South Asia (CASA) Region. He is the author of *Countering Violent Extremism by Winning Hearts and Minds* and a frequent presenter on the topics of Strategic Competition, Countering Violent Extremism (CVE), and Strategic Negotiations & Communication. Formerly, Dr. Farhadi served in senior positions for Afghanistan and has extensively advised the United States government and various other international organizations. Dr. Farhadi earned his Ph.D. in Economics from the University of Canberra, M.A. from New York University, and B.S. from East Carolina University.

Cyber and Great Power Competition in the Western Hemisphere

Alexander Crowther, Fabiana Perera, and Brian Fonseca

Abstract Great Power Competition is happening in every geographic region and across most domains. The People's Republic of China, Russian Federation, the United States, and other actors are engaged globally, regionally, and even locally. Not everyone competes everywhere, but there is no country in the world today that is not on a field of competition. The United States own neighborhood, the western hemisphere, has seen competition increase as the People's Republic of China, Iran, North Korea, and Russian Federation have all sought to engage to contest historical US hegemony over the region. Competition in the western hemisphere and in other regions is happening across all three major domains- land, sea, and air. It is also happening in the cyber domain. This chapter discusses great power competition in the cyber domain in the western hemisphere. It summarizes the interests and activities of US near-peers in the western hemisphere and presents an overview of their operations in cyberspace. The chapter argues that each of the US near-peers is pursuing different tools for different aims within the western hemisphere. However, in all cases advances from US near-competitors in the cyber domain are facilitated by weak governance of the defense sector, and weak capabilities in the cyber domain in the target countries.

Keywords Great power competition · Cyber · Cyber security · China · Cyberspace · United States · Western hemisphere

---

A. Crowther (✉) · B. Fonseca
Florida International University, Miami, FL 33199, USA
e-mail: gcrowthe@fiu.edu

B. Fonseca
e-mail: fonsecab@fiu.edu

F. Perera
National Defense University, Washington, D.C, USA
e-mail: fabiana.s.perera.civ@ndu.edu

A. Farhadi et al. (eds.), *The Great Power Competition Volume 3*,
https://doi.org/10.1007/978-3-031-04586-8_6

GPC in WHEM

The 2018 National Defense Strategy clearly identified "inter-state strategic competition" as the "primary national security concern" of the United States, displacing terrorism as the main focus. Subsequent policy documents have elaborated on what this means for the defense enterprise [40]. The Indo-Pacific Strategy Report released later that year aligns with this idea and singles out the "People's Republic of China, under the leadership of the Chinese Communist Party" as a competitor looking to reorder the region [61]. The US European Command's posture statement similarly references great power competition but highlights the Russian Federation as a "primary threat" and "long-term strategic competitor" [59]. More recently the Biden administration has appeared to move away from the use of the term "great power competition" to describe the relationship between the US and the People's Republic of China in favor of "strategic competition".

Though geographically distanced from these two competitors, our own hemisphere is also an arena in which US competitors have shown interest. US SOUTHCOM's 2020 posture statement referred to the western hemisphere as a "critical space" in great power competition [31]. US NORTHCOM's 2021 posture statement addresses the need to defend the homeland from our competitors who field long-range weapons and systems. Like the other strategy documents, it names the Russian Federation and the People's Republic of China as peer competitors.

Even as they are often addressed together under the heading of "peer competitors" or, more recently, "strategic competitors," the Russian Federation and the People's Republic of China differ significantly in their goals and the ways in which they attempt to achieve those goals. This is very true of the western hemisphere in which both competitors have increased their presence, but have done so in markedly different ways. Russia, under the leadership of Vladimir Putin, has shown a desire to build on its past engagement with the region to strengthen relations with Latin American and Caribbean states. The Russian Federation is pursuing two goals through reengagement. First, it seeks to boost its credentials as a global power. Second, it seeks to maintain a presence in the U.S.' near-abroad as a check against U.S. involvement in Russia's near-abroad (e.g., Georgia, Ukraine, the Baltic states).

Unlike Russia, PRC ties to the region are much more recent: Mexico and the Russian Federation established diplomatic relations in 1890, Mexico didn't establish diplomatic relations with the PRC until nearly a century later, in 1972. PRC ties to the region are also shallower: while there is a thriving Russian-Chinese interest in the region appears to be mainly driven by an interest in securing access to natural resources and markets [33, 62]. Both the People's Republic of China and the Russian Federation have been pragmatic in their approach to the region, displaying flexibility and ability to adapt quickly as they pursue economic and security objectives in Latin America [35, 54]. The two authoritarian states have sought to employ cyber tools including disinformation, cybercrime, electoral interference, and investment in critical infrastructure in different ways as they pursue their goals in the region.

Global Cyber Competition

The Russian Federation and the People's Republic of China operate in cyberspace around the globe and around the clock. They operate in the diplomatic, informational, military and economic areas. Although many in Latin America and Caribbean have seen cyber competition as an issue for Russia, the People's Republic of China, the European Union and the United States, they are discovering that, to paraphrase Leon Trotsky, "You may not be interested in cyber, but cyber is interested in you."

The People's Republic of China seeks to play the long game and ensure that Chinese interests are secured, including the Belt & Road Initiative (BRI) as well as access to natural resources and commodities. They seek to improve global perceptions and the influence of the People's Republic of China, advancing diplomatic claims including the South China Sea and recognition of Taiwan. Although they do not seek military supremacy, they do want to bolster the People's Republic of China's military capabilities, however their major area of competition is by advancing the People's Republic of China's economic interests, as increasing economic status will ensure a quiescent Chinese population and enable the other instruments of national power, i.e., diplomacy, information, and military power.

Since 2007, the Russian Federation has sought to overturn the current global system that they believe is designed to advantage the United States and partners to the detriment of other players such as the Russians. Their main strategic goal is regime survival and the morphing of the current system to a multipolar global system. They do not challenge the system overtly but seek opportunities to destabilize it. They seek to undermine international institutions by fracturing NATO and EU while undermining international law (humanitarian law, norms, etc.). They seek to fracture Western democracies through political warfare and seek to regain a buffer zone in Eastern Europe. Although the Western Hemisphere is not a strategic priority for them, they will seek to challenge the US in its near abroad with a goal to reduce US influence in the region.

Cybersecurity in the Western Hemisphere

In the western hemisphere internet access rates are higher than for other lower- and middle-income countries. While there is wide variation in internet connectivity rates across the region, in total there are nearly 400 million people with access to the internet. The rate of internet penetration varies greatly across the region. In The Bahamas (85%) and Chile (82%) internet penetration is equal or greater to the rate in the US (81.9%) [13].

The relatively high rates of internet connectivity could be good news for the region. Last year Latin American countries saw a significant growth in e-commerce as people struggled to get the things they needed while heeding the advice to stay home. Some countries in the region have been successful at harnessing internet connectivity to improve the delivery of government services through e-government. As with many other indicators of development in the region, however, there is significant intra-regional inequality in access to e-governance: whereas Argentina, Chile, and Uruguay rank among the 35 countries in the world with best e-government, Haiti, Honduras, and Cuba rank among the least developed (Table 1).

The rapid pace at which technology is expanding in the region is outpacing the state's capacity to regulate, coordinate, create awareness, and attract and cultivate online knowledge and skills. Though the region is keeping pace with others in its ability to enact legislation (about 86% of countries in the western hemisphere have some cybercrime legislation compared to 79% worldwide) very few cybercrimes are ever prosecuted in the region [63].

Governance shortcomings create vulnerabilities that can be exploited by authoritarian states seeking to compete with the United States. Testifying before Congress, US Southern Command commander Admiral Craig S. Faller recently identified cyber as one of the domains in which the US faces competition in Latin America.

The vulnerabilities that exist in the cyber domain in Latin America parallel other weaknesses in the region. Governance indicators point to the region's limitations with respect to government effectiveness, controlling corruption, and guaranteeing the rule of law. In cybersecurity, the region's biggest weakness is its poor governance. A recent report found that Latin America trails all other regions in the world, except the Commonwealth of Independent States, in its commitment to improving governance of cybersecurity, as demonstrated by a low commitment to cybersecurity policy and strategy, a dearth of cybersecurity indicators, and a lack of identification of a government agency responsible for this area.

Cyber Workforce in Latin America

As policymakers and strategists acknowledge cyberspace as a fourth war-fighting domain co-equal to the traditional domains of land, sea, and air, militaries across the world will have to recruit, retain, promote, and retire a workforce to operate in this domain [22]. The North Atlantic Treaty Organization took this step in 2016, two years after its members had agreed that a cyber attack could rise to the level of military assault, resulting in a triggering of the collective defense clause in the agreement [8]. The United States had recognized the importance of the cyber domain and created a sub-unified combatant command under US Strategic Command in 2010. It was elevated to full and independent command in 2018.

Latin American countries have also recognized the growing importance of the cyber domain. There is wide disparity in how the countries in the region have responded to the cyber domain: while some have established joint commands to respond to the threats, the Inter-American Development Bank (IADB) identified 21 countries in the region as being in the "initial" phase of building institutional capacity in cyber [49]. The outliers include Argentina, Brazil and Peru. Argentina established

Table 1 E-Governance Index for countries in Latin America United Nations. 2020. Available at https://www.un.org/development/desa/publications/publication/2020-united-nations-e-government-survey

Country name	E-Government index
Antigua and Barbuda	0.6055
Argentina	0.8279
Bahamas	0.7017
Barbados	0.7279
Bolivia	0.6129
Brazil	0.7677
Canada	0.842
Chile	0.8259
Colombia	0.7164
Costa Rica	0.7576
Cuba	0.4439
Dominica	0.6013
Dominican Republic	0.6782
Ecuador	0.7015
El Salvador	0.5697
Grenada	0.5812
Guatemala	0.5155
Guyana	0.4909
Haiti	0.2723
Honduras	0.4486
Jamaica	0.5392
Mexico	0.7291
Nicaragua	0.5139
Panama	0.6715
Paraguay	0.6487
Peru	0.7083
Saint Kitts and Nevis	0.6352
Saint Lucia	0.5444
Saint Vincent and the Grenadines	0.5605
Suriname	0.5154
Trinidad and Tobago	0.6785
United States of America	0.9297
Uruguay	0.85
Venezuela	0.5268

a cyber comm as part of the joint staff in 2014, Brazil established a cyber defense command in 2016 [3]. Peru did so in 2020 [52].

Beyond establishing organizational structures to operate in the cyber domain, countries attempting to be effective in this domain will have to recruit, train, and retain a cyber workforce. In the United State the National Initiative for Cybersecurity Education, a partnership between government, academia, and the private sector administered by the US Department of Commerce, warns that there is a large talent gap in cybersecurity. In the public sector specifically it estimates that the supply/demand ratio in this sector is about 1.7, meaning that there is one available professional for every 1.7 jobs available [26]. In the Armed Forces specifically there is a manpower requirement of 6,244 billets distributed across all services [21]. While the military has been able to successfully fill these, maintaining a mission-ready force in cyber will remain a challenge.

The challenge of recruiting, training and retaining a cyber force is not unique to the United States. The global cybersecurity workforce shortage means that by 2022 there will be nearly 2 million unfilled positions in this sector [28]. For the armed forces in the US and elsewhere, meeting the manpower demands of this domain will require big efforts in recruitment, training, and retention. In his nomination hearing US Secretary of Defense Lloyd Austin identified cyber as an area in which the US faces threats from the People's Republic of China, Russia, and Iran [60]. He elaborated on the importance of having a workforce that can respond to cyber threats promising that if confirmed he would If confirmed, "drive efforts to recruit, train, and retain our cyber workforce more effectively to better defend our networks" and "improve cybersecurity training, personal responsibility and awareness for the entire workforce."

Countries in Latin America are also facing challenges recruiting, training, and retaining a cyber workforce. The Inter-American Development Bank identifies only five countries in the region as having reached an "established" level of training in cybersecurity education: Argentina, Brazil, Colombia, Trinidad and Tobago, and Uruguay [48]. The vulnerabilities of cyberspace are well-known to the region. The 2020 Conference of Defense Ministers of the Americas, the highest-level gathering of defense officials in the hemisphere identified "concern over the growing threat of cyber attacks" as an area in which the region could work together and called for the creation of a new ad-hoc working group to explore how the region could best work together to face this threat [1].

In Latin America about 30% of hiring expect to increase their cyber workforce according to a report by the Center for Cyber Safety and Education [34]. This makes the region second only to North America in expected growth of the sector. Though most workers enter these jobs after working in another field (80% of cyber professionals in Latin America had a first career other than cybersecurity), they still require training and education to be able to transition into the cyber sector [34].

Partnering with Latin America as it works to expand and better its cyber workforce, particularly in the security and defense spheres, should be very important to the United. US competitors have sought to gain ground as providers of training and education for countries in the region. Already the People's Republic of China has established a Forum for Scientific and Technological Cooperation between the PRC and Latin America and the Caribbean that seeks to strengthen ties between the People's Republic of China and the region through sharing technology and education [11]. US Southern Command named both cyber and military training as areas in which the People's Republic of China has "increase[d] its self-serving activities to gain global influence and leverage [...] in the SOUTHCOM AOR." The statement highlighted the important role that security cooperation, particularly initiatives focused on capacity-building, plays in countering competitors in the region [31].

Capacity-building in the cyber-sector in Latin America will not only help to counter competitors' advances in this arena but could serve to create avenues for employment in the region, directly addressing a perpetual concern of Latin American leaders and citizens: economic underdevelopment. The cyber sector could be an area of employment growth for Latin America if its countries are able to find the right policies to recruit and retain professionals. Moreover, a strengthened cyber security infrastructure could reduce costly attacks in the region. In 2017 alone Latin America lost $97 billion to cyberattacks [20]. To put that figure in perspective, there are nearly a dozen countries in the region with GDP's smaller than that. Building a capable cyber workforce is economically important to the region and strategically important to the United States, perhaps even more so than strengthening capacity across the traditional domains.

State-Sponsored Cyber Operations in the Western Hemisphere

In addition to the criminals, terrorists and internal actors operating against states in the hemisphere, States operate as well. States or their proxies will operate in cyberspace against other states, international organizations, terrorists, businesses, and the populations of other countries. They also operate against criminals, proxies of others, and hacktivists who are operating against them.

The two major states operating in the Western Hemisphere are the People's Republic of China and Russia. Cuba and Venezuela operate on a limited basis in the area as well, mainly causing trouble for Colombia.

The Russian Federation and the People's Republic of China each use all of the instruments of national power: diplomacy, information, military, and economics. None of these instruments is separate from the others. As an example, the PRC uses diplomacy, intelligence operations and operations in the information environment in support of their economic goals. Each of these four areas is interlocked with the others; all are performed simultaneously in support of strategic goals. Though the instruments of national power are the same in both cases, Russia and China differ in how they deploy them and to what ends.

As argued above, the People's Republic of China has been argued to use instruments of power mainly in pursuit of economic objectives. Citing these aims it markets civilian and law enforcement technology to the region, such as the surveillance equipment purchased by Ecuador, for example. Though local groups there have expressed concern about the use of surveillance technology, Huawei, the manufacturer, has responded by saying that it provides technology but "does not get involved" in the public policy that would determine its use. The Russian Federation on the other hand is thought to be pursuing the advancement of ideas that oppose liberal democracy and the liberal world order. In the cybersphere this is pursued through a range of tools including perhaps most notably disinformation campaigns.

Diplomacy

Diplomatically, the Russians and Chinese seek international support for their policies. The Chinese seek support for their efforts such as the BRI as well as their stance vis-à-vis Taiwan and the South China Sea. The Taiwan issue is important in the Western Hemisphere in those nine of the 11 states that recognize the Republic of China are in the region: Belize, Guatemala, Haiti, Honduras, Nicaragua, Paraguay, Saint Kitts and Nevis, Saint Lucia and Saint Vincent and the Grenadines. These states are kept on side by Taiwan through lavish gifts and investments by Taiwanese companies. Despite this, all these countries are potentially in play, as shown by Panama switching sides in June 2017 and El Salvador recognizing the PRC in August of 2018. They will continue to work diplomatically to isolate Taiwan. This paper will do an in-depth analysis of Chinese digital diplomacy in the section that examines Chinese technological penetration in the hemisphere.

The People's Republic of China will also support those who oppose the U.S. in the hemisphere with two goals. They not only seek to frustrate the U.S. but also to ensure access to resources such as Venezuelan oil and a Nicaraguan canal dominated by the Chinese. However, the People's Republic of China will not work to destroy networks in the Hemisphere but work to replace the U.S. as the regional hegemon. They will also seek to build support for recognition of Chinese claims in the South China Sea as well as support for Chinese approaches to information & cyber in multilateral fora, such as the International Telecommunications Union.

The Russian Federation seeks to bolster partners in the area and to weaken US influence. As such, they will work to support Cuba, Venezuela, Nicaragua, and Bolivia in multilateral arenas. The Soviet Union enjoyed a modicum of success by supporting regimes that opposed the U.S. during the latter Cold War at a low cost to themselves; they seek to replicate that strategy which has a very low cost as compared to a potential payoff. They will also seek to gain support for Russia's international adventures such as the annexation of Crimea.

Information

Information is important in two ways: as data and as a tool that enables states to influence others. Both the People's Republic of China and the Russian Federation will conduct intelligence gathering and seek to influence states in the region. They will conduct these operations not only as a stand-alone but will also seek to enable diplomatic and economic activities through intelligence and influence operations.

The People's Republic of China will mainly conduct intelligence operations in support of their economic goals. Although the theft of intellectual property is the most well-known manifestation of this economic intelligence, the Chinese will conduct intelligence operations to determine what local markets look like. Not only will they seek to guarantee access to natural resources such as oil and lithium via contracts that are favorable to Chinese interests, but they will also seek to buy companies that will guarantee access to those resources but also seek to dominate local markets. They will conduct intelligence operations to determine which companies are weakest and how to use those vulnerabilities to Chinese advantage, to include taking over companies or destroying companies which compete against Chinese-aligned interests [14].

They will also conduct intelligence in support of diplomatic goals, accessing networks to determine the stance of a country vis-à-vis their goals as well as finding people who either sympathize with them or people who are vulnerable and can be subordinated to their goals.

The Chinese will also use influence operations to support their diplomatic and economic activities. These actions will increase support for and decrease resistance to Chinese activities. Some activities may be building a road, a railway, or a stadium. Others will include operations in the information environment such as external operations from Chinese sources such as Xinhua as well as operations from Chinese partners in the local countries.

The Russians will concentrate on intelligence operations, but only because the Hemisphere is a low priority. They will conduct widespread intelligence operations against governments in the region. They will conduct intelligence operations against those that support the United States such as Colombia and Chile as well as countries in the region who are influential such as Brazil and Mexico [47]. The Russian Federation will also conduct intelligence against their friends in the region to assist the regime in maintaining control as well as identifying actors within the various regimes who are pro- and anti-Russian.

As the Russian Federation has robust information capabilities, they will conduct operations in the information environment throughout the Hemisphere, but only as part of their strategy to provide low-cost irritation to the U.S. They will mainly conduct White and Grey Propaganda through their normal outlets such as *Actualidad RT* or RT (the former Russian Federation Today) in Spanish. This propaganda will support their global priorities such as gathering support for Russian activities in Crimea, Ukraine, and Syria as well as seeking to diminish support for the U.S., particularly in international fora [10].

Military

Neither the People's Republic of China nor the Russian Federation seeks to challenge the U.S. militarily, particularly in the Western Hemisphere. Their goal will be to support their friends militarily, to make new friends, and to open new markets for arms exports.

The main friend that they will support militarily is Venezuela. Unlike the 1980s, no one is interested in conducting military operations against Cuba or Nicaragua, and no one threatens other partners in South America. The two major threats to Venezuela are the United States and Colombia [46]. Although neither the U.S. nor Colombia want to intervene in Venezuela, aggressive rhetoric from the U.S. as well as a traditional enmity between Colombia and Venezuela lead the People's Republic of China and the Russian Federation to at least consider the possibility. The huge difference in capability between the extremely professional Colombian military and the internally focused and logistically incompetent Venezuelan military is also a danger sign. The Russian Federation has assisted Venezuela through arms exports such as SU-30 aircraft.

The Russian Federation will also monitor friendly networks to ensure that they maintain situational awareness of the military situation on the ground, as well as identifying potential military problems (i.e. coup plans) within Venezuela [45].

Russian activities will center on intelligence to determine the intent of the Colombians and the U.S. They will seek to enter Colombian and U.S. intelligence and command (C2) and control networks to conduct operational preparation of the environment so that they can slow down or take out C2 networks in case of conflict. They will also work to identify leaders of any potential military conflicts and conduct operational preparation of the environment with a goal of isolating those leaders in case of conflict.

The Chinese and Russians will also conduct intelligence operations against the more influential militaries in the region: Brazil, Chile, and Mexico. They will seek to enter the C2 networks "just in case" they might have to conduct offensive cyber operations against them in case of future conflicts.

Both the Chinese and the Russians will also work to bolster the offensive cyber operations capabilities of their partners, particularly Cuba and Venezuela.

Both the People's Republic of China and the Russian Federation will use diplomacy and cyber intelligence operations to see if they can convince states in the region to be a potential market for Russian arm sales.

Economic

The Chinese are truly a global operator economically. The Russians are just the opposite. One of their major lessons learned from the Minsk II sanctions by the EU and individual states such as the U.S. and Germany is that they need to de-link their economy from the global economy to limit the ability of the West to use sanctions to influence Russian activities. As such, the Russian Federation has few economic interests in the Western Hemisphere. They mainly seek locations where Russian elites can hide their illegally obtained resources.

The People's Republic of China seeks to guarantee access to resources and to open new markets. Much of their cyber activity will be economic espionage and influence operations, as described above [58]. They will seek to obtain favorable contracts to guarantee resources, take over local companies to guarantee resources, and seek to obtain monopolies in states to not only guarantee access to resources but to be able to coerce the state in question if necessary. Their cyber operations will be designed to support those goals: intelligence operations to determine opportunities, influence operations to decrease local resistance to Chinese operations, and operational preparation of the environment to increase Chinese ability to coerce states.

These economic goals are the most important Chinese goals in the Western Hemisphere. Much of their activity will be designed to support these goals. The Chinese understand that the economic instrument of national power is the one that enables the others and has identified the U.S. economy as the source of American power; they seek to replicate that situation in their own country. Their cyber operations in the Western Hemisphere will reflect this priority.

The US House of Representatives has published [64] a detailed list of Chinese engagement in Central America and Mexico, the Caribbean and South America while the Carnegie Endowment has published (2018) an analysis of Russian activity.

Chinese and Russian Network Penetration

When considering specific Chinese and Russian operations in the Western Hemisphere, states fall into one of three categories: large countries, friends of the People's Republic of China and/or Russian Federation, and other countries. Both the Russian Federation and the People's Republic of China have very good reasons to penetrate the various networks in most countries in the Hemisphere [32]. Based on their known targeting of networks in other regions around the world including North America, Europe, and Asia, it is possible to predict where they will target in WHEM.

The large countries include Argentina, Brazil, Canada, Chile, Colombia, Mexico, and the United States. Their friends include Cuba, Nicaragua, Venezuela, and sometimes Bolivia and Ecuador. Other countries attract attention based on their natural resources and critical infrastructure.

There are six basic types of networks that they will be in: intelligence (including unclassified data) [29], security, natural resources, information (influence, not data), critical infrastructure, and government. The Cyber and Infrastructure Security Agency (CISA), part of the Department of Homeland Security, says that "(Chinese) activities resulted in the theft of trade secrets, intellectual property, and other high-value information from …multiple foreign governments" [25] while the Russians perform "worldwide cyber exploitation of network infrastructure devices…by Russian state-sponsored cyber actors. Targets are primarily government and private-sector organizations, critical infrastructure providers, and the Internet service providers (ISPs) supporting these sectors" [24]. Indeed "Russian-attributed cyber actions were found to be present in 85 countries spanning a total of 6 continents and 16 world regions including Central America…Northern America…(and) South America" [23].

They will be in intelligence networks to see what their targets know. They will be in security networks to see what their targets' level of readiness is and what plans they have developed. The People's Republic of China is particularly interested in natural resources. Widespread use of mis- and disinformation demonstrates their interest in those networks. Penetration of critical infrastructure allows them to understand the state of critical infrastructure, to determine whether they can gain control of the critical infrastructure and figure out how to shut down their target's functions if necessary [53]. Last, they are interested in penetrating government networks to monitor government decision-making to understand what the process is and what decisions their targets are going to make.

The People's Republic of China, Russian Federation, and their fellow travelers Iran and North Korea have a proven history of penetrating all these types of networks around the world. Most proven examples come from North America and Europe because those areas are where most economic, internet and international political activity occur. Since these four actors perform these types of operations in North America and Europe, it is safe to believe that they are performing the same activities in other areas, including Latin America and the Caribbean.

Chinese, Russian, Iranian and North Korean interest in other countries is driven by their particular interests and political goals for each separate state. Some are strictly political, such as Russian and Chinese in who they can get to ideologically support their efforts or China's interest in who recognizes Taiwan versus the People's Republic of China. Other interests revolve around natural resources. Everyone has an interest in where they can make money, either legitimately through investments or buying and selling products where others (e.g. North Korea) are interested in where they can steal money. One example that shows the breadth of target countries and organizations is the FBI and Australian Signals Directorate (Thier version of the NSA) that reveals that the Avaddon ransomware campaign targeting academia, airlines, construction, energy, equipment, financial in a variety of countries including Brazil, Costa Rica, and Peru [6].

Large States

The Chinese and Russian are particularly interested in the large states in the hemisphere. Several are global players, many have natural resources that the People's Republic of China seeks to control, and all have critical infrastructure that the People's Republic of China would like to influence or purchase. Both the People's Republic of China and the Russian Federation have a proven track record of penetrating networks in the US. Canada, a NATO ally, is also a major target. They will seek to penetrate the intelligence networks of all these major countries to determine what they know. The People's Republic of China also has a penchant for gathering data from around the world to achieve their own version of big data. They believe that using AI will allow them to crunch that data to provide them with an advantage over other global players. The Russian Federation will additionally seek the information from intelligence networks to pass to their friends, particularly Cuba and Venezuela.

Both will seek to penetrate security networks in the large countries to determine vulnerabilities. Both states will seek to fully understand the command-and-control (C2) networks for large country security systems. Not only will that inform their operations around the world but will also potentially give them an advantage should any large country seek to use their security capacity against themselves or their friends [2]. Their ideal situation would be for them to be able to shut down the security networks of any potential adversary or the friends of any potential adversaries who would seek to assist them against the People's Republic of China in places like the South China Sea.

Large countries in the hemisphere have natural resources. The People's Republic of China understands their vulnerabilities because they must import most of their resources to feed their manufacturing base. Part of the Belt and Road Initiative (BRI) is providing Chinese investment to access natural resources. The People's Republic of China will seek to have permanent access to natural resources or will seek to control those resources themselves by purchasing natural resource companies. Penetrating networks of these companies will allow the People's Republic of China to identify vulnerabilities as well as give the People's Republic of China information that they can use to their advantage during contract negotiations for purchasing the resources or the companies themselves. Additionally, the People's Republic of China will be able to operate against networks of any companies that compete against Chinese influenced or owned companies.

As mentioned above, the People's Republic of China and the Russian Federation seek to use information operations to delegitimize the operations of others and legitimize their own. Large countries in the hemisphere are natural targets for both Chinese and Russian information operations. Penetration of information networks reveals their decision-making, allowing both the People's Republic of China and the Russian Federation to identify those who tend to support them as well as those who oppose them. Both will then develop cyber operations to the benefit of their fellow travelers and the detriment of those who oppose them. Even more insidious are information operations designed to undermine the political system of their targets and exacerbate existing fractures within societies. Information operation can cause the population to lose trust in both their actual government as well as their system of governing. Democracy requires the population to support the system, Both the People's Republic of China and the Russian Federation seek to delegitimize the whole concept of democracy and espousing their own authoritarian systems.

Both the People's Republic of China and the Russian Federation seek to penetrate critical infrastructure in large countries in the hemisphere [43]. As with natural resources and discussed above, this has an economic element. The People's Republic of China seeks to control critical infrastructure as much as natural resources and will penetrate all critical infrastructure in larger countries. One example of critical infrastructure operations in the Western Hemisphere was when "the Russian speaking cybercrime group, Revil" performed a ransomware attack on Brazilian based JBS, the world's largest meat processing company which shut down facilities in the United States, Canada and Australia.

Both the People's Republic of China and the Russian Federation penetrate government networks of large countries as part of current global competition. Knowing how governments make their decision is an advantage while monitoring decision-making in real time allows both countries to make moves internationally to support their points of view.

The People's Republic of China and the Russian Federation will penetrate every major network in the larger countries in the Western Hemisphere to gain economic and informational advantages.

The People's Republic of China and the Russian Federation have several partners around the world, mainly united in their opposition to what they perceive as a US-led global system that supports democracy, human rights, and free(ish) trade. This group includes those targeted by the US, allies, and partners as well as those who seek undemocratic rule.

Their Friends

Their partners in the hemisphere include Cuba, Venezuela, Nicaragua and sometimes Bolivia and Ecuador. Cuba has been a Russian client since the early 1960s, and the People's Republic of China has stepped up their diplomatic and economic activities with Cuba since the end of the Cold War [7]. Venezuela cozied up to the Russian Federation and the People's Republic of China after Hugo Chavez took over in 2000 and remains close [57]. Venezuela also has a very close relationship with Iran. Although Nicaragua the state is not inherently aligned with these actors, the Sandinista party has been, and the current president Daniel Ortega has been friends with the Russian Federation since the Sandinistas took control of Nicaragua in a revolution in 1979. Although Ortega was out of power between 1990 and 2007, he regained control of the country in an election and has maintained his grip on power since then. Although he does take pains to avoid antagonizing the US, he clearly aligns with the People's Republic of China (who has promised to build a trans isthmian canal through Nicaragua) and Russia. Bolivia and Ecuador have been close to both the People's Republic of China and the Russian Federation when their leftist parties have been in control. Bolivia was close to both when Evo Morales was the president 2006–2019. Although Morales was ousted in 2019, his Movement for Socialism (MAS) party is back in power and technocratic MAS President Luis Arce (who came to power in November 2020) has not specifically aligned yet, however there will be pressure from the more ideological elements of the MAS to continue their relationship with the People's Republic of China and Russian Federation. Former Ecuadorian president Rafael Correa aligned his country with Hugo Chavez and hence the People's Republic of China and Russian Federation. He served as President

2007–2017. The President between 2017–2021, Lenín Moreno (who was Correa's vice president 2007–2013), renewed relations and cooperation with the United States while conservative Guillermo Lasso took office in May 2021.

Both the People's Republic of China and the Russian Federation are solidly in the intelligence and security networks of their partners. Using the excuse of assisting their partners, both will dig in as far as possible to monitor the security situation in each country, because the security system is the only thing that keeps their friends in power. This will also allow them to influence the security system in each country to their advantage.

Their partners in WHEM usually do not control natural resources or major critical infrastructure assets and tend to be among the poorest countries in the region. The outlier is Venezuela who was well-off when Chavez took over, had large oil reserves, and was a major oil exporter [57]. Mismanagement has run the oil system into the ground which has given the People's Republic of China the opportunity to buy oil futures at good prices and allowed the Russians to assist *Petróleos de Venezuela SA* (SA is the equivalent of incorporated in the US) or PDVSA. The People's Republic of China will be tracking lithium in Bolivia while the Russian Federation will monitor nickel in Cuba and oil in Venezuela.

All their partners have learned from the People's Republic of China and the Russian Federation and make extensive use of information operations, both domestically and internationally. Aggressive use of the infosphere marks authoritarians everywhere. The Chinese provide inexpensive information infrastructure that supports authoritarians, which will be covered in detail in the following section about Chinese technology penetration in the hemisphere. Chinese and Russian help to their partners in developing domestic information systems will allow them unfettered access to domestic information infrastructure.

While the US, allies and partners seek to avoid eavesdropping on each other, both the People's Republic of China and the Russian Federation have a very strong motivation to monitor government decision-making inside their partners' governments. Because they have few friends, they cannot afford to give up the ones that they have. This will lead both states to penetrate every network in the governments of friendly states in the Western Hemisphere [19]. This will allow them to identify different factions within these states and support the ones that in turn support them, providing what to them would be a virtuous loop of supporting those who will support them. This was common practice during the Cold War where every major power had their own proxies and sought to keep them on side, updated for the twenty-first century by the inclusion of cyber operations.

Both the People's Republic of China and the Russian Federation are no doubt in the networks of other states throughout the hemisphere, based on their interest in each country. They will seek access to both the intelligence and security networks of anyone who is close to the US or anyone that could be a threat to them. They will also be in the networks for both natural resources and (to a lesser extent because there is less) critical infrastructure. This list would include countries like Jamaica (bauxite) and Trinidad & Tobago (natural gas). They will have a presence in information networks, but other countries are typically not a target of Chinese or Russian

Table 2 Chart #X graphically displays the three types of countries and the six types of networks that the Chinese and Russians have probably penetrated

	Intelligence	Security networks	Natural resources	Information	Critical infrastructure	Government network
Large countries	X	X	X	X	X	X
Their clients	X	X	X	X		X
Other countries	Selected	Selected	Selected		Selected	Selected

information operations except for those countries who still recognize Taiwan. These countries are also among the ones whose government networks the Chinese will seek to penetrate. Currently Guatemala, Honduras, Haiti, Paraguay, Nicaragua, St Vincent & the Grenadines, Saint Lucia, and Belize are among the 15 states that recognize Taiwan and not the People's Republic of China. Panama changed their recognition to the PRC in 2017 while the Dominican Republic and El Salvador switched in 2018. This is a major issue for the PRC who monitors this situation very closely. They are no doubt in each government decision-making network in the countries that still recognize Taiwan and must have been in the networks of the three that changed. As no one leaves a network that they have taken the time to penetrate, the People's Republic of China is most probably still inside the networks of the three states that changed their recognition most recently.

Panama is the smaller state that is probably the biggest target of the People's Republic of China. The Panama Canal is critical infrastructure; their relationship with the United States is very important, and they just switched recognition three years ago. That leads one to assume that the People's Republic of China has infiltrated every major network in the Republic of Panama and monitors Panamanian intelligence, security, economic, information and decision networks very closely.

Based on their track records elsewhere and their level of interest in various states within the Western Hemisphere, it is very likely that the People's Republic of China and the Russian Federation (and perhaps Iran and North Korea) have penetrated most important networks throughout the entire region as part of their strategic objectives. As individual countries become more important to the People's Republic of China and/or Russia, they will penetrate their networks as well.

Chinese Technology Penetration

Part of the growing cyber threat around the world is Chinese technology penetration. As with everywhere else around the world, the People's Republic of China is expanding its technological footprint throughout Latin America and the Caribbean. The spread of Chinese technologies over the last decade corresponds with its rise as a global economic and political power. The People's Republic of China's need for resources to sustain its growth and markets for Chinese goods has driven exponential increases in Chinese trade with the region, with total trade growing from US$8 billion in 1999 to US$307 billion by 2019. In recent years, Beijing has leveraged its economic and diplomatic access and large SEOs to spread a wide range of technologies—from laying fiber optic cables and information technology infrastructure to establishing safe cities and data centers.

There is a strategic and even aspirational intent to the People's Republic of China's technology penetration globally. Chinese engagement in Latin America and the Caribbean should be viewed as part of a global effort aimed at expanding Chinese influence, and in part, shaping a world conducive of its authoritarian political model. One way that Beijing is differentiating itself from its competitors is by leveraging Chinese technology companies to engage at local and provincial levels—often with security institutions—bypassing the need to work through central governments. Still, there are elements of opportunism that guide its global—and regional—engagement, and the spread of technology across the region varies.

The increase in Chinese technology penetration takes place at a time when Latin American and Caribbean countries are looking to the People's Republic of China as a means of diversifying their economic and diplomatic engagement beyond the United States and increasing their access to technologies critical to social, economic, and security development. Additionally, local and municipal governments are desperate for resources that central governments are unable to provide. COVID-19 is further incentivizing the region to find opportunities to enhance its connectivity to cyberspace—for educational, economic and public health reasons, among others. This paper explores the People's Republic of China's strategic expansion of technologies across Latin America and the challenges and opportunities for the U.S. and the region.

Chinese Global Technology Proliferation

With their 2013 *Made in China 2025* strategic development plan, the People's Republic of China made clear its aspirations as a global technological superpower. The plan underscores the importance of technological disruption across all economic sectors and emphasizes advanced information technology (IT) and automation as catalysts for industrial transformation and global economic growth. The People's Republic of China's IT and automation efforts focus on internet infrastructure, high speed computing, artificial intelligence (AI), internet of things (IoT) devices, and robotics. Many of these technological advancements have profound impacts and application in the domestic security spaces.

The strategy complements the People's Republic of China's 2013 Belt and Road Initiative (BRI). The BRI is an ambitious and still somewhat ambiguous foreign and economic policy strategy aimed at facilitating Chinese political and economic global integration around the world beyond just commodities and infrastructure. In 2017, the People's Republic of China's President Xi Jinping introduced the Digital Silk Road (DSR) as a third component to the BRI [27]. Xi asserted the centrality of the DSR and technology proliferation in the People's Republic of China growth strategy during the opening ceremony of the Belt and Road Forum for International Cooperation—a People's Republic of China-led forum designed to facilitate engagement with foreign governments and multilateral organizations [50].

The People's Republic of China also asserted its intentions to expand Chinese technologies to markets in Latin America and the Caribbean in its 2016 Policy Paper on Latin America and the Caribbean and during the 2018 meeting held between Chinese officials and the foreign ministers of the Forum of the Community of Latin American and Caribbean States (CELAC) [67]. In both, Beijing asserted information technology among a short list of priorities for cooperation in the region. Furthermore, technological innovation and expansion was used in describing a range of cooperative priorities from military exchanges to media, space and infrastructure cooperation.

The Digital Silk Road Expands to Latin America

The proliferation of Chinese technologies along the DSR in Latin America and the Caribbean comprises four general areas of expansion: digital infrastructure, advanced technologies, e-commerce, and digital diplomacy [30]. These four areas are interconnected, and combined they afford the People's Republic of China a comprehensive, multidisciplinary and expansive technological footprint that increases dependency of the People's Republic of China.

Digital Infrastructure

The Chinese use digital infrastructure as one of the most important aspects of the People's Republic of China's global technology penetration strategy. In addition to expanding Chinese influence, it also deepens nations' dependency on Chinese technologies. These digital infrastructure projects include cellular networks, data centers, smart/safe cities, surveillance technologies, and fiber optic cables.

Through massive Chinese state-owned enterprises such as Huawei and ZTE, the People's Republic of China is a lead in providing the region with 3rd, 4th, and 5th generation telecommunications networks. The People's Republic of China has made massive ICT investments in Mexico, Brazil, and Venezuela. In fact, in 2019 Huawei saw a 21% increase in its revenue from Latin America. Chinese companies have established telecommunications cables connecting Brazil to Africa and are working on a cable connecting Chile with the People's Republic of China. The adjacent map shows Chinese ICT penetration in the Caribbean, Central America and part of South America.

Chinese technology giants are rapidly increasing the transfer of surveillance technologies to the region. The notion of smart cities is a prominent concept in President Xi's vision of the DSR. A smart city consists of an array of surveillance capabilities ranging from surveillance cameras, facial and license plate recognition to command and control centers, data labs, IoT technologies, and real-time situation reporting and analysis [38]. Many components of these surveillance capabilities are spreading in LAC through dominant Chinese tech-firms such as Huawei, ZTE, Megvil, Dahua, Alibaba, and Hikvision, among others [5]. In Montevideo, the Uruguayan police installed approximately 2,100 video surveillance cameras donated by ZTE and Huawei across 347 locations [39]. In 2016, Huawei provided the city of Cochabamba in Bolivia with its Safe City Solution that included intelligent surveillance, visualized critical communication, surveillance cameras, and a converged command system [37]. Additionally, Guyana, Panama, and Mexico also procured and implemented components of Huawei's Safe City System throughout their respective countries to enhance law enforcement capabilities.

The Chinese state-owned enterprise, China National Electronics Import & Export Corporation (CEIEC) is another prominent actor in establishing smart defense systems in cities across LAC. According to CEIEC, the corporation specializes in delivering integrated-tailored-solutions for their customers to respond to challenges from sovereignty threat, natural disaster, social crisis, and economic development [12]. CEIEC systems are utilized by military, police, and fire organizations in Argentina, Bolivia, Ecuador, Peru and Venezuela. These systems are comprised of a network of regional response centers, servers, thermal and biometric surveillance cameras, and an automated platform designed to deliver video evidence to courts. CEIEC provides on-site engineering, technical and operational support to these centers. The establishment of physical infrastructure like data centers and telecommunications equipment is an important achievement for Huawei [18]. A data center is a physical facility that houses critical applications and massive amounts of data [36]. Per example, in Chile, Huawei formed a partnership with the University Federico Santa Maria (USM) to create a data center to store data from Chile's ALMA telescopes (66 12-m telescopes which provide detailed insights on star births and planet formation) for the next nine years [36]. Additional data centers were also established in Brazil and Argentina, and there are centers planned in other parts of Central America and the Caribbean.

Advanced Technologies

Developing advanced technologies is the People's Republic of China's second pillar and reinforces the importance of establishing a dominant Chinese-built digital infrastructure across Latin America to integrate Chinese-developed advanced technologies. As outlined in its strategic planning, advanced technologies encompass high speed computing (quantum), artificial intelligence (AI), biotechnology, and financial, space, surveillance technologies [16].

One notable example of Chinese advanced technologies is the development of the BeiDou System, the People's Republic of China's own satellite navigation system. The People's Republic of China launched its indigenous version of the GPS in early 2000s and it is already set to take over the U.S. GSP. The BeiDou System affords Chinese systems self-reliance in position, navigation and timing (PNT) and will allow the People's Liberation Army the ability to avoid using the prominent U.S. Department of Defense's GPS satellite navigation system. In 2014, the People's Republic of China was working with Brazil to establish a BeiDou monitoring station and in more recent years the People's Republic of China is working to expand its coverage over Latin America and the Caribbean. Widespread use of BeiDou would allow Chinese technologies such as surveillance monitoring to be fully integrated within Chinese telecommunications and GPS platforms and out of the reach of the U.S. and Latin American governments.

Another example is the People's Republic of China's recent collaboration with Cuba. In early 2020, the People's Republic of China's Hebei Foreign Studies University signed a series of agreements with Cuban universities to advance research and collaboration in AI technologies. As a product of the agreements, the People's Republic of China and Cuba established the International Institute of Research in Artificial Intelligence and the International Friendship Hospital—the latter argued to be an innovative and AI infused medical facility. Representatives stated that the collaborative ventures would pursue "high-impact" medical research with Cuban specialists that intersect with advanced technologies applicable to biotechnology, biomedicine, the environment, the quality of life of the elderly, smart cities, precision agriculture, and industry [55]. The center will also host research laboratories focused on AI, big data, Internet of Things, blockchain, cloud computing, data management and software development [41].

Of course, space research and quantum computing—both very much interconnected—are also important advanced tech spaces that the People's Republic of China is pursuing (with global reverberations). In 2014 the People's Republic of China and Argentina established a space agreement where the People's Republic of China located one of its China Deep Space Network ground stations in the Neuquen Province of Argentina. There is lots of speculation about the nature of the research, however, some speculate that it encompasses space technologies, including data acquisition. As far as quantum computing, the People's Republic of China recently claimed that it was in the lead as far as quantum computing technology is concerned. In December 2020, the People's Republic of China demonstrated that it created a machine that was a "trillion times faster than the next best supercomputer." Quantum computing is argued to be the next most important milestone in computing evolution and crucial to technological dominance in the coming decades [17].

E-Commerce

E-Commerce is the third pillar of Chinese technology penetration in Latin America and the Caribbean. During the 2018 the People's Republic of China—Latin America and Caribbean Summit in Montevideo, Uruguay, Chinese leadership emphasized the importance of e-commerce as a key component of its Belt and Road Initiative for the region. Since then, the region has seen an increase of Chinese tech companies such as Zhaogan.com and Alibaba promote e-commerce throughout Latin America and the Caribbean. E-commerce consists of the integration of digital and real economies to conduct transactions through digital platforms. The promotion and widespread use of cheap e-commerce activities through Chinese digital platforms further integrates Chinese and Latin America economies. There is also speculation that e-commerce could possibly strengthen the People's Republic of China's soft economy and advance their efforts to internationalize the Renminbi—the latter is something the People's Republic of China has been pursuing in the region since at last 2016.

For example, e-commerce has led to an explosion in agriculture commerce. Online supermarkets are bringing Latin America food products to Chinese middle-class tables. In 2018, total sales from Chinese online grocery sales amounted to more than US$51 billion. Alibaba, through various e-grocery platforms such as T-Mall, JD, Hema, Yiguo, and Suning are quickly capturing Latin America markets [66]. Still, Latin America is in nascent stages of embracing e-competence. COVID-19 is likely increasing the pace that Latin Americans are adopting online shopping and the People's Republic of China is ready to capture that increasing market space. The People's Republic of China is also a leader in digital payment methods.

Digital Diplomacy

The final pillar of the DSR is digital diplomacy. Beijing is engaging in digital diplomacy to "fill the developing world's digital infrastructure gap, to avoid dealing with the BRI's hard infrastructure quagmire, and pursue greater hard and soft power geopolitical ambitions [44]." The People's Republic of China leverages massive SEOs such as Huawei and ZTE to advance Chinese foreign policy objectives as part of ongoing business negotiations. The People's Republic of China also leverages its access to markets to promote its vision of internet governance. The People's Republic of China's preference is for a closed, state-controlled internet that firewalls its political system from ideas antithetical to its interest and subsequent survival. To that end, the People's Republic of China has relentlessly demonstrated its determination to champion digital diplomacy to shape the global and regional discussions on norms and principles of the internet. Chinese technology exports along the DSR paves the way for Beijing to garner support and authority in spreading principles consistent with its political model [15].

The People's Republic of China is also cultivating long term soft power. Often referred to as the People's Republic of China's Charm Offensive, the People's Republic of China is leveraging global media platforms in order to promote an image of itself that appeals to populations all around the world. According to Zhoa Huang, Chinese digital engagement serves to…(a) to spread official discourse and views in the diplomatic arena to strengthen the government's impact and (b) to accrue sympathy capital and increase attractiveness through strategic placement of imagery (e.g. static images, moving gifs, and video clips) and textual narrative" [65]. This is also known as "*Panda Engagement*." The People's Republic of China is increasing its media presence in Latin America and the Caribbean as well. There are several print and digital platforms that provide Beijing-filtered news directly to overseas Ethnic Chinese in Latin America as well as content for the broader Spanish and Portuguese speaking communities. For example, Xinhua, People's Daily and China's Radio International, three top Chinese state-sponsored outlets, provide continuous content in Spanish and Portuguese through various digital platforms.

While all the DSR pillars are complimentary, advancing digital infrastructure has been most salient in Latin America and the Caribbean. Chinese tech giants like Huawei and ZTE have been successful at securing digital infrastructure and building surveillance technologies and data centers around that infrastructure in countries such as Chile, Argentina, Venezuela, Brazil, Uruguay, Bolivia, Ecuador, Guyana, Panama, and Mexico.

Opportunities and Challenges for Latin America

Chinese technology proliferation comprises important opportunities and challenges for the region. On one hand, technologies increase the region's connectivity, enable better provisions of security, and infuse capital into the region. On the other hand, Chinese tech also compromises the region's digital sovereignty, creates physical cybersecurity vulnerabilities, offers intelligence collection opportunities for the People's Republic of China, and empowers autocratic governments. Ultimately, it is the regional leadership's responsibility to hedge the benefits against the challenges in order to advance the social, economic, political, and security interests of societies across Latin America and the Caribbean.

Improving the Region's Technological Footprint

The region is desperate for advanced technologies that help countries keep pace with the emerging digital economy and allows countries to innovate in critical areas such as economic development, governance, and security. Furthermore, COVID-19 is highlighting the technological gap across the region as communities struggle to pivot to online platforms for business, education and other vital services. The infusion of digital infrastructure provides technological capability for the region's growing internet user community. For example, social media users in Latin America grew in 2014–2016 by an average of 12.3%. Latin America comprises over 300 million social media users in 2020. Increased connectivity demands improvement in technological infrastructure. Additionally, the surge in technological sophistication also comes with upskilling Latin American workers required to operate in a more high-tech and connected workplace. Upskilling workers for technical jobs is vital to the future of work in Latin America and the Caribbean where there is still a large portion of the workforce that is unskilled and being displaced by machines.

Increase Security and Law Enforcement Capacity

Security and law enforcement institutions also benefit from greater technological sophistication. Surveillance technologies, big data analytics and data management, enhanced communications technologies, faster and more powerful computing capacity, and cybersecurity threat tools. The proliferation of surveillance technologies and data centers—with the right protections of human rights and civil liberties—in Latin America can provide security institutions enhanced capability. Surveillance technologies can act as important deterrents to common crime. Data centers, better computing power, and enhanced communications equipment allow law enforcement to run vehicle license plates, check suspect backgrounds, and communicate and coordinate activities more reliably—all in real time. These technologies also help governments target more pervasive threats from illicit trafficking organizations and domestic insurgences. Chinese tech is also useful for facial recognition, and biometrics and forensics investigative efforts. Finally, technology, used honestly, can improve transparency and accountability among security organizations.

Important Capital Inflows

Finally, Chinese technologies often come with important capital inflows vital to the economic health of the region. Although overall Chinese investment inflows appear to be slowing, access to capital remains vital for a region that has experienced economic stagnation over the last decade and is experiencing a downturn as a result of the global pandemic. In 2019, Chinese technology investors committed "$70.5 billion in technology investments, focusing on artificial intelligence, fintech, e-commerce, and mobility" around the world [42]. Latin American and Caribbean tech entrepreneurs can unlock tremendous resources partnering with Chinese technology monsters like Tencent, Alibaba, etc.

Overall, improving the region's technological footprint, enhancing security institutions capacity to combat threats, and access to capital inflows from Chinese investors are important to the social, political, economic, and security future of Latin America and the Caribbean. Scholars and observers argue that the limited technological penetration is holding back the region's development. Despite the benefits, there are still critical challenges that the region must hedge against. The region must contend with the loss of digital sovereignty, the vulnerabilities to physical cyber systems, the People's Republic of China's access to the region's data, prospects for Chinese intelligence gathering, and the empowering of autocratic governments like Venezuela.

Loss of Digital Sovereignty

One of the biggest challenges for the region lies in its ability to maintain digital sovereignty. The sheer range of Chinese digital infrastructure—including its advanced technologies—is creating tremendous dependency on the People's Republic of China for all technology needs, from routers and networks to data management software and cloud storage access. The loss of digital sovereignty, that is diminishing control a country has over its data, technology platforms, etc.—will undermine the independence of the country's governments and business leaders to make decisions antithetical to Chinese interests. It is akin to Russian influence over Ukrainian energy because Ukrainian gas flowed through pipelines that sat on Russian Federation territory. In several instances, the Russian Federation cut off Ukraine access to exact pressure on Ukrainian leaders. The People's Republic of China's control over its technology footprint in Latin America and the Caribbean could end in similar fate.

Intelligence and Data

The spread of Chinese technologies is challenging democratic institutions around the world, creating intelligence collection opportunities for Beijing. For now, there is substantial evidence that indicates the People's Republic of China is stealing American and western intellectual property in order to leapfrog global competition in important economic and security spaces such as technology innovation. Although the theft of intellectual property is the most often cited form of Chinese industrial espionage, the Chinese also undertake intelligence operations to obtain local market advantages. For example, a common Chinese tactic for bolstering economic reach involves the acquisition of local companies in foreign markets. In the past, Chinese state security services have assisted corporate takeovers by providing intelligence on the internal deliberations and potential vulnerabilities of local companies targeted for mergers or acquisitions. To date, however, there is little evidence that the People's Republic of China's access to surveillance or data in the region is being used for Chinese intelligence operations. Still, the threat remains that at some point in the future, the People's Republic of China could use their access to Chinese technologies around the world to inform decision making and gain advantage in global competition.

Empowering Autocratic Governments

Technology itself is neither autocratic or democratic. Technology can, however, serve to enable and embolden authoritarian regimes. In Latin America and the Caribbean, there are already several cases where autocratic leaders are using Chinese surveillance technologies to monitor political opposition. In Venezuela, the infamous "fatherland card" developed by ZTE was used to track voting patterns of citizens [56]. Still, Chinese surveillance technologies are still in infant stages and there are no credible open-source reports of Latin American governments misusing Chinese surveillance technologies, the region has a long history of misusing surveillance capabilities. In Colombia during the last decade, political leaders used wiretaps to monitor political opposition. Former Panamanian President Ricardo Martinelli was accused of the same thing when he deployed a spyware called Pegasus to spy on union leader and journalist [4]. There are countless other examples in Peru, Brazil, Argentina, Honduras, etc.

Critics are arguing that the concern over misuse of Chinese surveillance technologies by Latin American leaders is more critical now given the impacts of the global pandemic. Governments are debating using the technology to more closely monitor individuals and track vitals to stop the spread of COVID-19. For example, In Mexico two Uber drivers were fired because Mexican authorities alerted the company after using Chinese surveillance technologies to track a COVID-19 positive individual [9].

Finally, there are second, third order effects of Chinese technology engagement such as erosion of rule of law, inducing corruption, etc. that should also challenge governments' ability to embrace the opportunities. Cutting edge technologies and capital inflows are likely to entice corrupt public officials looking to maximize their personal wealth while in office. This is something common throughout Latin American history. Furthermore, Chinese business practices are agnostic to western ethics when it comes to conflicts of interest among government employees. There are various anecdotes about Chinese courting public officials with lavished trips to the People's Republic of China and other gifts to win lucrative contracts in Latin America.

View from the United States

Certainly, the view from Washington is complicated. On one hand, some in the American foreign policy establishment argue that Chinese engagement is good for both the region and U.S. interests—if done in a manner consistent with Western norms. They advocate for seeing Chinese engagement through the lens of competition and encourage facilitating Chinese engagement rather than contain it. They also seek to leverage Chinese engagement to help enhance the region's prosperity and help usher in a more secure and resilient Latin America. On the other hand, hawkish foreign policy professionals and public officials label Chinese technology proliferation a national security threat, and more importantly, advocate that view to U.S. partners in Latin America and the Caribbean. The policy approach for this group leans more towards containing and blocking Chinese technology proliferation rather than competing against it. They also argue that overall Chinese engagement, thus far, has done little to improve democratic governance and the social and economic well-being of the region. It is still too early to conclude whether the benefits outweigh the costs or vice versa for the U.S., but the implications for merit stronger assessments.

The potential gains for the U.S. are clear and consistent with the benefits to the region described above. The U.S. benefits from a stable, innovative, and prosperous hemisphere—providing those democratic values and liberal economic principles remain the preferred political and economic models. At the same time, Chinese technology penetration in Latin America can undermine U.S. strategic interests if it undermines the efficacy of democratic institutions and expands Chinese influence beyond the U.S. in vital economic, political, and security landscapes. This is happening in some places where Chinese digital penetration is advancing Chinese soft power and where Chinese tech is encroaching on citizens' civil liberties. The People's Republic of China's security engagement, and potential dependency on the People's Republic of China for technology provisions, is positioning itself as an alternative to U.S. security assistance and could influence the security calculations of countries in the region in the long-term, while also providing the People's Republic of China with important technological footholds and access in this hemisphere. Finally, Chinese intelligence capacity to collect against the U.S. public and private sectors is also a concern to the U.S.

Policy Recommendations

The U.S. should disaggregate the good from the bad and resist labeling all Chinese activities as a threat to the U.S. and the region's interests. The U.S. should also resist viewing Chinese technology proliferation as zero sum. As mentioned above, there are important potential benefits that the U.S. can leverage to advance its own interest in the region. To that end, the following are recommendations that should frame how the U.S. approaches the region [51].

Promote Democratic Governance and Rule of Law

Continue deepening U.S relationships across the hemisphere and encourage those partners in pressuring the People's Republic of China to operate according to the accepted international norms.

Reinforce and help strengthen the democratic institutions to build resiliency against the People's Republic of China's ability to leverage its engagement to bypass the rule of law or promote authoritarian alternatives to desperate leaders failing to deliver across the region. The U.S. has and should remain an important player in enhancing the governing capacity of our partners in the region.

Encourage Latin American and Caribbean countries to maintain integrity in their own laws and demand the People's Republic of China to adhere to those laws as it engages the region—especially labor and environmental policies.

Increase American Technology Proliferation in the Region

Provide funding to educational institutions to deliver technology training and education—in the areas of cybersecurity, AI, advanced computing, computation ethics, etc. Dominating the educational landscape around high-tech could serve as a firewall to some of the challenges above.

Invest in American technological innovation, including 5G, artificial intelligence, and quantum computing, and incentive U.S. companies to compete in Latin American and Caribbean markets abandoned by U.S. technology providers.

Create technology entanglements between U.S. and Latin American technology companies. In other words, incentive U.S. companies to partner, joint venture, and/or acquire Latin American technology start-ups. The U.S. can establish its own technological roots in the region, but with strong Latin American and Caribbean equity.

Advance American Soft Power & People to People Relationships

Increase American soft power and relationships by fostering global engagements using American universities, non-governmental organizations, private sectors, and other non-state actors. Build a "whole hemisphere" strategy.

Evaluate the usage of the 160 Bi-national Centers for Latin America (BNCs) currently operating in the region and explore ways to use them as a counter to Chinese propaganda and bolster U.S. image in the region.

Enable opportunities for deeper people-to-people relationships using non-governmental actors already abundant and rooted in the region.

Enhanced U.S. ability to compete in the information space. The U.S. could help expose and exploit the contradictions in Beijing's policy and the divisions that exist between the People's Republic of China and the region, such as cultural differences, political systems, and business practices, including over-promising commitments and labor practices (using of Chinese labor). The U.S. should help make the region aware of how the People's Republic of China is undercutting the region's long-term economic and political interests.

Enhance Security Cooperation Beyond Just Military

Increase U.S. counterintelligence operations at home and abroad to combat Chinese theft of intellectual property and its associated intelligence collection activities.

Institutionalize relationships beyond the region's militaries to include more substantive and lasting engagements/partnerships with national, provincial/state, and municipal law enforcement organizations. Our traditional military-to-military relationships are strong; however, the U.S. lacks robust domestic law enforcement cooperation with our Latin American partners. Chinese have outlined engagement with law enforcement entities as a priority—and that is where Chinese technology penetration is most dangerous.

Increase training in key areas transparency/anti-corruption and rule of law and continue focusing on developing long-term sustainable security relations with key partners (through tools such as International Military Education and Training).

Conclusion

The Russian Federation and the People's Republic of China are conducting cyber operations throughout the Western Hemisphere, seeking to obtain support for their global policies and operations. While the Russian Federation seeks to destabilize the global system to their own advantage, the People's Republic of China seeks to maintain the current system and replace the U.S. as the global hegemon. Although their daily cyber operations will not pose a clear and present danger to U.S. goals and activities in the region, the cumulative effect of their cyber operations will endanger U.S. influence in the region, could circumscribe the U.S. ability to operate in the region and, in the worst case, cause the U.S. to allocate resources to the area in case of a global conflagration. The U.S. needs to operate in the region as well as build partner cyber capacity throughout the area.

The People's Republic of China is expanding its technological footprint throughout Latin America and the Caribbean as part of its broader global strategy to lead the world as a technology superpower. It is also evident that Beijing is guiding and aligning its massive technology focused state-owned enterprises and state-influenced enterprises with the country's foreign and domestic policy objectives. Beijing is also differentiating itself from its competitors by leveraging Chinese technology companies to engage at local and provincial levels—often with security institutions—bypassing the need to work through central governments. This is an important distinction between the thrust of U.S. security engagement and how the Chinese are acquiring market share in the security landscape.

Still, it is important to note that there are elements of opportunism that guide its global—and regional—technology engagement, and the spread of technology across the region varies. It is also important to highlight that not all Chinese technology proliferation is bad. Increasing the technological capabilities of the region to compete in the twenty-first century, improving security capacity to tackle crime, and leveraging capital inflows from the People's Republic of China can improve the economic, social, political, and security wellbeing of the region. At the same time, the region must hedge against the potential threats such as the loss of digital sovereignty, government overreach and compromising of citizen human rights and civil liberties, and the associated intelligence vulnerabilities with Chinese safe cities and data centers. If managed right by Latin American leaders and U.S. policymakers, the cost–benefit of Chinese technologies could favor the benefits (Table 2).

References

1. (PDF) revista científica estudios en seguridad y defensa, volumen 15, número 30 | revista. CONFERENCIA DE MINISTROS DE DEFENSA DE LAS AMERICAS. (2020). Retrieved October 13, 2021, from https://www.academia.edu/49358355/Revista_Cient%C3%ADfica_Estudios_en_Seguridad_y_Defensa_Volumen_15_N%C3%BAmero_30.
2. By, U. P. I. (2021). United Press International https:// www.upi.com/UPI delivers the latest headlines from around the world: Top News, International, U. P., & UPI delivers the latest headlines from around the world: Top News. (2019, November 6). *U.S. security heads warn Russia, China, Iran will interfere in elections*. Gephardt Daily. Retrieved October 19, 2021, from https://gephardtdaily.com/national-international/u-s-security-heads-warn-russia-china-iran-will-interfere-in-elections/.
3. Agências Verde-Oliva. (2017). *Comando Conjunto na Defesa Cibernética*. Ministério da Defesa. Retrieved October 13, 2021, from https://www.gov.br/defesa/pt-br/centrais-de-conteudo/noticias/ultimas-noticias/comando-conjunto-na-defesa-cibernetica.
4. Alimonti, K. R. V. (2020, October 28). *Latin American governments must commit to surveillance transparency*. Electronic Frontier Foundation. Retrieved October 13, 2021, from https://www.eff.org/deeplinks/2020/10/latin-american-governments-must-commit-surveillance-transparency. (28 Oct 2020).
5. ASPI. (2021). *Mapping China's Tech Giants|ASPI|australian strategic ...* ASPI. Retrieved October 13, 2021, from https://chinatechmap.aspi.org.au/.

6. Australian Cyber Security Centre (ACSC) (2021) *2020–003: Ongoing campaign using Avaddon Ransomware*, from https://www.cyber.gov.au/sites/default/files/2021-05/2021-003%20Ongoing%20campaign%20using%20Avaddon%20Ransomware%20-%2020210508.pdf.
7. Bain, M. J. (2018). Russo-Cuban relations in the 1990s. *Diplomacy & Statecraft, 29*(2), 255–273. https://doi.org/10.1080/09592296.2018.1452431
8. Baldor, L. C. (2016). *Air, land, sea, cyber: NATO adds cyber to Operation Areas*. Phys.org. Retrieved October 13, 2021, from. https://phys.org/news/2016-06-air-sea-cyber-nato-areas.html. (14 Jun 2016).
9. Berri. (2020). *Covid-19 surveillance brings mixed results amid security concerns*. RSS. Retrieved October 13, 2021, from https://www.diplomaticourier.com/posts/covid-19-surveillance-brings-mixed-results-amid-security-concerns.
10. Biletska, T., Bratko, A., Ivashkov, Y., Kupriyenko, D., Rybshchun, O., & Torichnyi, V. (2021). Information and propaganda component of the Russian Federation Hybrid Aggression: Conclusions for developed democratic countries on the experience of Ukraine. *Trames. Journal of the Humanities and Social Sciences, 25*(3), 355, https://doi.org/10.3176/tr.2021.3.06.
11. *Bre El Fo R O Chin A -Celac*. China CELAC Forum. (2016). Retrieved October 13, 2021, from http://www.chinacelacforum.org/esp/ltjj_2/P020161207457618108481.pdf.
12. Ceiec. (2021). *About Ceiec*. Ceiec. Retrieved October 13, 2021, from https://www.ceiec.com/About/.
13. Census Bureau. (2017). *Digest of Education Statistics, 2017*. National Center for Education Statistics (NCES) Home Page, a part of the U.S. Department of Education. Retrieved October 13, 2021, from https://nces.ed.gov/programs/digest/d17/tables/dt17_702.60.asp.
14. Chang, R. C., lin, C. H., Chang, K. L., & Lai, C. M. (2021). *Dataset of propaganda techniques of the state-sponsored information operation of the People's Republic of China*. Dataset of Propaganda Techniques of the State-Sponsored Information Operation of the People's Republic of China. Retrieved October 19, 2021, from http://claws.cc.gatech.edu/mis2-papers/paper3-chang.pdf.
15. Cheney. (2019). [web log]. Retrieved October 13, 2021.
16. Chabra, T., Doshi, R., Hass, R., & Kimball, E. (2020, April 28). *Global China: Technology*. Brookings. Retrieved October 13, 2021, from https://www.brookings.edu/research/global-china-technology/.
17. *China claims quantum computing lead with Jiuzhang Photon Test*. South China Morning Post. (2020, December 4). Retrieved October 13, 2021, from https://www.scmp.com/news/china/science/article/3112649/china-claims-quantum-computing-lead-jiuzhang-photon-test.
18. Cisco. (2021). *What is a data center?* Cisco. Retrieved October 13, 2021, from https://www.cisco.com/c/en/us/solutions/data-center-virtualization/what-is-a-data-center.html#:~:text=Center%20and%20Virtualization-,What%20Is%20a%20Data%20Center,of%20shared%20applications%20and%20data. (5 Oct 2021).
19. Coha. (2011). *Arm sales, especially by Russia and China, continue to penetrate Latin America*. COHA. Retrieved October 19, 2021, from https://www.coha.org/arm-sales-especially-by-russia-and-china-continue-to-latin-america/. (3 Mar 2011).
20. Cote-Muñoz, N., & Laskai, L. (2019, August 15). *Is Latin America prepared for China's booming tech investments?* Americas Quarterly. Retrieved October 13, 2021, from https://www.americasquarterly.org/article/is-latin-america-prepared-for-chinas-booming-tech-investments/. (15 Aug 2019).
21. Crall, D. (2021). *United States senate - armed-services.senate.gov*. Armed Services Senate. Retrieved October 13, 2021, from https://www.armed-services.senate.gov/imo/media/doc/19-16_02-27-19.pdf.
22. Crowther, A. (2018, July 31). *The cyber domain*. The Cyber Defense Review. Retrieved October 13, 2021, from https://cyberdefensereview.army.mil/CDR-Content/Articles/Article-View/Article/1589056/the-cyber-domain/#:~:text=Both%20the%20Department%20of%20Defense,knowledge%2C%20influence%2C%20or%20activity. (31 July 2018).
23. Cunningham, C. (2020) *A Russian Federation information warfare primer*, from https://jsis.washington.edu/news/a-russian-federation-information-warfare-primer/.

24. Cyber and Infrastructure Security Agency (CISA). (2018). *Alert (TA18–106A) Russian state-sponsored cyber actors targeting network infrastructure devices*, from https://us-cert.cisa.gov/ncas/alerts/TA18-106A.
25. CISA. (2021). *China cyber threat overview and advisories*, from https://us-cert.cisa.gov/china.
26. Cyberseek. (2021). *Cybersecurity supply and demand heat map*. Cybersecurity Supply and Demand Heat Map. Retrieved October 13, 2021, from https://www.cyberseek.org/heatmap.html.
27. Deeks, R. (2020). *The digital silk road-China's $200 billion project*. BBC Science Focus Magazine. Retrieved October 13, 2021, from https://www.sciencefocus.com/future-technology/the-digital-silk-road-chinas-200-billion-project/. (15 Apr 2020).
28. *Digital Forensics & Cybersecurity Careers*. Methodist University. (2019, September 19). Retrieved October 13, 2021, from https://www.methodist.edu/dfcs/careers/.
29. Dorfman, Z. (2018) *Botched CIA communications system helped blow cover of Chinese agents*, from https://foreignpolicy.com/2018/08/15/botched-cia-communications-system-helped-blow-cover-chinese-agents-intelligence/.
30. Eurasia Group. (n.d.). *China's digital aid: The risks and rewards*. Council on Foreign Relations. Retrieved October 13, 2021, from https://www.cfr.org/china-digital-silk-road/.
31. Faller, C. (2020). *Posture statement of Admiral Craig S. Faller commander united states southern command*, from https://www.armed-services.senate.gov/imo/media/doc/Faller_01-30-20.pdf.
32. Faller, C. (2021). *Statement of admiral Craig S. Faller commander, United States Southern Command*, from https://www.southcom.mil/Portals/7/Documents/Posture%20Statements/SOUTHCOM%202021%20Posture%20Statement_FINAL.pdf?ver=qVZdqbYBi_-rPgtL2LzDkg%3d%3d.
33. Freitas da Rocha, F., & Bielschowsky, R. (2020). *China's quest for Natural Resources in Latin America*. Publication | Economic Commission for Latin America and the Caribbean. Retrieved October 13, 2021, from https://www.cepal.org/en/publications/44555-chinas-quest-natural-resources-latin-america. (19 Nov 2020).
34. Frost, S. (2017). *2017 global information security workforce study*. Center for Cyber Safety and Education. Retrieved October 13, 2021, from https://www.isc2.org/-/media/B7E003F79E1D4043A0E74A57D5B6F33E.ashx.
35. Giusti, S., & Penkova, T. (2008). From Ideology To Pragmatism. *World Affairs: The Journal of International Issues*, *12*(4), 14–53. Retrieved 2021, from https://www.jstor.org/stable/48505029?seq=1#metadata_info_tab_contents.
36. Huawei, S. (2016). *A telescope in Chile is generating massive amounts of data from the stars*. Business Insider. Retrieved October 13, 2021, from https://www.businessinsider.com/huawei-stargazers-are-data-mining-the-night-sky-2016-8. (28 Sept 2016).
37. Huawei. (2018). *Huawei smart city overview presentation*. Huawei Enterprise. Retrieved October 13, 2021, from https://e.huawei.com/en/material/industry/smartcity/02ad4d5ab608492ea24659ec667f04bd.
38. Huawei. (2018). *Visualized public safety lays a solid foundation for building a smart cochabamba*. Huawei Enterprise: New Value Together. Retrieved October 13, 2021, from https://e.huawei.com/topic/leading-new-ict-ru/visualized-public-safety-smart-bolivia-case.html.
39. Infodefensa. (2019). *Uruguay Blinda Con 1.000 Cámaras de Vigilancia La frontera con brasil*. Infodefensa. Retrieved October 13, 2021, from https://www.infodefensa.com/latam/2019/02/23/noticia-uruguay-blinda-camaras-vigilancia-frontera-brasil.html. (23 Feb 2019).
40. Mattis, J. (2018). *Summary of the 2018 national defense strategy*. DOD. Retrieved October 13, 2021, from https://dod.defense.gov/Portals/1/Documents/pubs/2018-National-Defense-Strategy-Summary.pdf.
41. Juventud, J. (2020). *Acuerdo cuba-china potenciará la inteligencia artificial*. Medium. Retrieved October 13, 2021, from https://medium.com/juventud-t%C3%A9cnica/acuerdo-cuba-china-potenciar%C3%A1-la-inteligenciaartificial-767006d80c33. (12 Feb 2020).

42. Kharpal, A. (2020). *Tencent pledges $70 billion investment in high-tech areas as Beijing pushes Digital Infrastructure.* CNBC. Retrieved October 13, 2021, from https://www.cnbc.com/2020/05/27/china-tech-giant-tencent-pledges-70-billion-investment-in-ai-cloud.html. (27 May 2020).
43. Liptak, K. (2021). *US blames China for hacks, opening new front in cyber offensive.* CNN. Retrieved October 19, 2021, from https://www.cnn.com/2021/07/19/politics/us-china-cyber-offensive/index.html. (20 July 2021).
44. Magnus, G., & Bretherton, H. (2020). *Will digital diplomacy cement the belt and road initiative's "common destiny"?* China Dialogues. Retrieved October 13, 2021, from https://blogs.lse.ac.uk/cff/2020/09/17/will-digital-diplomacy-cement-the-belt-and-road-initiatives-common-destiny/. (19 Nov 2020).
45. Malashenko, T. I. (2019). Russian military-technical cooperation in Latin America: Venezuela. *Ekonomicheskie i Sotsial'No-Gumanitarnye Issledovaniya, 3 (23)*, 136–144. https://doi.org/10.24151/2409-1073-2019-3-136-144.
46. Mena Report. (2020). *Inicio.* MPPRE. Retrieved October 19, 2021, from http://www.mppre.gob.ve/en/2020/05/03/usa-colombia-incursion-national-sovereignty/. (13 Oct 2020).
47. Mena Report. (2021). *FBI-DHS-CISA joint advisory on Russian Foreign Intelligence Service Cyber Operations.* Cybersecurity and Infrastructure Security Agency CISA. Retrieved October 19, 2021, from https://us-cert.cisa.gov/ncas/current-activity/2021/04/26/fbi-dhs-cisa-joint-advisory-russian-foreign-intelligence-service.
48. Moreno, L. (2021). *Ciberseguridad.* CiberSeguridad. Retrieved October 13, 2021, from https://observatoriociberseguridad.org/#/home.
49. Moreno, L., & Almagro, L. (2020). *Ciberseguridad.* CiberSeguridad. Retrieved October 13, 2021, from https://observatoriociberseguridad.org/#/home.
50. Net. (2017). *Full text of president Xi's speech at opening of Belt and road forum.* Xinhua. Retrieved October 13, 2021, from http://www.xinhuanet.com/english/2017-05/14/c_136282982.htm.
51. Office of the Clerk, U. S. H. of R. (2019). *Hearing: Dollar diplomacy or debt trap? examining China's role in the Western Hemisphere Subcommittee on Western Hemisphere, Civilian Security and Trade (Committee on Foreign Affairs).* Dollar Diplomacy or Debt Trap? Examining China's Role in the Western Hemisphere|Committee Repository|U.S. House of Representatives. Retrieved October 13, 2021, from https://docs.house.gov/Committee/Calendar/ByEvent.aspx?EventID=109459.
52. Pelcastre, J. (2020). *Fuerzas Armadas de Perú refuerzan defensa en ciberespacio.* Dilogo Americas. Retrieved October 13, 2021, from https://dialogo-americas.com/es/articles/fuerzas-armadas-de-peru-refuerzan-defensa-en-ciberespacio/#.YWb9_i2z1QI.
53. Pham, J. P. (2010). China's strategic penetration of Latin America: What it means for U.S. interests. *American Foreign Policy Interests, 32*(6), 363–381. https://doi.org/10.1080/10803920.2010.535762.
54. Piccone, T. (2020). *China and Latin America: A Pragmatic embrace.* Brookings. Retrieved October 13, 2021, from https://www.brookings.edu/research/china-and-latin-america-a-pragmatic-embrace/. (20 July 2020).
55. Publisher. (2020). *FIU Law Review, 14*(1). https://doi.org/10.25148/lawrev.14.1.3
56. Reports, S. (2018). *A new Venezuelan ID, created with China's ZTE, tracks citizen behavior.* Reuters. Retrieved October 13, 2021, from https://www.reuters.com/investigates/special-report/venezuela-zte/. (14 Nov 2018).
57. Reyes Vázquez, Á. (2021). The Sino-Venezuelan relationship (1974–2020). *Comillas Journal of International Relations*, (21), 030–052. https://doi.org/10.14422/cir.i21.y2021.002.
58. Saha, S. (2012). *CFIUS now made in China: Dueling national security review frameworks as a countermeasure to economic espionage in the age of globalization...* Retrieved October 19, 2021, from https://scholarlycommons.law.northwestern.edu/cgi/viewcontent.cgi?article=1030&context=njilb.
59. Scaparrotti, C. (2019). *USEUCOM 2019 Posture statement.* United States European Command. Retrieved October 13, 2021, from https://www.eucom.mil/article/39546/useucom-2019-posture-statement.

60. *Senate armed services committee advance policy questions for Mr. Lucas N. Polakowski nominee to be assistant secretary of defense for nuclear, chemical, and biological defense programs*. Senate. (2021). Retrieved October 13, 2021, from https://www.armed-services.senate.gov/imo/media/doc/Polakowski_APQs_08-06-20.pdf.
61. Shanahan, P. M. (2019). *Department of defense indo-pacific strategy report 2019*. Media Defense. Retrieved October 13, 2021, from https://media.defense.gov/2019/Jul/01/2002152311/-1/-1/1/DEPARTMENT-OF-DEFENSE-INDO-PACIFIC-STRATEGY-REPORT-2019.PDF.
62. Sullivan, M., & Lum, T. (2021). *China's engagement with Latin America and the Caribbean-Congress*. CRSREPORTS. Retrieved October 13, 2021, from https://crsreports.congress.gov/product/pdf/IF/IF10982.
63. UNCTAD. (2020). *Cybercrime legislation worldwide*. UNCTAD. Retrieved October 13, 2021, from https://unctad.org/page/cybercrime-legislation-worldwide.
64. US House of Representatives. (2021). *China regional snapshot: Mexico and Central America*, from https://gop-foreignaffairs.house.gov/mexico-and-central-america/.
65. Wang, R., & Huang, Z.A. (2020). *Panda engagement*. Taylor and Francis Online. Retrieved October 13, 2021, from https://doi.org/10.1080/01292986.2020.1725075.
66. Wei He, L., & Mu, H. (2019). *Farm to phone: How Latin America is feeding China with E-commerce*. Americas Quarterly. Retrieved October 13, 2021, from https://www.americasquarterly.org/article/farm-to-phone-how-latin-america-is-feeding-china-with-e-commerce/.
67. 杨希. (2018). *Full text of China's policy paper on Latin America and the Caribbean*. Full text of China's Policy Paper on Latin America and the Caribbean-China.org.cn. Retrieved October 13, 2021, from http://www.china.org.cn/world/2016-11/24/content_39777989.htm.

Dr. Alexander Crowther is a visiting research professor at the Jack D. Gordon Institute for Public Policy within the Steven J. Green School of International and Public Affairs at Florida International University (FIU) and is a non-resident Senior Fellow at the Center for European Policy Analysis in Washington DC.

Dr. Fabiana Sofia Perera is an Assistant Professor at the William J. Perry Center for Hemispheric Defense Studies. Prior to joining the Perry Center, Fabiana was a Rosenthal Fellow at the Office of the Secretary of Defense, Under Secretary for Policy, Western Hemisphere Affairs.

Brian Fonseca is director of the Jack D. Gordon Institute for Public Policy at FIU's Steven J. Green School of International and Public Affairs. He serves as the Institute's technical expert for national security and foreign policy. From 1997 to 2004, he served in the United States Marine Corps and facilitated the training of foreign military forces in both hostile theaters and during peacetime operations.

Operations in Cyberspace

When the Levee Breaks: A Global Trend of Cyber-Physical and Cyber-Operational Attacks Against Critical Infrastructure and Future Implications on the Great Power Competition

Steve Sin and Rhyner Washburn

Abstract Cyber-physical systems (CPSs) refer to systems that connect computers, communication channels, and physical devices. They lie at the heart of today's critical infrastructure. CPSs are currently one of the most targeted systems of adversarial actors operating in the cyber domain. Cyber-physical and cyber-operational attacks on critical infrastructures via attacks on CPSs have the potential to damage physical infrastructure assets and have widespread consequences for national security as well as the society. We analyzed 427 publicly reported cyber-physical and cyber-operational attacks conducted against critical infrastructures globally between January 1, 1992, and July 9, 2021. We find that of the attacks that can be attributed to an actor type, state actors (including state-affiliated and state-supported actors) were found to be the predominant actors that conduct cyber-physical attacks while state and non-state actors occupied approximately same ratio of attacks for cyber-operational attacks. We also find espionage to be the most statistically significant motivation for the state actors to conduct cyber-physical and/or cyber-operational attacks. Addition ally, we find rivalry between the attacker and the target to be the most statistically significant international security relevant variable. Finally, we provide an assessment of the implications of cyber-physical and cyber-operational attacks on critical infrastructure in the contexts of current **and future irregular warfare and great power competition.**

Keywords Cyber-attack · Cyber-physical · Cyber-operational · Cyber security · Cyber-physical system (CPS) · Critical infrastructure · Critical infrastructure protection · Industrial control systems (ICS) · Internet of things (IoT)

S. Sin (✉) · R. Washburn
University of Maryland, College Park, MD 20740, USA
e-mail: sinss@umd.edu

R. Washburn
e-mail: rwburn@umd.edu

A. Farhadi et al. (eds.), *The Great Power Competition Volume 3*,
https://doi.org/10.1007/978-3-031-04586-8_7

Introduction

Cyber-physical systems (CPSs) lie at the heart of modern society's critical infrastructure. Today, CPSs connect computers, sensors, communication devices, and actuators of the physical facilities and equipment via heterogeneous, open, system-of-systems, or hybrid configurations. As systems have become more computerized, they have also become more complex and interconnected, making disparate critical infrastructure sectors interdependent on each other [1]. Attacks against CPSs, therefore, do not simply disrupt or destroy a single equipment or facility but can potentially affect entire critical infrastructure sector or across different critical infrastructure sectors.

We have witnessed several cyber-physical and cyber-operational attacks against critical infrastructures since late twentieth century, with some recent examples being 2016 attack on Ukraine; 2017 global NotPetya attack that impacted companies such as Maersk; and 2021 Colonial Pipeline attack. Additionally, terrorist organisations such as Islamic State was suspected of attempting to 'hack' into the United States' electric grid [2] as well as attempting to launch cyber-attacks aimed at Britain's airports, hospitals, and electricity supply [3]. Successful cyber-physical and cyber-operational attacks on critical infrastructure would result in critical disruptions and high consequences. The grave consequences of cyber-physical and cyber-operational attacks against CPSs, therefore, makes security of these CPSs an utmost priority for any private industry and national government; furthermore, in the contexts of irregular warfare and great power competition, the need to secure the CPSs of our critical infrastructure against cyber-physical and cyber-operational attacks become even more urgent.

Although cyber-physical and cyber-operational attacks on critical infrastructure have been identified as one of the major security challenges for the foreseeable future, there has not been a systematic study of these attacks. One of the obstacles to doing so is that there has not been an aggregated collection of data on cyber-physical and cyber-operational attacks against critical infrastructure globally across critical infrastructure sectors. Lack of such aggregation of data has limited our ability to conduct systematic studies of the phenomenon, and our abilities to hypothesize and to gain a deeper understanding about the behaviors and motivations of the attackers.

To increase our ability to formulate theories and test hypotheses relevant to the behaviors and motivations of actors, we developed Significant Multi-domain Incidents against Critical Infrastructure (SMICI) dataset. The SMICI dataset, a first of its kind, collects and aggregates publicly available incident in-formation on cyber-physical and cyber-operational attacks on critical infrastructures across sectors globally. It collects on 38 individual variables per observation and, at the time of this writing, contains 427 attributed and unattributed cyber-physical and cyber-operational incidents carried out against critical infrastructures worldwide between January 1, 1992 and July 9, 2021.

Research Questions

Employing the SMICI dataset, we performed both qualitative and statistical analyses to address several research questions. The findings to these research questions will serve as the cornerstone of future research on cyber-physical and cyber-operational attacks on critical infrastructure. The findings will also allow us to glean implications these types of attacks will have in the future within the contexts of irregular warfare and great power competition. The research questions we address through our analyses are:

What are the trends of cyber-physical and cyber-operational attacks on critical infrastructure?

What is the distribution between cyber-physical and cyber-operational attacks?

Who are the primary actors that can be identified associated with cyber-physical and cyber-operational attacks on critical infrastructure?

What is the distribution between state (including state-affiliated and supported) and non-state actors for cyber-physical attacks?

What is the distribution between state (including state-affiliated and supported) and non-state actors for cyber-operational attacks?

What state actor is attributed most for cyber-physical attacks?

What state actor is attributed most for cyber-operational attacks?

What country's critical infrastructure is most targeted for cyber-physical and cyber-operational attacks by state actors?

What critical infrastructure sector is most targeted for cyber-physical and cyber-operational attacks by state actors?

The remainder of this chapter will be structured as follows: firstly, we define disruptive cyber-physical and disruptive cyber-operational attacks as well as providing the inclusion criteria for the SMICI dataset. Secondly, we provide a brief overview of the dataset's collection methodology, level of analysis, and variables included. Thirdly, we present the results of our analyses of the SMICI dataset. Finally, we conclude with a summary and implications of our findings.

Defining Cyber-Physical and Cyber-Operational Attacks

It was not too long ago that it was safe to assume a cyber-attack referred to an attack that began and ended in the cyber domain, with consequences of such an attack also confined largely to the cyber domain. Under this paradigm, a cyber-attack could affect essentially one or more of the three information security attributes, collectively known as the CIA triad [4]. Likewise, one could also assume that a physical attack referred to an attack that began in the physical world against a target in the physical world, with the consequences in the physical world. As information and communications technologies (ICTs) have advanced, and as these technologies have become ubiquitous, there are areas where these two spaces—cyber and physical—overlap.

For example, modern automobiles equipped with devices that rely on cellular services (such as OnStar or a WiFi hotspot) can become targets of cyber-attacks designed to cause physical malfunctions to the vehicles. In such an attack, the attacker's aim is to exploit the weaknesses of the ICT installed on the automobile to achieve his/her physical goal, such as causing the driver to lose control of the vehicle. The type of attack described immediately above is an example of a cyber-physical attack. Following this logic, if an attacker launched a cyber-attack against the industrial control system (ICS)[1] of a petroleum pipeline and caused the pipeline to rupture, then this would be a cyber-physical attack conducted against a critical infrastructure.

If the example of an attack on a petroleum pipeline described above is an example of a cyber-physical attack, then what if an attacker attacks the ICTs installed in the pipeline control room and causes the operators there not to be able to monitor the information that is being provided by the ICS or send instructions to the ICS to control the pipeline, resulting in the manual shutdown of the pipeline until the operators could once again receive information from and send instructions to the relevant ICS? Should such an attack be classified as a "traditional" cyber-attack, a cyber-physical attack, or something else? Strictly speaking, the attacker in this example did not attack the ICS, but the attacker did cause a disruption in the operation of the petroleum pipeline in the physical world. Since such an attack does not directly cause physical damage or disruption, it is difficult to categorize it as a cyber-physical attack. On the other hand, the attack did cause the operators to shut down the pipeline, which resulted in the disruption of pipeline operations. To account for attacks in this vein, we adopt the term cyber-operational attacks.

For the SMICI dataset, we include both cyber-physical and cyber-operational attacks on critical infrastructure in order to create a more comprehensive dataset. More specifically, to be considered for inclusion in the dataset, an incident must be a disruptive cyber-physical or a disruptive cyber-operational attack conducted against (a) critical infrastructure sector(s) as defined in the Presidential Policy Directive 21 (PPD-21), dated February 12, 2013.[2][5] We provide definitions of these two attack categories as well as the full inclusion criteria for the dataset in the sections below.

---

[1] We define ICS as the aggregation of various operational technology systems, devices, and process (e.g. HMIs, PLCs, RTUs, SCADA, DCUs, etc.); therefore, malware that targets operational technology is reported as ICS malware in the SMICI dataset. To date, ICS malware is exceptionally rare and only five types of ICS malware have been publicly reported as involved in disruptive cyber-physical incidents: Stuxnet, Havex, BlackEnergy (BE3 in particular), Industroyer/CRASHOVERRIDE, and Triton/Trisis. Havex and BlackEnergy (BE2 AND BE3) interacted with ICS systems, but did not directly cause cyber-physical disruptions. The original design of BlackEnergy malware was for cybercrime.

[2] PPD-21: Critical Infrastructure Security and Resilience, identifies 16 critical infrastructure sectors whose assets, systems, and networks, whether physical or virtual, are considered so vital to the United States that their incapacitation or destruction would have a debilitating effect on security, national economic security, national public health or safety, or any combination thereof. The 16 sectors identified are: Chemical, Commercial Facilities, Communications, Critical Manufacturing, Dams, Defense Industrial Base, Emergency Services, Energy, Financial Services, Food and Agriculture, Government Facilities, Healthcare and Public Health, Information Technology, Nuclear Reactors, Materials, and Waste, Transportation Systems, and Water and Wastewater Systems.

Definition of Disruptive Cyber-Physical Attacks

We defined disruptive cyber-physical attacks as an incident where a threat actor—state or non-state—executes malicious action(s) in the cyber domain that have a disruptive kinetic effect(s) in the physical domain. For this definition, we focus on the fact that the threat actor is able to cause disruption to operational technology by bridging the information technology and operational technology gap. In general, this type of incident occurs when a threat actor targeting critical infrastructure compromises the ICS.

From the Dataset:

Something Must be in the Water: Cyber-Physical Attack against Oldsmar Water Treatment Facility

Though cyber-physical incidents are rarer than cyber-operational incidents, their potential impacts can be much higher than cyber-operational incidents. Unlike cyber-operational incidents, however, cyber-physical incidents are those where threat actors execute attacks that directly affect the physical domain kinetically by exploiting vulnerabilities present in the industrial control systems (ICS) and causing disruption, degradation, and/or destruction of the physical systems. The rarity of these incidents is signified by the difficulty to gain access to ICS assets that directly leads to a physical effect. As such, these incidents are almost exclusively conducted either by state actors or by non-state actors (who quite often are considered to be insider threats). In the case of the latter, insider threats may have access to ICS assets directly and do not need to be technically sophisticated. Occasionally, these actors exploit access of software used by the target entity to conduct lateral movement in the network. That is allegedly the case with the incident that occurred at the Oldsmar Water Treatment Facility in the Tampa Bay area of Florida in 2021.

In the morning of February 5, 2021 (approximately 8 AM local time), an unidentified actor gained initial access to Oldsmar's network. A plant operator noticed the access on his workstation but was not concerned because the plant, as well as other Oldsmar city facilities, uses TeamViewer to give employees access to workstations remotely. In this instance, the plant operator assumed it was his supervisor who regularly used TeamViewer to access the plant network and workstations. Five and half hours later, the same unknown actor that accessed the system earlier in the day gained access to the Supervisory Control and Data Acquisition (SCADA) controls (an ICS system) at the Oldsmar plant. The plant operator noticed the mouse move across the screen of his workstation and initiating access to the software that monitors and controls water treatment systems. Over the next approximately five minutes, the unknown actor used the software to increase the sodium hydroxide (lye) level in the water from 100 parts per million (PPM) to 11,100 PPM. After increasing the sodium hydroxide level, the unknown actor exited the system. Fortunately, the plant operator who was monitoring what was happening worked quickly to bring the sodium hydroxide level back down to a safe level of 100 PPM.

While this incident did not result in any physical damage to the critical infrastructure affected or loss of life, it does serve to clearly illustrate the components of a cyber-physical attack on critical infrastructures. Firstly, the start point of the incident was in the cyber domain—the unidentified threat actor gained access to an online network remotely. Secondly, critical infrastructure was compromised through cyber means—the unidentified threat actor gained control of an ICS environment remotely and manipulated ICS control software remotely. Thirdly, the incident resulted in physical consequences—the unidentified threat actor manipulated the ICS controls via cyber means to change the sodium hydroxide level in the water.

As for attribution of this incident, based on what has been revealed about the incident, authorities believe the attack was probably carried out by an insider. Testifying before the U.S. Congress, former Cybersecurity and Infrastructure Security Agency (CISA) Director Chris Krebs said, "It's possible that this was an insider or a disgruntled employee. It's also possible that it's a foreign actor… But we should not jump to a conclusion that it's a sophisticated adversary."

Finally, this attack has not been attributed to any specific threat actor, it does illustrate the potential consequences of cyber-physical attacks on critical infrastructure. The fact that this incident occurred at a water treatment facility also causes some pause. In general, when one thinks about a cyber-physical attack on critical infrastructure, most people imagine rolling blackouts caused by electric grids being taken offline; or pipelines rupturing due to over-pressurization; or an explosion at a steel mill due to loss of its furnace controls. This incident, however, reminds us that cyber-physical attacks do not need to have the "Hollywood feel", and that cyber-physical attacks against critical infrastructure can be "quiet" with potentially enormous consequences regardless of the specific industrial sector being targeted.

Definition of Disruptive Cyber-Operational Attacks

We define disruptive cyber-operational attacks as an incident where a threat actor executes malicious actions through the cyber domain that have a disruptive effect in the physical domain without having direct action(s) against ICS. Rather, these attacks are designed to disrupt the information technology systems that are connected to the ICS or internet-of-things (IoT) systems and devices.[3] For the SMICI dataset, we included espionage incidents as cyber-operational incidents because remediation of systems after detection disrupts operations, and much of the cyber espionage in critical infrastructure is interpretable as either: (a) conducting reconnaissance for intelligence preparation of the battlefield, or (b) stealing intellectual property for economic gain.

[3] Ransomware and wiper malware are particularly effective in these incidents; however, sophisticated use of IoT malware such as BrickerBot and Mirai demonstrate the adaptability of the malware development and the vulnerability of our critical infrastructure to such threats.

From the Dataset:

Oil Runs Through It: Cyber-Operational Attack against the Colonial Pipeline

As our world has grown to be more connected, so too has our critical infrastructure. The interconnectedness of sectors such as energy and the transportation sectors illustrate a complex and vulnerable element of modern society. The disruption of one can affect and magnify disruption of the other. Such was the case with the Darkside ransomware attack against Colonial Pipeline, which began on Friday, May 7, 2021. A threat actor, who had exfiltrated 100 GB of sensitive data a day prior to the attack, deployed a ransomware known as the Darkside onto the Colonial Pipeline's network deployed a ransomware, known as the Darkside, onto Colonial Pipeline's network. The ransomware, primarily affecting Colonial Pipeline's billing systems, did not compromise any of the company's operational technology. With its billing systems affected, Colonial Pipeline, however, was unable to bill its customers appropriately. This forced the company to halt its pipeline operations, stopping the flow of petroleum for the majority of the East Coast of the United States. Although Colonial Pipeline paid the ransom the same day as the attack and received the key to unencrypt its system, it took additional five days (May 12, 2021) to restore fully its pipeline operations. During that five-day period, panic buying severely depleted service stations up and down the East Coast, and the effect lingered on beyond resumption of full operations. For example, an estimated 30% of service stations in Florida; 53% in Virginia; 72% in North Carolina; and 86% in the District of Columbia were "dry" as of May 14, 2021.

The cascading effect of ransomware affecting the billing systems of an oil pipeline company forced the company to shut down operations and disrupted petroleum product distribution to over 120 million people. Although this attack was believed to be carried out by a non-state actor, the incident demonstrated—to the United States and its adversaries alike—the interdependencies and accompanying vulnerabilities of modern critical infrastructure.

Inclusion Criteria

For an incident to be included in the SMICI dataset, an incident must meet all three of the following criteria:

The attack must have originated from the cyber domain.

The attack must target a critical infrastructure sector as defined by Presidential Policy Directive 21 (PPD-21), dated February 12, 2013.

The attack must be a disruptive cyber-physical incident OR a disruptive cyber-operational incident.

Dataset Overview

The SMICI dataset relies strictly on publicly available sources. Data is collected manually using several pre-designated search strings that we specifically developed for this research effort. Sources considered for the data collection effort include news reports, professional magazine articles, peer reviewed journal articles, industry briefs and reports, government reports, and credible social media sources.

We designed the dataset in an incident-target dyad format, meaning that every observation in the dataset provides one incident to one target pair.

For each observation, we record 38 individual variables. These variables are classified into seven variable categories. Table 1, provides a list of these seven variable categories as well as individual variables contained within each variable category.[4]

Findings

Overwhelming number of attacks identified in the SMICI dataset were cyber-operational attacks, accounting for approximately 94% of the identified incidents.[5] Cyber-physical attacks against critical infrastructure, on the other hand, accounted for approximately 6% of the identified incidents included in the SMICI dataset (see Fig. 1).

One of the most challenging tasks to determine when examining any cyber attack is the attribution of the threat actor. Determining the threat actors that committed cyber-physical and/or cyber-operational attacks against critical infrastructure is not different—attribution is extremely difficult, especially utilizing open-source materials. For the SMICI dataset, we were able to attribute 76 incidents, which is approximately 18% of the incidents included in the dataset. Of the incidents that were able to be attributed to either a state or a non-state actor, approximately 59% (49 incidents) were attributed to state actors while approximately 41% (29 incidents) were attributed to non-state actors (see Fig. 2).

Overall Cyber-Physical and Cyber-Operational Attack Trends

Cyber-physical and cyber-operational incidents against critical infrastructures have exponentially increased in 2012 compared to previous two decades, and again in 2017. Starting in 2017, number of incidents increase exponentially each year. Prior to 2012, the number of cyber-physical and cyber-operational incidents remained in

[4] For a full description and coding rules for each variable contained in the SMICI dataset, please consult the SMICI Codebook v1.2, dated July 28, 2021 [8]. A copy of the codebook can be obtained by emailing your request to: infostart@umd.edu.

[5] Of the 427 incidents included in the SMICI dataset, we were able to collect enough information to identify 376 incidents (approximately 88.06%) as either cyber-operational or cyber-physical attacks. We were not able to determine whether the remaining 51 incidents were cyber-operational or cyber-physical incidents, only that they occurred and were publicly reported.

Table 1 Variable categories and variables contained in the SMICI dataset

Variable groups	Variables
Actor variables	Attribution
	Actor type (State / Non-State)
	Threat actor name
	Motive
	Threat actor notes
	Polity of actor
Target variables	Rivalry with actor
	Involved in international conflict
	Currently engaged in conflict with actor
	Experiencing domestic conflict
	Polity of country where target is located
	GDP per capita of the targeted country
	Quality of overall infrastructure
	Connectedness of the targeted country
	Targeted country has adopted information security standards
	Critical infrastructure sector targeted/affected
	Critical infrastructure sub-sector targeted/affected
	Specific target of attack
Geographic variables	Target region
	Target sub-region
	Target intermediate region
	Target country
	Target State-Region (US Only)
	Target State-Division (US Only)
	Target State (US Only)
Malware/technical variables	Malware type
	Malware name/family
	Malware notes
Temporal variables	Execution date(s)
	Discovery date(s)

(continued)

Table 1 (continued)

Variable groups	Variables
	Disclosure date(s)
Impact variables	Attack type
	Financial impact
	Impact notes
Miscellaneous variables and sources	Confidence level
	Incident summary
	Analyst notes
	Sources

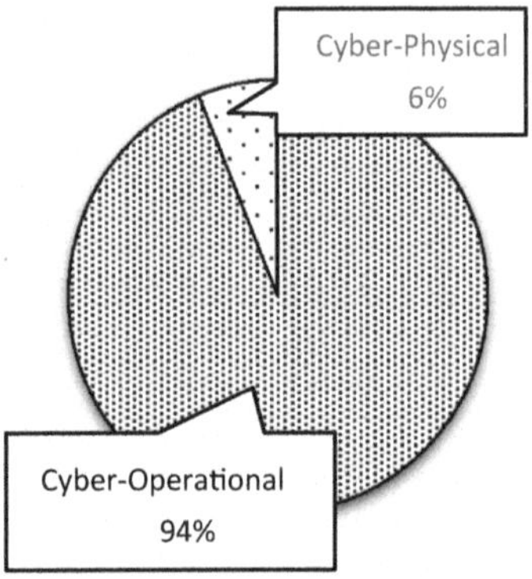

Fig. 1 SMICI distribution of incident type

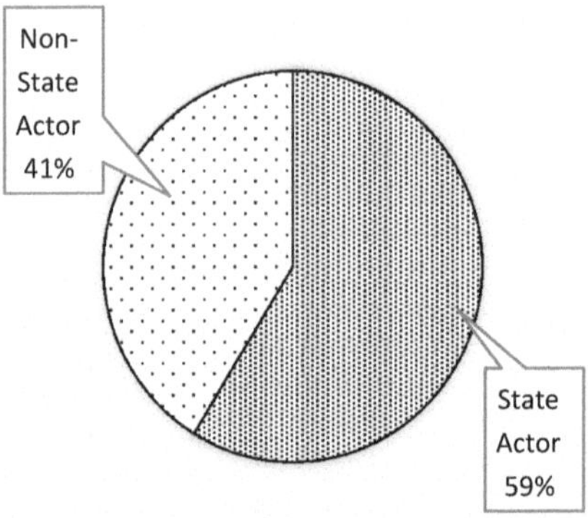

Fig. 2 Distribution of incidents by threat actor type

single digits, with 7 cyber-operational attacks in 2003 being the greatest number of attacks observed in a single year. In 2012, the number of attacks doubles the number of attacks observed in 2003, with 14 total attacks. Number of attacks return to previous pattern between 2013 and 2015. In 2016, the number of attacks again hits double digits with 16 total attacks. In 2017, however, the number of attacks increases by 2 1/2 times over 2016, with 41 total incidents. This trend continues with

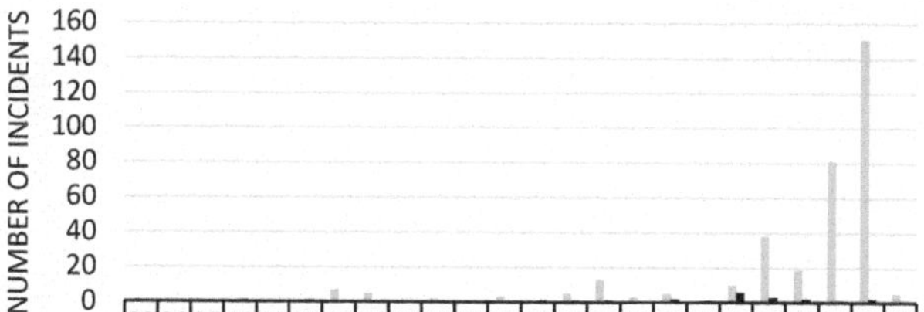

	92	96	97	00	01	02	03	04	05	06	08	09	10	11	12	13	14	15	16	17	18	19	20	21
Cyber-Operational	1	1	1		1		7	5	1	2	1	3		5	13	3	5		10	38	19	81	15	5
Cyber-Physical				1		1							1	1	1		2	1	6	3	2	1	2	1

Fig. 3 Total identified cyber-operational and cyber-physical incidents by year

82 total attacks in 2019 (twice as many at 2017) and with 153 total attacks in 2020 (approximately 1.8 times as many as 2019) (see Fig. 3).

We believe there are two primary reasons for this observed trend. Firstly, attacks on critical infrastructure have indeed increased globally over the period covered by the SMICI dataset due to advancements in technologies and increased connectedness of the critical infrastructures. This is especially true of the previous five years. Secondly, the trend is reflective of available public information on cyber-attacks on critical infrastructure. Locating data on these incidents prior to 2010s can be extremely difficult given that these incidents were not well recorded and that reporting of these incidents did not become routine until early 2010s.

Threat Actors and Attack Types

Of the incidents that we were able to identify the threat actor types, state actors (including state-affiliated and state-supported entities) account for approximately 59% of the incidents while non-state actors account for the remaining approximately 41% of the incidents (see Fig. 2). Since this volume is primarily interested in the implications of these attacks in the contexts of current and future irregular warfare and great power competition, we will focus our discussions primarily on state actor behaviors.

Of all the incidents identified as carried out by state actors, Russia and its affiliated entities accounted for approximately 52% of the incidents while North Korea accounted for approximately 20% of the incidents, with Iran following closely behind accounting for 19% of the incidents. Other state actors identified were the United States, China, and Israel, accounting for, respectively, 5% and 2% each (see Fig. 4).

Examining the types of attacks carried out by the state actors, we find that approximately 79% of the attacks were cyber-operational attacks and approximately 21% were cyber-physical attacks (see Fig. 5). This contrasts quite a bit with non-state actors. For non-state actors, cyber-operational attacks accounted for approximately 90% of their attacks while cyber-physical attacks only accounted for approximately 10% of the total attacks (see Fig. 6).

For both cyber-operational and cyber-physical attacks, Russia and its affiliated entities were most active, accounting for approximately 50- and 80% of the attacks,

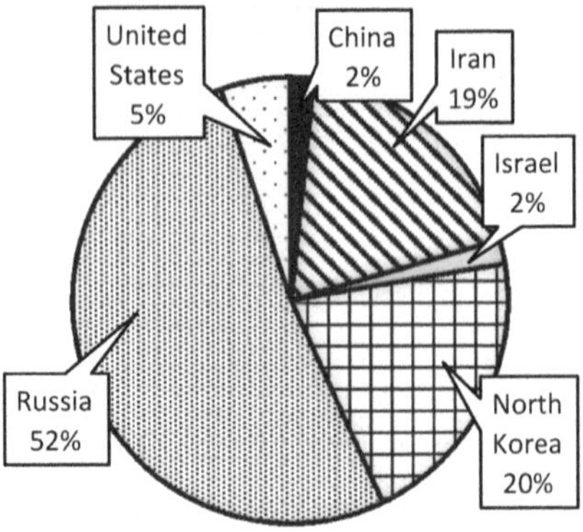

Fig. 4 Percentage of incidents with identified state actor

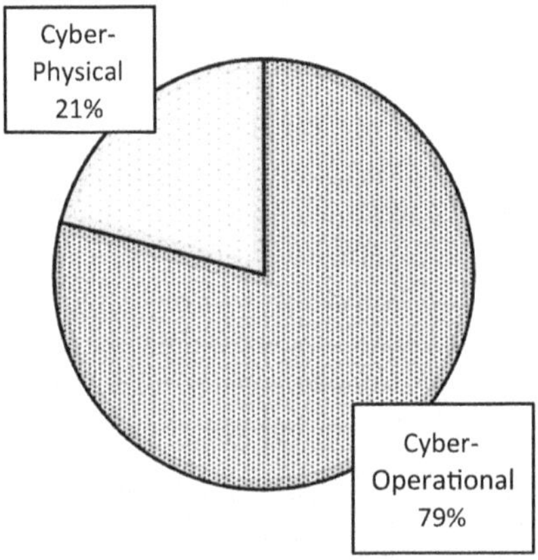

Fig. 5 Distribution of incident types carried out by state actors

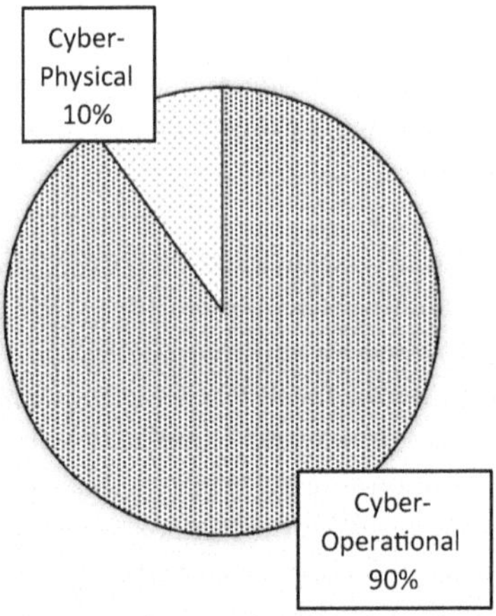

Fig. 6 Distribution of cyber-physical and cyber-operational incidents attributed to state actors by country

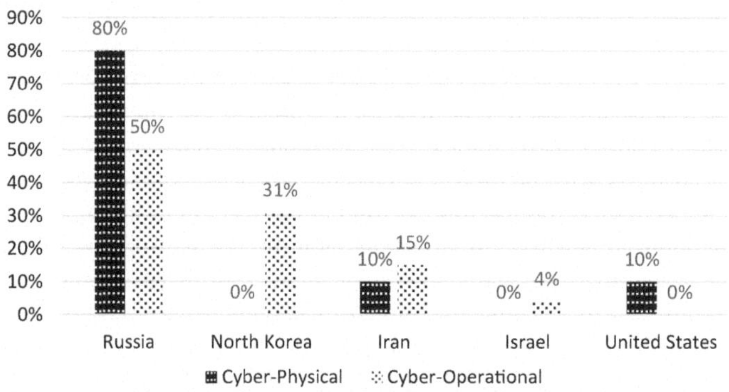

Fig. 7 Distribution of incident types carried out by non-state actors

respectively. Iran and the United States were found to be the only other state actors that have carried out cyber-physical attacks, with 10% each. For cyber-operational incidents, North Korea was found to be the second most active state actor conducting such attacks, with approximately 31% of the cyber-operational incidents being attributed to it. Iran and Israel were found to have the next most number of cyber-operational incidents attributed to them, with approximately 15% and 4% of the total, respectively (see Fig. 7) China, although the dataset records one incident that is attributed to it, was not included in this portion of the analysis because we were not able to clearly determine whether the incident was a cyber-operational or a cyber-physical incident.

An interesting finding about the relationship between threat actors (both state and non-state) and the attack types (cyber-physical and cyber-operational) is that there is not a statistically significant relationship between the two variables. In other words, according to the data, the fact that there have been more cyber-operational than cyber-physical attacks (by both state and non-state actors) was more likely due to attackers' motivations, goals, and capabilities rather than due to their preferences of one type of attack over the other.

These findings suggest while all actors engaging in cyber domain originated attacks against critical infrastructures are more likely to conduct cyber-operational attacks than cyber-physical attacks, cyber-physical attacks against infrastructures still remain the domain of state actors. These findings further suggest although advancements in technologies have increased the vulnerabilities of critical infrastructures and decreased barrier to entry to the field for non-state actors, the level of resources and technical capabilities required to carry out successful cyber-physical attacks still remain well outside of most non-state actors' abilities.

Motivations

While it is important to have a firm understanding of the trends of cyber domain originated attacks against critical infrastructures, it is much more important to understand the motivations behind such attacks. Each incident included in the SMICI dataset was coded for one of six motivations behind the attack—financial gain, destruction, espionage, hacktivism, proof of concept, and unknown/indeterminate.

For an incident to be coded as Financial Gain for motivation, the research team must be able to identify clearly that the threat actor(s)' primary or sole motivation in executing an attack was for financial gain. To be coded as Destruction, the research team must be able to identify clearly that the threat actor(s) primary goal was to disrupt, degrade, and/or destroy the critical infrastructure under attack rather than holding victim's data hostage for ransom or exfiltrating data to sell elsewhere. Incidents were coded as Espionage when the threat actor(s)' sole motivation was to exfiltrate as much data from the victim as possible. An Advanced Persistent Threat is an example of a threat actor whose motivation would be coded ordinarily as Espionage. If the threat actor(s) was/were determined to have carried out an attack to garner public support, to influence public opinion, and to make a political statement about a specific issue, those incidents were coded as Hacktivism. Incidents that were coded as Proof-of-Concept were not attacks in the general sense. Often times, these incidents were related to academic experiments that had "gone loose"; a grey hat hacking disrupting a target to demonstrate a security vulnerability; or security firms conducting a penetration test on a live target with appropriate permission. Finally, when the research team was not able to identify clearly the threat actor(s)' motivation(s), those incidents were coded as unknown/indeterminate.

We found that motivation for approximately 47% of the attacks carried out by state actors was destruction. Approximately 31% of the time, state actors carried out cyber-operational and/or cyber-physical attacks on critical infrastructures for espionage purposes (or to cover up an ongoing/past espionage efforts). Approximately 16% of the time, state actors conducted attacks for financial gain. Finally, we were not able to determine the motivations for approximately 6% of the cases while there were no incidents where the motivations were either hacktivism or proof-of-concept for attacks carried out by state actors (see Fig. 8).

Examining more closely state actors' motivations behind the attacks, we find that North Korea was attributed to all of the attacks where financial gain was the motive. North Korea was also attributed to approximately 21% of attacks for espionage purposes. Furthermore, we find that Russia was attributed for the greatest number of attacks for both destruction and espionage as the motivation. Russia accounted for approximately 67% of attacks for both destruction and for espionage. China only accounted for approximately 7% of attacks with espionage as the motivation. Iran, on the other hand, accounted for approximately 21% of attacks for destruction and approximately 7% of attacks for espionage (see Fig. 9).

Analyzing the critical infrastructure sectors targeted based on motivation; we find that the energy sector has been targeted most for destruction and espionage. We also find that the critical manufacturing sector was most targeted for financial

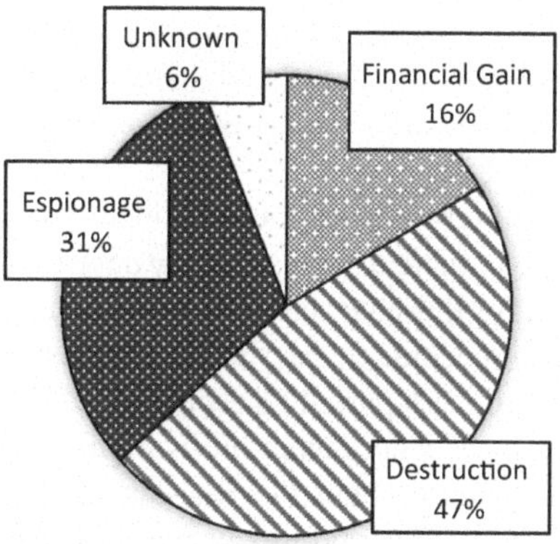

Fig. 8 Distribution of state actor motivations behind the attacks

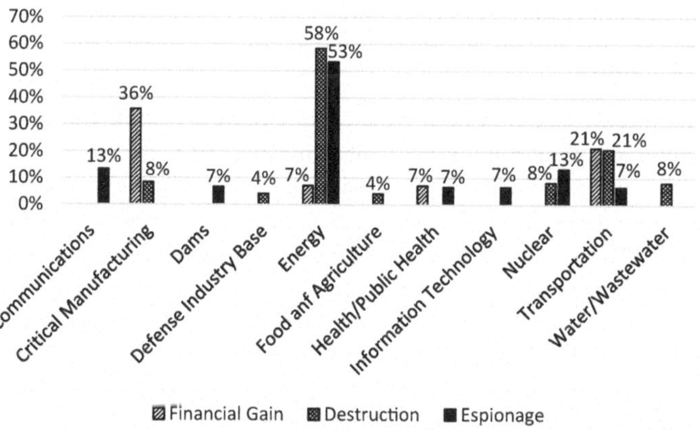

Fig. 9 Distribution of state actor attack motivations by critical infrastructure sector

gain. Probing further, we find that Russia accounts for approximately 73-peercent of the attacks against the energy sector motivated by destruction. We also find that approximately 63% of these attacks were aimed at the energy sector in Ukraine and that the attacks took place between 2015 and 2017. The dates of these attacks coincide with the height of active Russian military involvement in Eastern Ukraine. Russia's cyber-operational and cyber-physical attacks against the Ukrainian energy sector with the intent to disrupt, degrade, and destroy during an ongoing armed conflict is a clear example of Russia operating in the information environment in direct support of its ground forces within the context of irregular warfare.

The energy sector has also been targeted the most for espionage purposes. Of the attacks against the energy sector for espionage purposes, Russia is attributed for 83% of the incidents. Russia's primary target for this purpose has been the United States, accounting for 60% of its attacks. Russia's other targets in this area have been the

United Kingdom and Ireland, each accounting for 20% of Russian attacks. Based on the dataset, the United States is the only other state actor that was attributed to an incident against an energy infrastructure for espionage purposes. The target of the United States' activity was Russia.

Finally, although the dataset includes more incidents identified as motivated by destruction than espionage, statistical analyses of the dataset show espionage to be the most impactful motivation for the state actors. In other words, statistically, it is more likely that a state actor would decide to carry out a cyber-operational/physical attack against a critical infrastructure as a part of its espionage effort than for other reasons. Indeed, there are circumstances where other motivations—such as destruction—are driving the decision. Our results about the Russian attack against the Ukrainian energy sector discussed above illustrate this exact point. Overall, what the data is telling us is that while the state actors generally prefer to employ these methods as a part of their espionage efforts, but they will certainly employ them—especially our adversaries—as tools of irregular warfare if these methods can facilitate the achievement of their goals.

Extension of Politics

"…war is not merely an act of policy but a true political instrument, a continuation of political intercourse, carried on with other means" [6].

The fact that approximately 47% of all attacks targeting critical infrastructures attributed to state actors were for destruction purposes, coupled with the fact that approximately 35% of all attacks carried out for destruction purposes were against targets located in a country where the attacker was actively engaged in conflict against, suggests there is a need to examine more closely the international relations context surrounding these attacks. Since these examinations were fundamentally interstate conflict analyses, we selected two international conflict variables that have been shown as having the largest impact on the decision for states to enter into conflict against each other—Rivalry and International Conflict (or External Distraction) [7]—to test whether international security contexts had any effect on state-actors carrying out cyber-operational/cyber-physical attacks targeting critical infrastructure. Additionally, since a state actor could decide to carry out a cyber-operational/-physical attack targeting critical infrastructures of another state actor who it is already engaged in active conflict against, we also included Conflict against Attacker in our analyses.

For the *Rivalry* variable, the information was derived from the New Rivalry Dataset version 5.10 (NRD v.4.10) [8]. The NRD v5.10 records rivalries between dyads (two countries) from 1816 thru 2001 and it categorizes the rivalries into isolated, proto-rivalry, and enduring rivalry. For the SMICI dataset, the *Rivalry* variable only recognizes enduring rivalries from NRD v5.10 for the two state actors to be considered "rivals." Since the NRD v 5.10 only includes data up to the end of 2001, additional coding was done for the rivalry variable for incidents that occurred between 2002 and 2021 following the same coding rules set forth in NRD v5.10.

For the *International Conflict* variable, the information was derived from the *External Distraction* variable contained in the Distractionary Theory of Interstate Crisis Initiation Datase [7]. For the SMICI dataset, a state of where the targeted

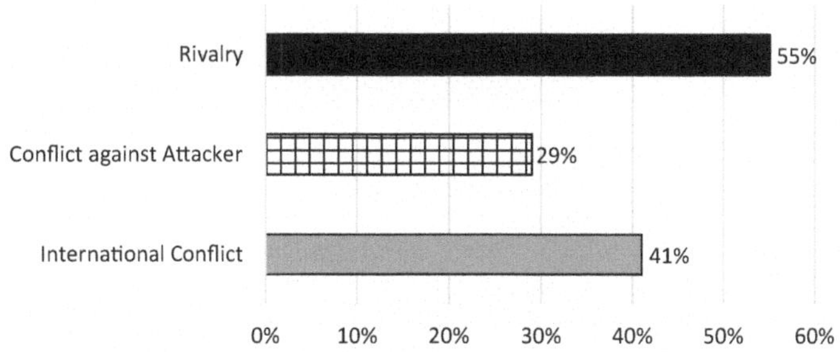

Fig. 10 International relations contexts at the time of attack

critical infrastructure is located is considered to be involved in *International Conflict* if that state is already engaged in an interstate crisis or conflict prior to becoming a target of the state actor carrying out the cyber-operational/-physical attack. Since the variable is only interested in whether the state where the target is located was involved in any other interstate crisis and/or conflict at the time of the observed attack, the variable is operationalized as a binary variable (yes/no).

To account for situations where cyber-operational/-physical attacks on a target occurred while the state actor initiating the attack and the state where the target is located are already engaged in a conflict with each other, the SMICI dataset also includes *Conflict against Attacker* variable to capture this condition, which the *International Conflict* variable does not capture.

Analyzing the dataset within the context of these international relations context, we find that approximately 55% of all attacks attributed to state actors were conducted against targets located in states with enduring rivalry. We also find that approximately 29% of all attacks occurred against targets located in states that are engaged in active conflict against the attacking state actor(s). Finally, we. find that approximately 41% of all attacks occurred against targets located in states that are already engaged in some form of international conflict with state(s) that is not the attacking state actor at the time of the attack (see Fig. 10).

Statistical analyses of incidents attributed to state actors in relation to the three variables discussed above revealed *Rivalry* to be the most impactful variable for state actors to carry out cyber-operational/-physical attacks against a target. We find that rivalry between the state actor and the state where the target is located increases the probability of a cyber-operational/-physical attack by approximately 41%. Interestingly, both *International Conflict* and *Conflict against Attacker* were found not to be statistically significant. What the statistical analyses are telling us is that a target that is in a state (let's call it State A) that happens to be a rival of the potential attacker (let's call it State B) will experience a higher volume of attacks emanating from State B regardless of other factors included in the international relations context analyses.

If (some) state actors indeed view cyber-operational/-physical attacks on critical infrastructures as mere execution of their desired policies with other means, we should expect a differing distributions of international relations context factors in

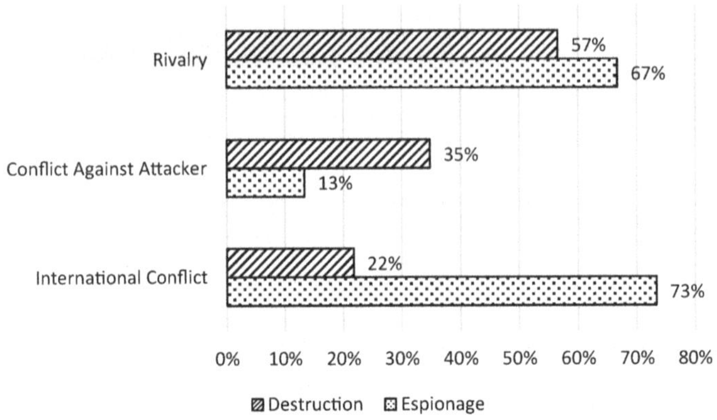

Fig. 11 Relationship between attack motivations and international relations contexts

relation to target selection based on the attackers' motivations. The analyses indeed found this to be the case. As expected, Rivalry between the states remained as the primary driver for both destruction and espionage motivated attacks. The results for *Conflict against Attacker* variable was quite striking, however. We found that only approximately 13% of attacks for espionage purposes were aimed at targets located inside states actively engaged in conflict with the attackers while approximately 34% of attacks aimed at destruction targeted critical infrastructures located inside states actively engaged in conflict with the attackers. These results illustrate that while state actors are more likely to employ cyber-operational/-physical attacks against their rivals than non-rivals in general, states are much more likely to employ these methods when the states being targeted are already engaged in active conflict(s) with them if their goal is to destroy, disrupt, and/or degrade the targeted critical infrastructure, negatively affecting the targeted state's capabilities during conflict (see Fig. 11).

Finally, approximately 73% of the attacks motivated by espionage targeted states involved in international conflicts with other states while only approximately 22% of the attacks motivated by destructions targeted states involved in international conflicts (see Fig. 11). A closer examination of the data revealed, however, this result was because the United States accounted for approximately 53% of the state where targeted critical infrastructures are located that were targeted for espionage purposes (and the United States has been involved in continuous international conflict since 2001). When incidents involving targets located in the United States were removed, the number of incidents involving targets located in states involved in international conflict decreased to approximately 20% for attacks with espionage as the motive—which was at parity with the attacks motivated by destruction.

Overall, analyses of our data indicate that state actors have already employed cyber-operational/-physical attacks targeting critical infrastructures of their adversaries as a policy implementation and facilitation tool—during both "hot" (e.g., Russia's attacks on Ukraine) and "cool" (e.g., Israel's attacks on Iran) conflicts as

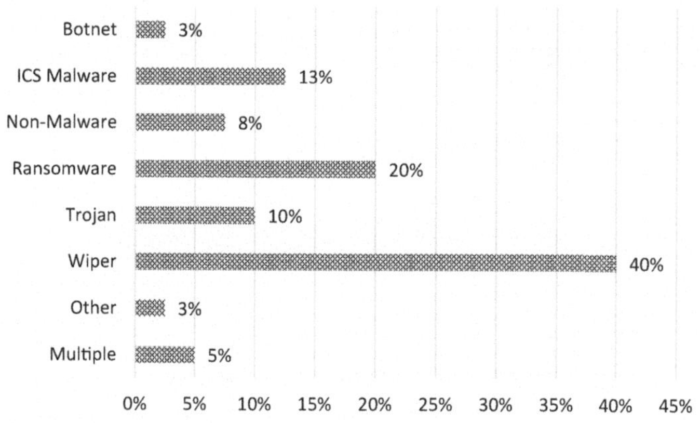

Fig. 12 State actor employed malwares

well as in the context of espionage. Furthermore, results indicate context surrounding the target (i.e., who is the target, what is the relationship between the target and the attacker, and what is the target up to) and the goal of the attacker are two most important drivers of state actor attributed cyber-operational/cyber-physical attacks targeting critical infrastructure.

Most Employed Malwares

The dataset was able to identify six (6) specific types of malwares state actors employed to conduct cyber-operational and/or cyber-physical attacks against critical infrastructures. The data shows state actors employed wiper malware the most (approximately 40%), followed by ransomware (approximately 20%), followed by ICS Malware (approximately 13%) (see Fig. 12).

We were also able to ascertain different types of malware state actors employed based on attack types. For cyber-operational attacks, state actors employed wiper malware the most, accounting for approximately 60% of the attacks. Ransomware was used the second most, accounting for approximately 32% of the attacks. Finally, ICS malware was used for approximately 8% of the attacks (see Fig. 13).

For cyber-physical attacks, state actors employed trojan malware the most, accounting for approximately 40% of the attacks. State actors employed ICS malware the second most for cyber-physical attacks, accounting for approximately 30% of the attacks. Finally, for approximately 20% of the attacks, state actors employed multiple malwares in conducting their attacks (see Fig. 14).

When conducting attacks for destruction, state actors employed wiper malware the most (approximately 68% of the attacks). ICS malware and non-malware—or fileless malware—were employed the most when conducting attacks for espionage purposes (approximately 18% each).

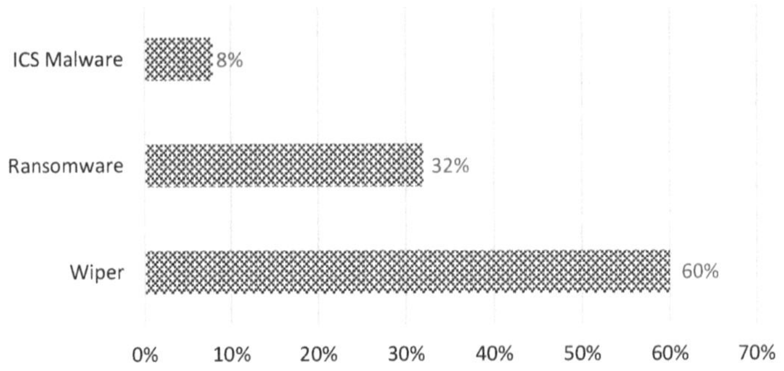

Fig. 13 State actor employed malwares for cyber-operational attacks

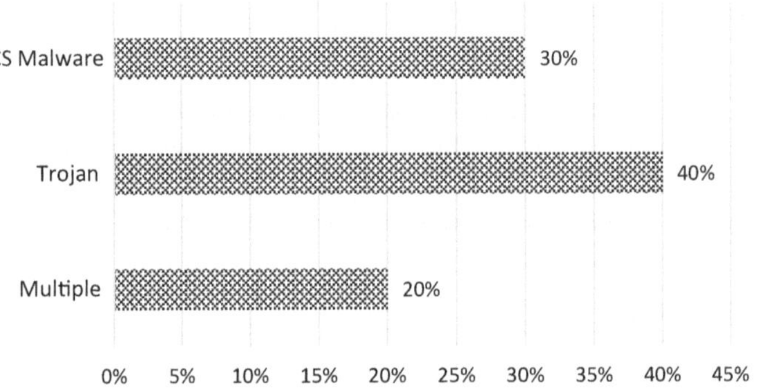

Fig. 14 Top three state actor employed malwares for cyber-physical attacks

Conclusion

Although cyber-physical and cyber-operational attacks on critical infrastructure have been identified as one of the major security challenges for the foreseeable future, there has not been a systematic study of these attacks. One of the obstacles to doing so is that there has not been an aggregated collection of data on cyber-physical and cyber-operational attacks against critical infrastructure globally across critical infrastructure sectors. Lack of such aggregation of data has limited our ability to conduct systematic studies of the phenomenon, and our abilities to hypothesize and to gain a deeper understanding about the behaviors and motivations of the attackers.

We developed Significant Multi-domain Incidents against Critical Infrastructure (SMICI) dataset in order to begin building our ability to formulate theories and test hypotheses relevant to the behaviors and motivations of threat actors. This first of its kind dataset collects and aggregates publicly available incident information on cyber-operational/-physical attacks targeting critical infrastructure across sectors globally. The dataset collects on 38 individual variables per observation and, at the time of this writing, contains 427 attributed and unattributed cyber-physical and cyber0operational incidents carried out against critical infrastructures worldwide between January 1, 1992 and July 9, 2021.

Utilizing both qualitative and statistical analyses, we were able to draw some conclusions on the trends, actors, attack types, motivations, international context, and malwares used for cyber-physical and cyber-operational attacks. First, we found that cyber-physical and cyber-operational incidents against critical infrastructures have exponentially increased in 2012 compared to previous two decades, and again in 2017. Starting in 2017, number of incidents increase exponentially each year. Prior to 2012, the number of cyber-physical and cyber-operational incidents remained in single digits, with 7 cyber-operational attacks in 2003 being the highest number of attacks observed in a single year. In 2012, the number of attacks doubles the number of attacks observed in 2003, with 14 total attacks. Number of attacks return to previous pattern between 2013 and 2015. In 2016, the number of attacks again hits double digits with 16 total attacks. In 2017, however, the number of attacks increases by 2 1/2 times over 2016, with 41 total incidents. This trend continues with 82 total attacks in 2019 (twice as many at 2017) and with 153 total attacks in 2020 (approximately 1.8 times as many as 2019).

Second, overwhelming number of attacks identified in the SMICI dataset were cyber-operational attacks, accounting for approximately 94% of the identified incidents. Cyber-physical attacks against critical infrastructure, on the other hand, accounted for approximately 6% of the identified incidents included in the SMICI dataset. Of the incidents that were able to be attributed to either a state or a non-state actor, approximately 59% were attributed to state actors while approximately 41% were attributed to non-state actors. Finally, of all the incidents identified as carried out by state actors, Russia and its affiliated entities accounted for approximately 52% of the incidents (with approximately 80% of cyber-physical attacks and approximately 50% of cyber operational attacks attributed to it) while North Korea accounted for approximately 20% of the incidents, with Iran following closely behind accounting for 19% of the incidents.

Third, destruction was found to be the motivation behind the attacks carried out by state actors for approximately 47% of the incidents. Approximately 31% of the time, state actors carried out cyber-operational/-physical attacks on critical infrastructures for espionage purposes (or to cover up an ongoing or past espionage efforts). Finally, approximately 16% of the time, state actors conducted attacks for financial gain. Incidentally, Russia was attributed for the greatest number of attacks for destruction and espionage purposes, accounting for approximately 67% of attacks for both destruction and espionage motivation categories. North Korea, perhaps unsurprisingly, was the only state actor attributed to have conducted cyber-operational attacks targeting critical infrastructure for financial gain purposes.

Analyzing the critical infrastructure sectors targeted based on motivation; we find that the energy sector has been targeted most for destruction and espionage. Probing further, we find that Russia accounts for approximately 73% of the attacks against the energy sector motivated by destruction. We also find that approximately 63% of these attacks were aimed at the energy sector in Ukraine and that the attacks took place between 2015 and 2017. The dates of these attacks coincide with the height of active Russian military involvement in Eastern Ukraine.

Fourth, we found that approximately 55% of all attacks attributed to state actors were conducted against targets located in states with enduring rivalry. We also determined that approximately 29% of all attacks occurred against targets located in states that are engaged in active conflict against the attacking state actor(s). Additionally, we found that approximately 57% of attacks conducted for destruction purposes were carried out against rival states; furthermore, approximately 34% of attacks were carried out against states who were already in active conflict with attacking state(s). Ultimately, context surrounding the target (i.e., who is the target, what is the relationship between the target and the attacker, and what is the target up to) and the goal of the attacker are two most important drivers of state actor attributed cyber-operational/cyber-physical attacks targeting critical infrastructure.

Finally, state actors employed wiper malware the most to conduct cyber-operational attacks as well as to conduct attacks for destruction (approximately 60- and 68% respectively). For cyber-physical attacks, state actors employed trojan malware the most, accounting for approximately 40% of the attacks. Additionally, ICS malware and non-malware—or fileless malware—were found to be the most employed malwares for state actors when conducting attacks for espionage purposes (approximately 18% each).

Great Power Competition Implications

Cyber-physical and cyber-operational attacks by state actors targeting critical infrastructure of the adversary in support of their national strategic objectives is here and now (and has been a reality for some time), and it will only become more prominent in the future. Our analyses clearly show that state actors have already employed cyber-operational and cyber-physical attacks targeting critical infrastructures of their adversaries during both "hot" and "cool" conflicts as well as during times of apparent peace. For example, Russia has already demonstrated its willingness and capability to conduct cyber-operational as well as cyber-physical attacks against critical infrastructures of its adversaries in support of its irregular warfare operations when it attacked Ukraine's energy infrastructure between 2015 and 2017. It has also demonstrated its willingness to conduct attacks against the United States energy infrastructure as a part of its espionage efforts. Likewise, China has demonstrated its willingness and ability to conduct cyber-operational attack against Canada's information technology company to compromise SCADA administration tools and exfiltrate information. These examples are but a few cases that illustrate our strategic adversary and competitor's willingness to incorporate cyber-operational and cyber-physical attacks targeting critical infrastructures into their operations in the information environment in support of their strategic and operational goals.

Additionally, as the context surrounding the target (i.e., who is the target, what is the relationship between the target and the attacker, and what is the target up to) and the goal of the attacker have been shown as the two most important drivers of state actor's decision to whether or not to conduct cyber-operational/cyber-physical attacks targeting critical infrastructure, it would only be prudent for the United States to anticipate attacks from its primary strategic adversary and competitor will increase—in both frequency and in sophistication—in the future. Finally, the United States must be the leader and a champion in creating an international norms and rules of engagement for operations in the information environment that provides clear boundaries of acceptable behaviors as well as clear repercussions for those who violate them. Furthermore, the United States should adopt a policy that clearly articulates a credible threat should a state actor or a state affiliated/supported entity conducts an attack targeting critical infrastructure located in the United States.

Looking Forward

Development of the SMICI dataset and the analyses of the data contained in the dataset provided us with the first comprehensive look into the global trends and patterns of cyber-physical and cyber-operational attacks targeting critical infrastructures. Through this effort, we were able to begin addressing several fundamental research questions that will serve as the cornerstone of future research on this topic. To continue our contribution to further advancement in this area, we will continue our SMICI dataset collection effort, improve data granularity, and expand the dataset temporally in both directions. Our hope is that this dataset will provide the foundation for future studies into the phenomenon of cyber-physical and cyber-operational attacks against critical infrastructures for scholars, industry practitioners, and policy makers alike.

References

1. Friedberg, I., McLaughlin, K., Smith, P., Laverty, D., & Sezer, S. (2017). STPA-SafeSec: Safety and security analysis for cyber-physical systems. *Journal of Information Security and Applications, 34*(2), 183–196.
2. Pagliery, J. (2015). ISIS is attacking the U.S. energy grid (and failing). *CNN Money*. https://money.cnn.com/2015/10/15/technology/isis-energy-grid/index.html.
3. Bloom, D. (2015). ISIS trying to launch deadly cyber attack on airports, hospitals, and national grid, warns George Osborne. *The Mirror*. https://www.mirror.co.uk/news/uk-news/isis-trying-launch-deadly-cyber-6845517.
4. Loukas, G. (2015). *Cyber-Physical Attacks: A Growing Invisible Threat*. Waltham, MA: Butterworth-Heinemann.
5. The White House. (2013). *Presidential Policy Directive 21: Critical Infrastructure Security and Resilience*. Washington, D.C.: The White House.
6. Von Clausewitz, C. (1832). *Vom Kriege*. Berlin: Dümmlers. English Edition: Howard, M., Paret, P. (1976). *On War* (trans: Paret, P.). Princeton, NJ: Princeton University Press.
7. Sin, S. S. (2017). While the enemy if preoccupied: a distractionary theory of interstate crisis initiation [Doctoral Dissertation, University at Albany, State University of New York].
8. Sin, S. S., & Washburn, R. T. (2019). Significant Multi-domain Incidents against Critical Infrastructure Dataset. Codebook v1.2, START. College Park, MD: START, July 28, 2021. https://www.start.umd.edu/pubs/START_UWT_SignificantMultiDomainIncidentsAgainstCriticalInfrastructure_Dec2019.pdf.
9. Klein, J. P., Goertz, G., & Diehl, P. E. (2006). The new rivalry dataset: Procedures and patterns. *Journal of Peace Research, 43*(3), 331–348.

Dr. Steve Sin is the Director of the Unconventional Weapons and Technology Division (UWT) of START, where he manages large research projects, explores new avenues for research, and establishes collaborative research relationships. Dr. Sin specializes in a broad range of international security, terrorism, and homeland security challenges. His expertise includes radiological and nuclear (RN) terrorism; illicit trafficking of RN materials; terrorist use of cyber domain; emerging technology; emergency preparedness and management; intelligence/counterintelligence operations, analysis, and exploitation; Northeast Asia regional security; and counter-terrorism training, exercise, and curriculum development.

Mr. Rhyner Washburn is the Cyber Intelligence Researcher of the Unconventional Weapons and Technology Division (UWT) of the National Consortium for the Study of Terrorism and Responses to Terrorism (START), headquartered at the University of Maryland, where he manages and conducts research projects spanning across a broad range of international security, terrorism, and cybersecurity topics. His expertise includes election security; terrorist use of cyber domain; multi-domain operations; cyber-physical nexus; insider threat investigations; mass surveillance; cybercrime; emerging technologies; influence operations; and Northeast Asia regional security.

Alternate Reality—The Use of Disinformation to Normalize Extremism

Elie Alhajjar

Abstract Disinformation is becoming abundant in our civic society, and it poses a national cyber threat to our democracy. While some conspiracy theories are born spontaneously (in the heat of the moment), others are premeditated by near-peer nation-states such as Russia, China, and their proxies. In this chapter, it is argued that our enemies are using game-like cognitive tactics to reach a large number of people within the United States and radicalize their beliefs. By turning random events and unrelated pieces of information into an alternative reality in which the vulnerable population is getting immersed, a new cognitive state is being established that can basically justify any extreme thought one might have.

Keywords Information · Disinformation · Apophenia · Gamification · 5G

Introduction

Misinformation is false information that is spread either by mistake or with intent to mislead. When there is intent to mislead, it is called disinformation. Misinformation is everywhere online and has the potential to cause substantial harm to individuals and society. While both misinformation and disinformation can deceive audiences, the distinction is that disinformation is intentionally, maliciously deceptive.

The term misinformation refers to information that is false or inaccurate, and is often spread widely between individuals and groups, regardless of an intent to deceive. Examples are abundant, such as the claim that the new 5G technology causes cancer (or even COVID-19) despite there being no scientific evidence to support it. The main idea behind this false claim is that 5G radio waves are harmful to the brain and cause health issues such as autism and cancer. However, experts have debunked this concern, explaining that 5G radio waves cannot damage the DNA in our cells, nor can they even penetrate past the skin, which acts as a protective barrier.

E. Alhajjar (✉)
United States Military Academy, West Point, NY 10996, USA
e-mail: alhajjar@westpoint.edu

A. Farhadi et al. (eds.), *The Great Power Competition Volume 3*,
https://doi.org/10.1007/978-3-031-04586-8_8

In contrast, disinformation is created to be deliberately deceptive. The European Commission defines disinformation as verifiably false or misleading information that is created, presented and disseminated for economic gain or to intentionally deceive the public, and may cause public harm. Disinformation does not include reporting errors, satire and parody, or clearly identified partisan news and commentary [5]. The aim of disinformation is to distract and divide, to plant seeds of doubt by distorting and falsifying facts, thus confusing people and weakening their faith in institutions and established political processes.

Disinformation erodes trust in institutions and in digital/traditional media as well as harms democracies by hampering the ability of citizens to make informed decisions. More importantly, disinformation often supports radical and extremist ideas and activities. It impairs freedom of expression that encompasses respect for media freedom and pluralism, as well as the right of citizens to hold opinions and to receive and impart information and ideas without interference by public authorities and regardless of frontiers.

Mass online disinformation campaigns are being widely used by a range of domestic and foreign actors to sow distrust and create societal tensions, with serious potential consequences. Furthermore, disinformation campaigns by third countries are part of hybrid threats to internal security, including election processes in combination with cyberattacks. For example, Russian military doctrine explicitly recognizes information warfare as one of its domains. The spread of disinformation also affects policy-making processes by skewing public opinion and diminishing trust in science and empirical evidence. Domestic and foreign actors can use disinformation to manipulate policy, societal debates and behavior in areas such as climate change, migration, public security, health, and finance.

There are many factors that play a role in the spread of disinformation. First, economic insecurity, rising extremism, and cultural shifts generate anxiety and provide a breeding ground for disinformation campaigns to foster societal tensions, polarization, and distrust. Second, the spread of disinformation takes place in the context of a media sector undergoing profound transformation. The rise of active platforms in the media sector has deeply affected journalists and professional news media outlets, which are still generally seeking to adapt their business models and find new ways to monetize content. Third, social networking technologies are manipulated to spread disinformation through the following cycle: creation, amplification through social/online media, and dissemination by users.

While disinformation is inexpensive, and often economically profitable, it has the dire effect of shifting people's mindsets on critical societal issues by guiding them step by step into making cognitively sound conclusions. The goal of this chapter is to shed light on how disinformation can be used as a cognitive weapon by our enemies to fool citizens into believing they are solving some type of mystery by following a set of clues to eventually get to their own synopsis. After a short introduction, the concept of apophenia is defined and explained in Sect. 2. Section 3 deals with the idea of gamification, namely using boards, simulations, challenges, etc. to lay out a cognitive map leading to a premeditated result. Section 4 ties the two concepts and expands on ways they could be exploited by malicious actors. The work concludes in Sect. 5 in which some mitigation measures are mentioned to orient the way forward.

Apophenia

Apophenia refers to the human tendency to see patterns and meaning in random information. The term was coined in 1958 by German neurologist Conrad [2], who was studying the unmotivated seeing of connections in patients with schizophrenia. Statisticians refer to apophenia as patternicity or a type-I error. Generally, there are four subcategories of apophenia: pareidolia, gambler's fallacy, clustering illusion, and confirmation bias. Below is a short explanation of each of them.

Pareidolia is a type of apophenia that occurs specifically with visual stimuli. People with this tendency most often see human faces in inanimate objects. Some examples of pareidolia include seeing a face in a slice of toast or seeing the shape of a bunny in a random mass of clouds. Second, people who regularly gamble often fall prey to the gambler's fallacy. They may perceive patterns or meaning in random numbers, often interpreting the pattern as an indication of an oncoming win. Third, a clustering illusion occurs when looking at large amounts of data; humans tend to see patterns or trends in data even when it is entirely random. Finally, confirmation bias is a psychological phenomenon in which a person tests a hypothesis under the assumption that it's true. This form of apophenia can lead to overemphasizing data that confirms a hypothesis and explaining away information that disproves it.

Examples of apophenia appear in a lot of instances in our daily lives, among which are conspiracy theories, the inkblot test, and the man in the moon to name just a few. Conspiracy theories are the most common example of apophenia: people seeing meaningful patterns in events or information that are likely completely unrelated such as UFO cover-ups, Bigfoot conspiracies, paranormal experiences, etc. The Rorschach inkblot test is a psychological test in which doctors show patients random inkblots and ask them to interpret meaning from them [10]. Swiss psychiatrist Hermann Rorschach designed these tests to gain insight into patients' minds through their pattern recognition. A popular example is the "The Man in the Moon": from the United States to Europe, people worldwide have looked up at the moon and claimed to have seen a face!

Of particular importance here are conspiracy theories, due to the many tragic events that were triggered by them. Conspiracy theories can be found among radical violent extremist groups; they can increase individuals' feelings of anxiety, fuel distrust toward institutions, and promote feelings of powerlessness, all of which may increase one's propensity for radical violent extremism [11]. Such theories play a role in the process of radicalization, as they promote beliefs that are designed to specifically target identified groups (e.g., Jews, Westerners, Government) against whom proponents of the theories wish to nurture society-wide grievances. Indeed, belief in conspiracy theories is linked to an overall distrust of mainstream narratives as well as the institutions that govern society.

People's beliefs in different conspiracy theories persist regardless of the specifics of these conspiracies, many of which contradict each other! Due to this closed, self-referential thought processes, people who believe in one conspiracy theory are more likely to believe in others. A main factor in the spread of conspiracy theories is the Internet. Malicious actors have exploited the Internet as an effective means of peddling their message, based on their ubiquity and repetition on the net. It has become common knowledge now that the use of the Internet and social media sites makes the sharing of conspiracy theories immediate and pervasive.

While some conspiracy theories never pick up any momentum, others may land on fertile grounds and lead to catastrophic consequences. One example of such incidents is the genocide in Christchurch, New Zealand, on 15 March 2019. In the aftermath of the tragedy in which fifty-one people were murdered and forty-nine were injured, it turned out that the killer Brenton Tarrant was brainwashed by one of these conspiracy theories. Detailing his rationale [14], he wrote online: "This crisis of mass immigration and sub-replacement fertility is an assault on the European people that, if not combated, will ultimately result in the complete racial and cultural replacement of the European people."

Gamification

Gamification is defined as the application of game design principles in non-gaming contexts [8]. At a time when much of what happens daily is mediated by digital technologies and social media, many firms are turning traditional processes into deeper, more engaging game-like experiences for many of their customers as well as their employees. This concept has wide applications such as healthcare, sustainability, government, transportation, education, etc.

The term gamification started to attract widespread attention in non-gaming contexts only a decade ago, mainly due to three recent developments [3, 9]. First, game designers have invested significantly in studies to better understand what makes a computer game engaging and successful. This has led to a number of theories and lessons about the design and management of gaming experiences, and to frameworks about incentives that motivate individuals to play. Second, the pervasiveness of social media and mobile-based and web-based technologies has changed how individuals and organizations participate in, share, co-create, discuss, and modify any type of experience. Third, firms are continually looking for new and impactful ways to better connect with, learn from, and influence the behaviors of employees and customers.

Gamification can change a person's behavior because it taps into motivational drivers of human behavior in two connected ways: reinforcements and emotions. First, both positive and negative reinforcements encourage repetition of behaviors. These approaches have long been used in psychology to explain a range of human behaviors as well as behavior modification. They also claim that behavior changes can be motivated either through extrinsic or intrinsic reinforcements. That is, while external factors such as money or fame can certainly motivate human behavior, emotions are also powerful motivators for behavior change [6]. Second, behaviors which lead to satisfying outcomes are more likely to lead to repeated or ongoing behavior changes while ones with unsatisfying outcomes are far less likely to be sustained [13].

Successful gamification involves the repetition of desired outcomes. Through the motivational mechanisms of reinforcements and emotions, desired outcomes become automatic behavioral processes or habits. Habits are formed through providing cues that elicit behaviors and then rewarding the behavior, thus forming a behavioral loop that requires less and less cognitive resources as the desired behavior is repeatedly reinforced [4]. Gamification can produce desired behavior change through the formation of habits by reinforcing the reward and emotional response of the individuals participating in the experience, thus requiring fewer cognitive resources each time the desired activity is reproduced.

All parties involved in gamified experiences can be described using two fundamental dimensions: variations in participation and connection with the gamified environment [7]. Player participation describes the extent to which the individual actively contributes to the experience, while player connection describes the type of environmental relationship (absorption vs. immersion) that unites the individual with the experience. In absorption, the experience unfolds before the person and occupies the person's mind, whereas in immersion, a person becomes part of the experience itself, either physically or virtually.

In a generic gamified experience, there are usually four types of people involved, namely players, designers, spectators, and observers whose respective roles are explained briefly. Players are those who compete in the game; they are the real performers in the experience. Designers are the decision makers who develop and design, as well as often manage and maintain, the gamified experience. Spectators are those individuals who do not directly compete but whose presence will influence how the gamified experience works. Observers are outside individuals passively involved and absorbed in the experience. While they have no direct impact on the gamified experience, their presence impacts the popularity of the experience.

Gamification dynamics are the types of player behavior that emerge as players partake in the experience. These dynamics describe in-game behaviors and the strategic actions and interactions that emerge during play. It is worth mentioning that gamification dynamics are difficult to predict and thus can lead to unintended behaviors and outcomes, which can be positive or negative in nature. This creates a challenge for designers to anticipate the types of dynamics that can emerge and to develop the mechanics of the experience appropriately.

Exploitation

In this section, the two concepts of apophenia and gamification are combined to demonstrate how malicious state and non-state actors can use them to smoothly radicalize sectors of the population. This will be done through a real-life example of particular interest and relevance: the QAnon movement (or any similar ones like FBIanon and CIAanon), in which people are manipulated into believing certain premeditated theories, all through a series of fake perceptions and gamified instances.

QAnon is the umbrella term for a set of internet conspiracy theories that allege that the world is run by a cabal of Satan-worshiping pedophiles. QAnon followers believe that this cabal includes top Democrats like President Joseph R. Biden Jr., Hillary Clinton, Barack Obama and George Soros, as well as a number of entertainers and Hollywood celebrities like Oprah Winfrey, Tom Hanks and Ellen DeGeneres and religious figures including Pope Francis and the Dalai Lama. Many of them also believe that, in addition to molesting children, members of this group kill and eat their victims to extract a life-extending chemical called adrenochrome.

QAnon grows on the wild misinterpretation of random data, presented in a suggestive fashion in a setting designed to help the users come to the intended misunderstanding. It is best described as "guided apophenia", since the masterminds behind it have already planned the desired conclusions and are constantly getting the player lost by pointing out unrelated random events and creating a meaning for them that fits the propaganda message. It is like a breadcrumb trail away from actual solutions and towards a dangerous psychological rush.

As Reed Berkowitz explains it: "it works very well because when you figure it out yourself, you own it! You experience the thrill of discovery, the excitement of the rabbit hole, the acceptance of a community that loves and respects you. Because you were convinced to connect the dots yourself, you can see the absolute logic of it; this is the conclusion you arrived at." [1]. In the gamified setting introduced in the previous section, the American citizen is the player, some enemy is the designer, previous victims are spectators who already bought in on the conspiracy theory, and gullible surrounding persons are the observers. When players arrive at the "correct" answers they are showered with adoration, respect, and social credit.

A major difference between QAnon and an actual game is that the former is almost pure political propaganda. QAnon uses the oldest trope of all mystery fiction: a mysterious stranger shows up and drops a strange clue leading to long-hidden secrets which his clues, and the player's detecting power, can reveal. The fictional mysterious stranger already knows, but instead of telling the answer in the first ten minutes, they give hard to follow, ambiguous clues. This is what creates the whole plot, the sense of mystery, and everything entertaining that is to follow. This is the breadcrumb trail out of the forest!

Solving puzzles in general is a special way of learning as it encodes information into the brain in a different way than other learning. Puzzles and knowledge gained through a person's own efforts are incredibly rewarding and also come with a hit of dopamine, the brain's pleasure drug, as a reward. Finding creative solutions to difficult problems is a fundamental aspect of human culture. This process often occurs with a unique phenomenal experience, the Aha!-moment, referring to the moment of transition from being completely dark about the solution to suddenly "seeing" it. This phenomenological experience brings a sense of ease, is intrinsically pleasurable, and is accompanied by a feeling of certainty about the solution [12]. Recently, a group of neuroscientists investigated insightful problem-solving processes using fMRI (Functional Magnetic Resonance Imaging) and showed that structures of the dopaminergic midbrain are associated with solutions per se and especially with highly insightful solutions [15]. This implies that the Aha!-Moment is a special form of fast retrieval, combination, and encoding process in the brain.

Thinking again about the spectators and the observers, it is remarkable that solving puzzles together is a great way to form community and to join community. Everyone has something to focus on, a shared interest, and something to do; it is a bonding mechanism. The player is having fun looking around and solving riddles reinforced by dopamine rushes and the feeling of being involved in something that no one else knows. They are also now part of a growing, active, and friendly community. Everything they see and hear creates further doubt about people and things they already doubt. Every new theory is easy to understand, compelling, and arrived at through their own ability to reason. Now to doubt certain ideas is to doubt themselves, their ability to see the truth, their community, and maybe the world around them. This mixture of guided apophenia and a gamified experience is creating an alternate reality that is very psychologically satisfying and seems to be supported by make-believe events.

Fake news can not be dispelled easily; however, they can be easily spread. It is very hard to prove something does not exist, e.g., aliens scientifically could exist so it is unlikely to prove that they do not. A random idea could start very innocently, then this same idea gets force-multiplied in thousands of sites, YouTube videos, ads, comments, Instagram influencers, tweets, etc. Now that people are fully on board with such scenarios, they can continue the game for themselves with very few cues. All the designers have to do is to drop a conspiracy theory on the internet, and within weeks, it is impossible to contain much like a flood, a wildfire or a volcano.

At the core, the difference between apophenia and science is just the scientific process and the reliance on proof. Scientists may not want to let go of a good theory that just isn't panning out, since the feeling of correctness is always over-powering. This is why they have peer-reviewers, colleagues who replicate their results and test their solutions. On the other hand, the proof corresponding to events is more apophenia. The answer is simply whatever feels the best, makes the most sense, and helps the story. It is whatever fuels the propaganda and reinforces the conclusions of the apophenia and central narrative. The more people engage in this gamified environment, the more it is likely to become a religion, a cult, a political movement, or anything else. Hence, these tactics are becoming the indoctrination campaigns of the twenty-first century.

Conclusion

In the past few years, there has been a consistent trend of malicious actors using misinformation to promote extreme views and polarize local debates. Given the increasingly sophisticated nature of disinformation activities and the difficulties of independent researchers to access relevant data from the platforms, a conclusive assessment of the scope and impact of disinformation campaigns will require a concerted effort by civil society, academia, public actors and online platforms.

The goal of this chapter is not to provide a literature review on the state-of-the-art of the disinformation research field, but rather to emphasize on a specific vulnerability that can be exploited by enemies of the United States of America. The human element, or the human-in-the-loop, is a topic of interest to the cyber community in general and there is no shortage of themes and venues that discuss its details. The particular focus here is on the dangerous combination of two different pillars, namely guided apophenia and gamification, and their effects on the cognitive behavior of citizens. With the current widespread of gamified environments and the human nature of perception, everyone is vulnerable to being fooled into premeditated scenarios and fabricated stories.

Moving forward, it is imperative to start taking steps towards a coordinated approach to tackle disinformation. First, the government agencies' capabilities to detect, analyze and expose disinformation should be strengthened by stepping up the financial and human resources therein. Second, as online platforms play a key role in the spread of disinformation, these entities should be urged to step up their efforts in the fight against disinformation. The main platforms, including Facebook, Google, Twitter, along with software companies and bodies representing the advertising industry should commit to improve the transparency, accountability and trustworthiness of their services. Third, empowering citizens and civil society is crucial to counter disinformation and foster societal resilience. This can be accomplished by raising awareness of disinformation across the country, providing multilingual communication to millions of citizens through social media and through communication campaigns, distributing myth busting materials from both academic and institutional expertise, scaling up media literacy initiatives at regional and national levels, and supporting new technologies for content verification and fact checking.

Since the Facebook/Cambridge Analytica case and the revelations about election interference around the world, protecting the integrity of elections became a timely challenge. As cyber threats keep increasing in number and magnitude, it is becoming more and more urgent to ensure stronger preparedness and coordination in the fight against disinformation. There is no room for complacency in this context. Disinformation is an evolving threat that requires continuous research to update the policy toolbox in line with new trends and practices. Malicious sources, both within and outside the country, are constantly using new tactics, opting increasingly for smaller-scale local operations that are less likely to be detected and exposed. However, the objective remains the same: dividing the US society and undermining the trust of citizens in democratic processes and institutions.

References

1. Berkowitz, R. (2020). *A Game Designer's Analysis of Qanon*. https://medium.com. Accessed 30 September 2021.
2. Conrad, K. (1958). *Die beginnende Schizophrenie*. Thieme Verlag.
3. Deterding, S., Dixon, D., Khaled, R. & Nacke, L. (2011). *From game design elements to gamefulness: Defining "gamification"*. Proceedings of the 15th International Academic MindTrek Conference. pp. 9–15. https://doi.org/10.1145/2181037.2181040.
4. Duhigg, C. (2012). *The Power of Habit: Why we do what we do in Life and Business*. Random House LLC.
5. European Commission (2018) *Tackling Online Disinformation: A European Approach*. https://ec.europa.eu. Accessed 27 September 2020.
6. Higgins, E. T. (2006). Value from hedonic experience and engagement. *Psychological Review, 113*(3), 439–460.
7. Pine, B. J., II., & Gilmore, J. H. (1998). Welcome to the experience economy. *Harvard Business Review, 76*(4), 97–105.
8. Robson, K., Plangger, K., Kietzmann, J., McCarthy, I., & Pitt, L. (2015). Is it all a game? Understanding the principles of gamification. *Business Horizons., 58*(4), 411–420. https://doi.org/10.1016/j.bushor.2015.03.006.
9. Roozenbeek, J., & van der Linden, S. (2019). Fake news game confers psychological resistance against online misinformation. *Palgrave Commun, 5*, 65. https://doi.org/10.1057/s41599-019-0279-9.
10. Rorschach, H. (1924). *Manual for Rorschach Ink-blot Test*. Stoelting.
11. Rousis, G. J., Richard, F. D., & Wang, D. Y. D. (2020). The truth is out there: The prevalence of conspiracy theory use by radical violent extremist organizations. *Terrorism and Political Violence*. https://doi.org/10.1080/09546553.2020.1835654.
12. Shen, W., Yuan, Y., Liu, C., & Luo, J. (2016). In search of the 'Aha!' experience: Elucidating the emotionality of insight problem-solving. *British Journal of Psychology, 107*(2), 281–298.
13. Skinner, B. F. (1938). *The Behavior of Organisms: An Experimental Analysis*. Appleton-Century.
14. Tarrant, B. (2020). *The Great Replacement*. http://www.documentingreality.com. Accessed 28 September 2021.
15. Tik, M., Sladky, R., Luft, C. D. B., Willinger, D., Hoffman, A., Banissy, M. J., Bhattacharya, J., & Windischberger, C. (2018). Ultra-high-field fMRI insights on insight: Neural correlates of the Aha!-moment. *Human Brain Mapping, 39*, 1–12. https://doi.org/10.1002/hbm.24073.

Dr. Elie Alhajjar is an experienced Research Scientist with a demonstrated history of working in the higher education industry and the US government. Skilled in Management, Network Science, Leadership, and Data Analysis. Strong professional with a Doctor of Philos-ophy (Ph.D.) in Mathematics from George Mason University. Dr. Alhajar is a member of the Intelligent Cyber-Systems and Analytics Research Laboratory (ICSARL), where he works in the fields of network science and cybersecurity.

The Commercialization of Influence Operations

Sean Ryan, Ian Conway, and Kathleen Cassedy

Abstract This research establishes a clear threat to U.S. national security through the information domain. Influence exercised through this domain impacts domestic stability and internal political processes in the United States. Scenarios presented demonstrate deliberate activities that focused on influencing the U.S. and allies. Scenarios demonstrated actual intent by Russia, China, and regional actors like Iran, to impact the perceptions of target audiences in U.S., Europe, and Africa working by, with, and through commercial platforms to both obfuscate attribution and corrupt objectivity in reporting. Going further, this research connects activities by Great Power competitors, namely Russia and China, to the subversion of commercial enterprises for the primary purpose of influencing media outlets that provide information to the American (and global) public, and to lever the global business environment to their advantage. Exploiting commercial relationships built legally through overt and clandestine corporate ventures is a primary vehicle for exerting influence in Great Power Competition. Loopholes in antiquated U.S. laws and policies allow such exploitation to go unchecked and often unobserved. High net worth individuals, multinational corporations, sovereign wealth funds, and NGOs are tied to political support leading to subversion and corruption through the direct use of finance and digital information. Finally, recommendations are made to tighten loopholes in U.S. policies and statutes to mitigate foreign exploitation of information aimed at distracting and disrupting the United States in Great Power Competition for global influence.

Keywords Global influence · Great powers · Influence operations · Social media · Russia · China · United States

---

S. Ryan (✉)
West Liberty University, West Liberty, WV 26074, USA
e-mail: Sean1ryan@aol.com

I. Conway · K. Cassedy
Arlington, Virginia, USA

A. Farhadi et al. (eds.), *The Great Power Competition Volume 3*,
https://doi.org/10.1007/978-3-031-04586-8_9

Introduction

Influence shapes everyday life in countless ways, large and small. We see influence in conversations, in politics, in playground interactions, and in advertising. Traditionally, influence as practiced by nation states takes many forms, including threats, coercion, diplomacy, economic reward, and storytelling. In the subversion toolkit, influence is one of the most powerful weapons that exists. Influence can shape perceptions about things or actions in the past, present, or future.

Since the invention of the printing press, newspapers have informed and influenced large and/or widely distributed populations. Information, however, is not synonymous with influence. With the advent of modern social media over the past two decades, Internet-based influence operations have arguably supplanted newspapers as the influence operators' medium of choice. Facebook, Reddit, Snapchat, TikTok, Weibo, and WeChat convey news, information, and rumor at the speed of the Internet,

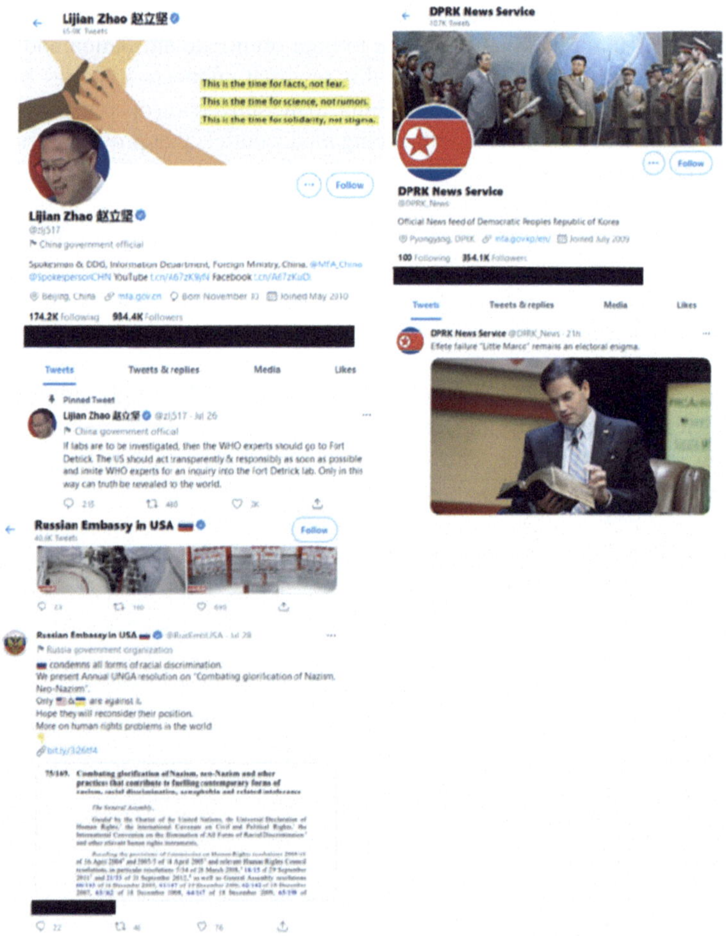

potentially reaching hundreds of millions of people in seconds. Breaking news no longer belongs to broadcast television or radio—if it's happening now, you will probably hear about it first on social media.

And then there's Twitter.

Since 9/11, the US military focused its technological superiority on turning its forces into an unstoppable machine to destroy and disrupt enemies largely through a cycle called F3EAD (Find-Fix-Finish-Exploit-Analyze and Disseminate). After the World Trade Centers fell and the Pentagon was attacked, conventional state-on-state combat capabilities became a lower priority than outfitting and training the hunter/killer teams that were sent to root out Islamic extremists around the globe. Our traditional adversaries in today's Great Power Competition–in particular, Russia and China—took note of the US's redirected attention, and used the opportunity to slowly and quietly set global operating conditions more to their advantages. Russia used the first decade of this century to tackle their near-abroad opportunities—Crimea and Ukraine—while experimenting with new approaches to their modern warfare doctrine. China used the same period to refocus and retrain its forces to conduct modern full-spectrum warfare, particularly leveraging the informational and economic elements of national power. In parallel, both nations and other adversaries observed ways in which the numerically and technologically overwhelmed Islamic terrorists were able to use a commercial digital backbone to maintain virtual operating space when physical havens were attacked, and fighters were killed. They used the Internet to recruit followers, raise funds, conduct command, control and communications, coordinate activities, and taunt the United States. These lessons in how to frustrate, circumvent, and counter a force without combat superiority did not go unlearned.

In this chapter on the commercialization of influence operations, two arguments emerge.

(1) Modern influence operations are conducted disproportionately not only on commercial platforms, but also by and through private sector entities and non-state actors.
(2) The U.S. lacks a comprehensive strategy for dealing with complex influence operations.

The shortcomings of U.S. capabilities include dealing with the proliferation of influence actors, in identifying and tracking said actors, or in countering or interceding in detrimental foreign influence operations. The chapter concludes with a discussion of potential actions that the United States could implement to pursue a sustainable competitive advantage in the quest for global power and influence.

Background

Historically, the United States has wrestled with balancing various aspects of strategy. Principle among these were finding the optimal balance between domestic and national security interests. Every President and Congress, tasked with managing budgets, were confronted with deciding how much of the federal budget should be allocated to defense spending and what specific roles and missions the Department of Defense should fill. Further, the United States faced a particular challenge with the need to synchronize the activities performed by the various agencies that comprise the U.S. Government. Today, threats to national security are increasing through both direct and indirect means; consequently, the need to synchronize U.S. capabilities and activities is arguably greater than ever before.

Whereas the U.S.' primary global adversaries, China and Russia, manage quasi-centrally controlled strategies, the United States struggles with developing, implementing, and managing a coherent strategy across the multitude of government agencies in the U. S. government. Liddell Hart would call an overarching strategy a Grand Strategy [1].[1] The U.S. addressed the need for an overarching strategy most notably three times since World War Two.

Hahn points out that as WWII ended the national security paradigm shifted. Harry Truman's presidency was the first to confront the need for a grand strategy in 1947 for new security situation. The United States held the largest multi-domain military in the world with the ability to project power over the sea, air, and land. Our primary rival, Russia, had a large Army, but little power projection capability at that time. While Truman sought to contain military spending, the strategy pursued by his administration initiated the Marshall Plan, committed to the World Bank and International Monetary Fund, involved greater international engagement, and focused broadly on increasing American influence.

George Kennan (a career diplomat who authored the now notorious Department of State Policy Planning Staff Memorandum titled "*The Inauguration of Organized Political Warfare*,"), and Paul Nitze (who led the Policy Planning Staff at Department of State in the development of National Security Council Memorandum 68) were instrumental in architecting and evolving the U.S. Containment Strategy. While Kennan and Nitze disagreed on Stalin's intentions, both arrived at the conclusion that it would take combined political, economic, and military efforts to shape the environment to contain Soviet expansion and avoid Soviet military action [2–4].[2,3,4] Both realized that it would take a position of combined relative strength to manage containment.

Later as the direct hostilities in Korea wound down, Eisenhower confronted new national security threats, as China had fallen to Mao and Russia emerged to lead

[1] Ripley, R.B. and Lindsay, J.M. eds., 1997. US foreign policy after the Cold War. University of Pittsburgh Press.

[2] X, (George Kennan) (1947). The Sources of Soviet Conflict. Foreign Affairs. July 1947. Retrieved from https://www.foreignaffairs.com/articles/russian-federation/1947-07-01/sources-soviet-conduct.

[3] Nitze, P., (1950). United States Objectives and Programs for National Security. NSC 68. Retrieved from https://history.state.gov/milestones/1945-1952/NSC68.

[4] PPD-23, (1948)., Report by the Policy Planning Staff. Foreign Relations Of The United States, 1948, General; The United Nations, Volume I, Part 2. Retrieved from https://history.state.gov/historicaldocuments/frus1948v01p2/d4.

the Communist Eastern Block with expanded power projection abilities, including nuclear weapons. Mao clearly demonstrated a willingness to confront the U.S. in Korea. In this context, Eisenhower recognized the long-term nature of the threats to the United States. Eisenhower characterized the basic problem confronting the U.S. as meeting the Soviet threat to U.S. security and "in so doing, to avoid seriously weakening the U.S. economy or undermining our fundamental values and institutions [1, 5]."[5,6]

Eisenhower's New Look policy continued a focus on containment. Avoiding direct military conflict, the New Look policy focused on nuclear deterrence, alliances, psychological and covert operations and negotiations [1].[7] Through the New Look, his administration focused on building productive allies while deterring Russia strategically.

Eisenhower created the United States Information Agency in 1953 to manage public diplomacy to external populations [8].[8] Seese and Ryan highlighted the greater complexity of contemporary communications when compared with those during the Cold War [9].[9] Still, Eisenhower clearly recognized the need to insulate the U.S. population from foreign messaging while simultaneously broadcasting our own messages.

While the U.S. may have tackled the need for an updated grand strategy after the end of the Cold War, without a clear adversary, U.S. National Security strategies lost focus. Clinton focused efforts on economic issues, which encouraged globalization and led to NAFTA among other international trade agreements. The Republican congress mitigated Clinton's efforts somewhat [1].[10] Little in the way of a coherent grand strategy emerges during the 1990s, however. U.S. military actions in Panama, Somalia, the former Yugoslavia, and arguably even Kuwait, suggest a tendency toward interventionism, rather than a purposeful overarching grand strategy. The past two decades seem to build on the interventionist approach as the U.S. pursued a global game of whack-a-mole chasing extremists in an indecisive process focused on high value targets. It is not especially difficult to make the case that tactics and strategy became conflated, culminating in the two becoming virtually indistinguishable.

The time to acknowledge the current threats and adopt an appropriate and pragmatic grand strategy is overdue. Tactics should support this strategy, not define it. Such a strategy must capitalize on U.S. strengths, mitigate our weaknesses, and insulate our population from foreign information operations and cyber-attacks. We must be clear-eyed and candid about the threats we face and avoid dismissing elements of

---

[5] FRUS, 1952–1954. 1984. See U.S. Department of State. Papers Relating to the Foreign Relations of the United States, 1952–1954. Washington, D.C., GPO.

[6] Ripley, R.B. and Lindsay, J.M. eds., 1997. US foreign policy after the Cold War. University of Pittsburgh Press.

[7] Ripley and Lindsay, eds., (1997).

[8] Cull, N.J., 2009. The cold war and the United States information agency: American propaganda and public diplomacy, 1945–1989. Naval War College Review, 62(2), p.14.

[9] Seese G., and Ryan, S., (2021). Great Power Competition, 2nd Ed., Springer. Pending publication.

[10] Ripley and Lindsay, eds., (1997).

these threats we find politically or economically inconvenient. These inconveniences are what our adversaries seek to exploit.

While current national strategy documents term this threat variously as Great Power Competition, strategic competition, competition below armed conflict, gray zone conflict, etc.—the aforementioned statesman George Kennan called it Political Warfare. While all of these terms have their merits, the threat is perhaps best defined simply as subversion. Specifically, the sustained subversion campaign targeting America's instruments of national power including—and especially—the economy and the information environment [8].[11] Critically, in the authors' research, these subversion campaigns almost invariably target and/or leverage the private sector. Subversion lives in the gaps and seams between government policy and private sector freedoms.

Developing an effective information mechanism focused on countering foreign subversion and building resilience in our domestic population must be a priority. Clear policies that build on American values is an essential element. While there is no question that there must be both offensive and defensive components to this strategy, we must not compromise our values by playing this game by our adversaries' rules. Surrendering the moral high ground will undermine the strategy itself. It is important to acknowledge that a strategy built on ideology will only lead to division, as ideology is intrinsically binary. Conversely, values-based decisions and strategy leaves room for dialogue and compromise.

Today, as globalization presents opportunities and threats in multiple domains, the key challenge for policy makers and strategists is to address threats and seize opportunities not in a single domain, but rather holistically. If the United States wishes to maintain the "global order" that has brought relative prosperity and peace to so many, a new strategy is imperative.

Situation Today

In the twenty-first century's global strategic competition with an increasingly multipolar world, the United States' greatest competitors are Russia and China. In the absence of a strategy to effectively counter the subversion campaigns of these competitors, an increasingly larger group of regional powers are ascendant, including India, Turkey, the Gulf States (led economically and practically by Saudi Arabia—or arguably Israel), Japan, Brazil, and even possibly Argentina according to Zeihan [9].[12]

---

[11] Conway, I., Cassedy, K., Ryan, S., and Danish, C., (2019). Reconsidering Great Power Competition Below Armed Conflict. The Cipher Brief. December 13, 2019. Retrieved from https://www.thecipherbrief.com/reconsidering-great-power-competition-below-armed-conflict.

[12] Zeihan, P., 2020. Disunited nations: the scramble for power in an ungoverned world. HarperCollins.

In Europe, Germany and France currently dominate economically and geopolitically. The European banking system in particular plays a crucial role in influence operations proliferated by malign actors around the globe, as client privacy is ingrained in the banking culture. Authoritarian regimes, ultra-high net worth individuals (UHNWIs), and malicious non-state actors all move personal wealth and operational funds throughout the global banking system, often emerging "clean" in Western private banks [10].[13]

In Southwest Asia, Iran continues to function as the Middle East's boogie man, confounding Western sanctions, refusing to go away quietly, and generally keeping the world on its toes with "will they or won't they?" guessing games about the state of their nuclear weapons program. Despite Iran's status in the West as a pariah state, the country exerts considerable influence over other regional players and generally keeps its neighborhood slightly off balance and wary of the potential to exert outsized influence if Iran either negotiates its way out of sanctions or acquires a functioning nuclear weapons program.

In short, this still-young century comes with a period of global redistribution of power and relationships. The world order that looked to the U.S. as its global superpower is crumbling, and foes both old and new are seizing the opportunity to remake the world more to their liking. Multipolarity is the order of the day, and strategic competition is the game. In this time and place, your relative size matters far less than your agency. And agency is available for the taking to individuals, businesses, groups, and nations with the will to execute, in the Age of the Internet.

Large or small, what do these actors have in common? A sophisticated understanding of, and capability to employ, large scale information and influence operations, based almost exclusively in the commercial sector and targeting select civilian populations around the world. While the U.S. birthed many if not most modern commercial influence operation capabilities, as a nation the U.S. lags behind its global competitors in controlling the cognitive high ground in the information battlespace. Observing information and influence activities allows the examination of the effects of the global outsourcing of influence operations, the under-utilized domestic influence capabilities, and the domestic policy challenges presented by the privatization of influence operations.

Russia

Modern Russian military and security strategy places heavy emphasis on what it terms the Initial Period of War (IPW) [11].[14] Russian military strategists believe that the IPW sets all the terms and conditions for conventional kinetic conflict, and that the IPW may last days, weeks, months, or years. Within the IPW, information

[13] International Consortium of Investigative Journalists; *Global banks defy U.S. crackdowns by serving oligarchs, criminals and terrorists.* International Consortium of Journalists. September 20, 2020. https://www.icij.org/investigations/fincen-files/global-banks-defy-u-s-crackdowns-by-serving-oligarchs-criminals-and-terrorists/.

[14] Thomas (2017). The Evolving Nature of Russia's Way of War. Military Review July–August 2017. Army University Press. Retrieved from: https://www.armyupress.army.mil/Journals/Military-Review/English-Edition-Archives/July-August-2017/Thomas-Russias-Way-of-War/.

dominance at the strategic, operational, and tactical levels is viewed as critical to success. In short, in current Russian military doctrine, if one does not control the information space during the IPW, one cannot win the conventional war.

This view provides a different lens with which to view Russian information operations this century, under the leadership of Vladimir Putin. While Russia (Soviet and post-Soviet) has long been acknowledged as master of the information craft—particularly its role in political warfare—the past two decades have seen a number of Russian information innovations. Chief amongst these has been the outsourcing of influence operations to non-governmental entities, which in turn operate as commercial entities, leveraging the Internet as a global operating space without needing to leave the terra firma of Mother Russia.

As Żochowski Darczewska wrote in 2017,

> ...active measures are now being perpetrated through layers of proxies – front organizations using non-governmental and corporate platforms established in a wide variety of countries, Russian expats, and funded through with both Russian government money and private money laundered in tax havens...or quasi-legitimate business ventures in the West [12].[15]

Outsourcing and operating commercially allowed these entities to access sophisticated western digital marketing and social media tools remotely, acquiring capabilities only as needed and appearing for all the world like routine business operations. This method of operation allowed Russian information efforts of the past decade to go largely undetected in the west, despite a scale and scope that the U.S. is only beginning to understand.

2016 Presidential Election and The Brexit Movement

The fifth report from the Senate Select Committee on Intelligence about the Russian interference in the 2016 U.S. Presidential Election illustrates Russian information operations via outsourcing. This report states that,

> ...highly evolved tools used to shape popular sentiment were utilized ... commercially available services - many of which are based overseas - rely on an array of personal information to build targeted messaging profiles. Russia applied these same technologies and methodologies to its campaign during the 2016 election and, in doing so, conducted foreign influence operations against the United States with a speed, precision, and scale not previously seen. The commoditization of these influence capabilities by for-profit firms working in the political and particularly electoral space, coupled with deeply concerning foreign government and intelligence service ties to some organizations, were troubling enough to warrant additional Committee scrutiny [13].[16]

[15] Darczewska, Żochowski; "Active Measures: Russia's Key Export." *Point of View* Issue 64. Warsaw: Centre for Eastern Studies. June 2017.

[16] Report of the Select Committee on Intelligence United States Senate on Russian Active Measures Campaigns and Interference in the 2016 U.S. Election Volume 5: Counterintelligence Threats and Vulnerabilities.

The Senate Select Committee on Intelligence states later in this document that they had "*…identified an international marketplace for digital services to shape popular sentiment and electoral outcomes* [13]."[17]

While the committee has been careful with their language (they do not explicitly assert that Russia did or did not engage in outsourcing influence operations to the private sector during the 2016 election), they leave no ambiguity whatsoever as to the availability of sophisticated influence tools and services in the private sector, and the ties these companies maintain to foreign intelligence services.

Critically, the firm most widely implicated in media scandals and both congressional and parliamentary oversight bodies was the UK's Cambridge Analytica. While ties to Russia are sketchy at best (and the British newspaper *The Guardian* really led the way here through the exemplary investigative reporting of Carole Cadwalladr), they include—at a minimum—a Russian data scientist, the Russian firm Lukoil [14],[18] and alleged meetings between pro-Brexit funders and Russian embassy staff [15],[19] among others.

Russia, or Non-state Actors? or, Both?

Irrespective of Russian ties to Cambridge Analytica, this firm absolutely worked for multiple British pro-leave political organizations concurrently (a violation of British electoral law), and in the U.S., first the Ted Cruz campaign and then Donald Trump's campaign. Testimony to the British Parliament revealed that Cambridge Analytica had provided influence as a service to political campaigns in more than two dozen countries spanning the globe [16].[20] While originally a UK company affiliated with the parent entity SCL Group, Cambridge Analytica came into its own with the backing of American hedge fund billionaire Robert Mercer, and the leadership of his daughter Rebekah and Steve Bannon. Bannon was a former investment banker, media executive, and later Chief Strategist to President Trump.

With or without Russian money (and we may never know the answer to this question), UK- and U.S.-based wealth and talent were largely behind the effectiveness of this company. In this way, influence operations bearing many of the Kremlin's

[17] Report of the Select Committee on Intelligence United States Senate on Russian Active Measures Campaigns and Interference in the 2016 U.S. Election Volume 5: Counterintelligence Threats and Vulnerabilities.

[18] Cadwalladr and Graham-Harrison (2018). Cambridge Analytica: links to Moscow oil firm and St Petersburg university. March 17, 2018. Retrieved from: https://www.theguardian.com/news/2018/mar/17/cambridge-academic-trawling-facebook-had-links-to-russian-university.

[19] Cadwalladr and Jukes (2018). Arron Banks 'met Russian officials multiple times before Brexit vote.' June 9, 2018. Retrieved from: https://www.theguardian.com/politics/2018/jun/09/arron-banks-russia-brexit-meeting.

[20] House of Commons Digital, Culture, Media and Sport Committee. (2019) Disinformation and 'fake news': Final Report. Retrieved from: https://publications.parliament.uk/pa/cm201719/cmselect/cmcumeds/1791/179102.htm.

hallmarks are now employed by western political parties and are being turned against their domestic political rivals. This cycle feeds 24/7 cable news outlets and Silicon Valley's social media platforms, which are squarely in the business of marketing outrage for advertising revenue. All of this resulting in national disunity, ideological fratricide, and the perpetuation of an information-based civil war.

By way of example, when President Trump was suspended from Twitter, there was a groundswell amongst his base to abandon Twitter and shift to the platform Parler. Parler was later found to be using Russian infrastructure [17].[21] Further, when Amazon Web Services (AWS) then terminated Parler as a client, the filings of the subsequent Parler v AWS lawsuit revealed that the ultimate beneficial owner (UBO) of Parler was NDM Ascendent LLC, owned by Rebekah Mercer of Cambridge Analytica [18].[22] Cambridge Analytica later became Emerdata, largely run out of Abu Dhabi in the UAE, where the firm doubtless found no shortage of clientele [19].[23] As the online publication *Lawfare* states:

> Private actors are adapting their online disinformation campaigns to avoid detection as social media companies continue to target them with takedowns and suspensions, and governments will likely continue to work through these private firms because they benefit from the indirect approach [20].[24]

Case study: Bonanza Media

The online investigative collective known as Bellingcat published a lengthy report in late 2020, in which they expose another investigative platform—Bonanza Media—as likely being funded, guided, and coordinated with individuals sympathetic to (if not directed by) Russia's GRU and the Russian state. The report details how Bonanza Media was used to create and amplify disinformation regarding the official investigation into the downing of MH17 over Ukraine in 2014. The Bellingcat report presents strong evidence that Bonanza Media was purposely built by Russian controlled elements, specifically to present alternative narratives about the crash that would cast doubt on Russian involvement in the crash. The Bonanza Media team used Kickstarter to raise funds for a documentary; leaked documents via their platform that appear to be from the GRU; and more recently, used YouTube to proliferate

[21] Menn, Li and Culliford (2021). Parler partially reappears with support from Russian technology firm (Reuters, January 18, 2021) Retrieved from: https://finance.yahoo.com/news/parler-partially-reappears-support-russian-234401397.html.

[22] Sanders (2019). See You in Which Court? (Substack) Retrieved from: https://wildwest.substack.com/p/see-you-in-which-court.

[23] Siegelman (2018). Chart: Emerdata Limited — the new Cambridge Analytica/SCL Group? (Medium, March 25, 2018). Retrieved from: https://wsiegelman.medium.com/chart-emerdata-limited-the-new-cambridge-analytica-scl-group-63283f47670d.

[24] Grossman and Ramali (2020). Outsourcing Disinformation. (Lawfare December 13, 2020) Retrieved from: https://www.lawfareblog.com/outsourcing-disinformation.

COVID conspiracy-related videos. The entire platform appears to exemplify modern Russian information use of commercial entities and the leveraging of digital media and the open Internet to amplify disinformation narratives.

Derived from the Bellingcat Investigative Report on Bonanza Media [21][25]

China

China's use of predatory lending for economic exploitation is becoming legendary. Their loans are strategically placed. In Africa, for example, China loaned more than $20 billion to Angola. For context, Angola is Africa's second largest oil exporter [22].[26] Angola is China's largest destination for loans on the African continent [23].[27] China's loan program with Angola involves several effects that favor Chinese strategy in terms of influence and access to oil.

This scenario begs several questions. Why would a large oil-exporting nation need to borrow billions of dollars? Why would China make loans of this nature? Reuters (2021) also reported that China granted Angola relief from some of their debt service obligations. What does China gain from granting debt relief? The South China Morning Post addressed part of this last question for Chinese lending institutions in terms of financial risk.

The answers to these strategic business questions is fairly straight forward. Such staggering debt provides China with significant influence in a major oil producing nation. The strategic advantages go much further, however. Angola continues to receive financial loans from the International Monetary Fund. Reuters and SCMP acknowledged that Angola is also the largest recipient of IMF loans in sub-Saharan Africa [22, 23].[28,29]

The IMF already made loans totaling more than $2.5 billion to Angola (Reuters, 2021). It is not a secret that the U.S. is the largest donor to the IMF. This correlates to an interesting cost–benefit relationship between China and the United States. Since the U.S. contributes nearly three times as much as China to the IMF, the U.S. incurs nearly three times the cost as China when the IMF loans Angola money (IMF, 2021). The IMF loans are going, at least in part, to serving Angolan debt to Chinese

---

[25] Bellingcat Investigation Team; *The GRU's MH17 Disinformation Operations Part 1: The Bonanza Media Project*. Bellingcat. November 12, 2020. https://www.bellingcat.com/news/uk-and-europe/2020/11/12/the-grus-mh17-disinformation-operations-part-1-the-bonanza-media-project/.

[26] Arnold and Strohecker (2021). Angola gets breathing space from Chinese creditors says finance minister. Reuters. January 11, 2021. Retrieved from: https://www.reuters.com/article/us-angola-economy/angola-gets-breathing-space-from-chinese-creditors-says-finance-minister-idUSKBN29G18X.

[27] Nyabiage, J., (2021). Lack of detail in Angola's debt deals with China could hide future risk. South China Morning Post. January 18, 2021. Retrieved from https://www.scmp.com/news/china/diplomacy/article/3118140/lack-detail-angolas-debt-deals-china-could-hide-future-risk.

[28] Arnold and Strohecker (2021).

[29] Nyabiage, J., (2021).

lenders. Consequently, it is reasonable to conclude that China is gaining influence in Angola, and in turn, the U.S. is indirectly funding the influence operation for the Chinese government. Further, China gains USD-valuated currency to further its foreign exchange gains from decades of trade imbalances and a largely insular yuan market.

There is one additional strategic advantage for China in this debt scenario. Chen [24] reported that China adopted long-term strategic objectives focused on becoming a leading world superpower. This objective is competitive in nature relative to the U.S. and Russia. Increasing influence in developing nations with strategic resources like oil can only advance China toward the goals adopted during the 19th Party Congress [24].[30]

Pakistan's misinformation campaign over their support to the Taliban created a unique opportunity for China. The loss of U.S. funding when the Trump Administration revealed Islamabad's duplicity opened the door to Chinese money [25].[31] Commercial ventures under auspices of programs such as the Pakistan-China Economic Corridor and the One-Belt-One-Road (OBOR) Initiative leaves Pakistan $24 billion in debt to China[26].[32] This represents 27% of Pakistan's $90 billion in external debt.

Pakistan presented China with a unique opportunity. Financial investment in Pakistan enabled Pakistan to protract the American involvement in Afghanistan and damage perceptions of the United States globally. Beyond supporting Pakistan in a surrogate role in Afghanistan, China also leveraged misinformation and financial debt to seize control of a warm water port. China effectively has control over the recently developed port of Gwadar [27].[33] This provides China with strategic advantage relative to the United States in the Great Power Competition, as well as improving China's strategic position with India.

Debt leverage is clearly a commonly used tool in China's influence toolkit. Where else is China's use of influence expanding into the commercial domain? In the media, new and traditional. China seeks to control the narrative and is willing to invest national treasure—and pressure Chinese and local nationals in global media fields—to see that China is portrayed in the most positive light possible. This includes in fictional works, where China gatekeeps access to its citizens' entertainment dollars

[30] Chen, D. (2017) China Has a New Grand Strategy and the West Should Be Ready. The Diplomat. Retrieved from: https://thediplomat.com/2017/10/china-has-a-new-grand-strategy-and-the-west-should-be-ready/.

[31] Ward, A., (2018). Why Trump Cut Millions in Military Aid to Pakistan. Vox. September 4, 2018. Retrieved from https://www.vox.com/2018/9/4/17818396/pakistan-aid-military-trump-pompeo-afghanistan.

[32] Haqqani, H., (2021). Islamabad will come to regret aiding the Taliban's Resurgence. Foreign Affairs. July, 22, 2021. Retrieved from: https://www.foreignaffairs.com/articles/afghanistan/2021-07-22/pakistans-pyrrhic-victory-afghanistan?utm_medium=promo_email&utm_source=special_send&utm_campaign=mktg_actives_we_all_lost_afghanistan&utm_content=20210816&utm_term=all-actives.

[33] Chellaney, B., (2011). China's Strategic Subterfuge. Great Power Competition in the Indian Ocean. Stagecraft and Statecraft. https://chellaney.net/2011/06/10/618/.

in exchange for softening or changing storylines or characters the government deems unsuitable for Chinese audiences [28].[34] China will also invest in or outright acquire media outlets (traditional or online) in locations outside China where it desires to shape local populations' perceptions of China. Their approach all comes back to controlling the narrative, and the narrative in the world of today is driven through online media and entertainment.

Case study: China and foreign media

Significant focus was placed in the last few years on China's One Belt, One Road initiative as an example China's modern influence operations. Less discussed—but no less prolific—are China's global investments in traditional news, media and entertainment platforms. China takes a multi-pronged approach, investing in or outright acquiring movie production and distribution firms, new media platforms around the globe, and traditional news companies with both online and paper distribution. Through these commercial investments, China gains leverage in shaping how China is portrayed and in shaping what narratives are amplified in which regions. In Africa, for example, China expanded overt Chinese media entities' presence, such as the Xinhua News Agency, and also invested in local and regional media. Kenya in particular serves as a base of operations for Chinese news and media companies on the African continent. China's approach to African-owned media is slightly different; they may offer badly needed operational funds, advertising purchases, and/or free content to local entities. Nothing is free, however; African journalists reported being pressured by Chinese investors to stop writing stories portraying China in a negative light. Media companies report Chinese companies and government outposts threatening to withdraw funding or advertising if sufficient positive stories about China are not published.

Derived from "Experts warn of China's growing media influence in Africa" published in *Deutsche-Welle by* Christopher Mwakideu. January 29, 2021 [29][35]

Commercialization of news and media presents China with unique opportunities to control and influence narratives. China is now positioned to influence positive narratives and to quash any message in news or media that might be viewed as negative in nature to China's image. Threats, openly stated or not, to postpone or cancel commercial contracts must factor into the decision calculus of media moguls at

[34] Robinson, Joanna; "Did You Catch All the Ways Hollywood Pandered to China This Year?"; *Vanity Fair*. August 5, 2016. https://www.vanityfair.com/hollywood/2016/08/did-you-catch-the-ways-hollywood-pandered-to-china-this-year.

[35] Mwakideu, Christopher; "Experts warn of China's growing media influence in Africa"; *Deutsche-Welle (DW)*. January 29, 2021. https://www.dw.com/en/experts-warn-of-chinas-growing-media-influence-in-africa/a-56385420.

Disney, Viacom, NewsCorp, GE, Time Warner, and others [30, 31].[36,37] In the protectionist perspective, China is also uniquely positioned to use its influence to prevent the release of information related to their extensive loans. This correspondingly creates compound favorable conditions for increasing China's influence.

China is also ramping up direct overt rhetoric regarding responsible partnerships and casting themselves as the premier example of a responsible partner [32].[38] Following a 2020 meeting of the Quad foreign ministers (Australia, Japan, India, and U.S.), Chinese Foreign Minister Yi issued harsh inflammatory language forecasting security concerns and risks to the Indo-Pacific region due to the Quad alliance. Portraying the Quad as a vehicle to start another cold war, China is leveraging both informational and economic power through the Regional Comprehensive Economic Partnership to portray itself in a positive light, while attempting to portray the Quad participants as risky partners to other Indo-Pacific countries.

Most recently, Chinese state media fabricated a Swiss scientist who claimed to be under "enormous pressure" from the U.S. to validate the COVID-19 "lab-leak theory." Under subsequent pressure from the Swiss embassy, the Chinese removed these stories [33].[39] This example provides excellent evidence of how far China will go to convey and protect a positive image. When unable to quash negative news media releases, China will fabricate and disseminate disinformation in attempts to delegitimize negative reports.

Iran

Brooking and Kianpour impressively detail Iran's modern influence operations in the Atlantic Council, centered heavily on social media usage.

As a whole, Iran's digital influence operations represent a continuation of public diplomacy, albeit conducted through misleading websites and social media sock-puppets. Iran broadcasts a fairly consistent message to many different audiences: in Africa, in Southeast Asia, in Europe, in North America, and, most notably, in Latin America and the Middle East. The aim of these efforts is to "tell Iran's story," the same as any Western government broadcaster might strive to do. The difference is

[36] Lutz, A., Business Insider. These 6 Corporations could control 90 percent of the media in America. June 14, 2012. https://www.businessinsider.com/these-6-corporations-control-90-of-the-media-in-america-2012-6.

[37] Seese, G., and Ryan, S., Great Power Competition Conference 2. Counter Intuitiveness of Influence Science, Social Judgment Theory, and Strategic Messaging in Great Power Competition (Pending publication). December 2020.

[38] Rudd, K., (2021). Why the Quad Alarms China. Its Success Poses a Major Threat to Beijing's Ambitions. Foreign Affairs. August 6, 2021. Retrieved from https://www.foreignaffairs.com/articles/united-states/2021-08-06/why-quad-alarms-china?utm_medium=newsletters&utm_source=twofa&utm_campaign=Why%20the%20Quad%20Alarms%20China%20&utm_content=20210813&utm_term=FA%20This%20Week%20-%2011 2017.

[39] https://asia.nikkei.com/Spotlight/Coronavirus/US-China-fight-over-COVID-origin-escalates-with-dash-of-fake-news.

that, as an international pariah, Iran must pursue this work through more clandestine means [34].[40]

Broadly, Iran's foreign influence efforts evidence a level of routinization that distinguish it from Russia, China, Saudi Arabia, or any other nation that has built a digital influence apparatus. This is a result of the early integration of digital manipulation into Iranian government and military functions. In 2009, Ayatollah Khamenei stated that "content promotion" was "the most effective international weapon" against foreign adversaries. In 2011, the head of the IRIB bragged that he had developed seven cyber battalions of "media experts and specialists," supposedly consisting of 8,400 members. Tehran's IRGC headquarters has trained thousands of recruits in "content production," teaching them social media strategy and graphic design. Given this heavy and longstanding emphasis, it is likely that many elements of the Iranian state—the IRIB, the office of the supreme leader, the intelligence services, the IRGC and associated militias, and the regular Iranian military—each employ their own Internet operatives [34].[41]

Iran sees information and influence as central to what it calls "soft war," and views the Republic itself as the target of a multinational battle using information. While the U.S. government and private U.S. social media companies have sought to control, shut down, or otherwise expose Iranian influence operations against the West, so too has Iran identified media platforms and organizations and prohibited them from operating within Iran [34].[42] This is possible because the Iranian government waged a campaign over the last 10–15 years to control Internet access for its citizens, and thereby, the flow of information from non-official sources into the country. Like its fellow authoritarian nations (Russia, China, and North Korea), Iran aggressively surveils its domestic communications platforms and uses the findings to threaten, detain, and even eliminate dissenting or even simply more liberal Iranian voices [35, 36].[43,44] This latter type of influence operation is used abroad as well, as evidenced in the below vignette.

However, Iran may also be involved in using U.S.-based commercial media entities to conduct influence operations. While there has been no publicly available proof, investigative journalists cite both concerning financial opacity and several suspicious correlations between Iran and Minnesota-incorporated Mint Press News. Reporters Rosie Gray and Jessica Testa assessed Mint Press as follows in 2013.

---

[40] Brooking, Emerson T., and Kianpour, Suzanne; *Iranian digital influence efforts: Guerrilla broadcasting for the twenty-first century*"; the Atlantic Council. February 11, 2020. https://www.atlanticcouncil.org/in-depth-research-reports/report/iranian-digital-influence-efforts-guerrilla-broadcasting-for-the-twenty-first-century/.

[41] Ibid.

[42] Ibid.

[43] BBC News; "Iranian blogger Sattar Beheshti dies after arrest"; *BBC News*. November 8, 2012. https://www.bbc.com/news/world-middle-east-20259533.

[44] Deutsche-Welle; "Ruhollan Zan: Iran executes blogger who inspired protest"; DW.com, English version. December 12, 2020. https://www.dw.com/en/ruhollah-zam-iran-executes-blogger-who-inspired-protests/a-55914201.

Interviews with former employees and people familiar with the inner workings of Mint Press, and an examination of public records, paint a portrait of a dysfunctional outlet where employees are left in the dark about the site's sources of funding and are alienated from the Muhawesh family that runs it: Mnar, the editor-in-chief, her brother-in-law and managing editor Muhammad Muhawesh, and her father-in-law Odeh Muhawesh, 54, a Minneapolis businessman born in Jordan. They also reveal an agenda that lines up, from its sympathy with the Syrian regime to its hostility to Sunni Saudi Arabia, with that of the Islamic Republic of Iran, where Odeh Muhawesh studied under an ayatollah for five years after the Islamic Revolution, and where he visited as recently as this summer [37].[45,46]

Case study: Iran's use of U.S. private investigators for surveillance

As this chapter was being written, the U.S. Department of Justice announced charges against an Iranian intelligence officer and members of an Iranian intelligence network. The charges include multiple counts of conspiracy in a plot to kidnap an Iranian-born journalist who is a U.S. citizen and resident. The indictment includes details of the Iranian intelligence personnel using U.S. private investigators to conduct surveillance against the journalist, who has been an outspoken critic of the Iranian government and its human rights abuses.

> On multiple occasions in 2020 and 2021, as part of the plot to kidnap Victim-1, FARAHANI and his network procured the services of private investigators to surveil, photograph, and video record Victim-1 and Victim-1's household members in Brooklyn. The extensive surveillance that FARAHANI's network procured included requests for days' worth of surveillance at Victim-1's home and the surrounding area, videos and photographs of Victim-1's family and associates, surveillance of Victim-1 outside Victim-1's residence, and the installation of and access to a live high-definition video feed depicting Victim-1's home. The network repeatedly insisted on high-quality photographs and video recordings of Victim-1 and Victim-1's household members; a large volume of content; pictures of visitors and objects around the house; and depictions of Victim-1's body language. The network procured the surveillance by misrepresenting their identities and the purpose of the surveillance to the investigators, and laundered money into the United States from Iran in order to pay for the surveillance, photos, and video recordings of Victim-1.

If the defendants are found guilty, this case will demonstrate clear attempts by Iran to subvert both the U.S.'s Constitutionally guaranteed freedom of the press and Iran's deft use of the commercial private investigation industry (using false identities to procure services) as part of an effort to censor an American journalist.

---

[45] Gray, R., and Testa, J., (2013). The Inside Story Of One Website's Defense Of Assad. A small Minnesota news outlet caused a storm when it ran a story claiming Syria's rebels carried out a chemical attack near Damascus in August. A look inside the murky world of Mint Press News. October 1, 2013, Retrieved from https://www.buzzfeednews.com/article/rosiegray/the-inside-story-of-one-websites-defense-of-assad.

[46] Ibid.

Derived from filings in the Southern District of New York [38].[47]

The Gulf States

The Gulf States—particularly Saudi Arabia, the United Arab Emirates, and Qatar—are in a league of their own when it comes to contemporary social media-based influence operations. Bot farms, troll farms, sockpuppets, pseudo-news fronts, purpose-built consultancies, lobbying, narrative spinning, mercenaries—all these influence tools and more are employed by the Gulf States through outsourced commercial relationships. Players in this region wage campaigns domestically, against and for each other, and abroad on a scale that's only beginning to be understood. A regional cultural orientation (and enormous oil revenue-derived sovereign cash reserves) towards employing "guest workers" to perform menial work and jobs that Gulf state citizens consider beneath them means that the Gulf is teeming with private sector entities lining up to benefit from the free-flowing oil money. This attitude extends into influence operations, making the Gulf region ripe with opportunity for enterprising companies and individuals with software, expertise, military training, and a willingness to put revenue ahead of their countries' of origin strategies [39, 40].,[48,49]

Case study: Qatar Blockade and the FIFA 2022 World Cup

Surrounding the award to Qatar as the host of the 2022 FIFA World Cup (which may have been the result of Qatari influence operations), neighboring Gulf States—led by Saudi Arabia—launched an international influence operation to try to persuade FIFA to revoke Qatar's hosting status and instead award it to either Saudi Arabia or the UAE. In response, Qatar launched its own international counter-influence effort. The *New York Times* reported that this battle for influence and information dominance between the Gulf States lasted at least six years and involved a London-based think tank, a multi-year blockade, news stories amplified by the BBC, multi-million dollar lobbying and counter-lobbying efforts in the U.S., and accusations on all sides of bribery, malfeasance, surveillance, hacking and even terrorism support. All of these operations, in support of the larger influence campaigns, involved private PR

[47] U.S. Attorney's Office, Southern District of New York press release. "Manhattan U.S. Attorney Announces Kidnapping Conspiracy Charges Against An Iranian Intelligence Officer And Members Of An Iranian Intelligence Network"; U.S. Department of Justice, Southern District of New York; July 13, 2021. https://www.justice.gov/usao-sdny/pr/manhattan-us-attorney-announces-kidnapping-conspiracy-charges-against-iranian.

[48] Wood, Josh; "Outsourcing war: How foreigners and mercenaries power UAE's military"; *Middle East Eye*; July 13, 2018. https://www.middleeasteye.net/news/outsourcing-war-how-foreigners-and-mercenaries-power-uaes-military.

[49] Salisu, Mohammed; "Hiring of Forces and Security in the Arab Gult: The Case of Private Security Companies in Qatar"; *Chinese Business Review*; July–September 2019, Vol. 18, No. 3, 10–18. http://www.davidpublisher.com/Public/uploads/Contribute/5df2f989a60c4.pdf.

firms, commissioned consulting reports, commercial cyber capabilities, and private media organizations. In the U.S. alone, Saudi Arabia and Qatar alone spent upwards of $50 million combined in 2017 on lobbying and counter-lobbying efforts. While centered in this case are the prestige and substantial national economic benefits of hosting duties for the World Cup, this decade-long influence campaign is emblematic in the global strategic competition environment of rising regional powers choosing to outsource their influence fights rather than conduct them using government assets alone. It enabled these competitors to expand their reach to global audiences and take advantage of the best commercial capabilities that their money can buy.

Derived from New York Times articles on UAE and Qatar outsourcing military trends [41, 42],[50,51]

Non-state actors: terrorists, MNCs, NGOs, UHNWIs, and sovereign wealth funds

Finally, in an era when the most effective and pervasive influence operations are either wholly outsourced to (or at least conducted through) commercial entities and platforms, no discussion of the operating environment would be complete without touching on the role of non-state actors in this practice. Social media platforms and the Internet made it possible for anyone with a mobile device and an Internet connection to broadcast anything they wish. Social media started as a form of communication and entertainment, but it was quickly adopted by companies and entities around the globe. The development and proliferation of these tools in the private sector means that capabilities once only available to governments are now available to anyone, anywhere, anytime.

Terrorists—Certainly, from the early days of the Global War on Terror, Internet-based spaces (hidden websites, members-only web fora, chat rooms, and information repositories) were a boon to al Qaeda and its affiliated offshoots and partners. Online, they found they could tell their story directly to followers and potential recruits everywhere, without having to go through the onerous task of getting news sources on their side. They were also able to fund-raise, provide training materials, and conduct communications and guidance using commercial encryption tools and even steganography. And al Qaeda used this access to grow its influence, both with followers and with its targets. Who doesn't remember the shock and horror in Western audiences when the live beheading of journalist Daniel Pearl was released onto the Internet? One could even argue that 9/11 was, at its heart, an influence operation, since its

[50] Montague, James and Panja, Tariq; "Ahead of Qatar World Cup, a Gulf Feud Plays Out in the Shadows"; *New York Times*. February 1, 2019. https://www.nytimes.com/2019/02/01/sports/world-cup-2022-qatar.html.

[51] Palumbo, Morgan and Draper, Jessica; "How Saudis, Qataris and Emiratis took Washington"; *Asia Times*. June 10, 2020. https://asiatimes.com/2020/06/how-saudis-qataris-and-emiratis-took-washington/.

intended effect was not simply devastation at the physical attack locations, but the pervasive imagery conveying the ultimate message: the U.S. was not invulnerable to attack on its own soil.

Multinational Corporations (MNCs)—Successful MNCs use influence operations every single day. We simply know them as advertising. What is advertising if not an attempt to influence the target audience to buy something, believe something, go somewhere, participate in an activity, etc.? But influence operations can also be seen amongst MNCs in how they interact with governments and local populations in different parts of the world. Sophisticated MNCs use public relations to spin narratives about new projects, business developments, employment, environmental issues. MNCs also conduct less overt influence operations, including competitive intelligence, commercial investigations, and even corporate espionage. When revenue and reputation are at stake, many less-scrupulous MNCs set their ethics bar quite low. For instance, the conduct in the Niger delta of Royal Dutch Shell and its alleged used of bribery, threats, and corruption to damage the environment, seize land, and extract natural resources at the expense of the local citizenry [43][52] is a classic example of full-spectrum influence operations as conducted by MNCs.

Ultra-High Net Worth Individuals (UHNWIs)—Prior to the 2016 Elections in the U.S., few people may have ever heard of Robert Mercer outside of the world of finance. A trained computer scientist who moved into artificial intelligence and trading algorithms for hedge funds, Mercer made billions of dollars as the co-CEO of hedge fund Renaissance Technologies. When the Cambridge Analytica scandal broke on both sides of the Atlantic, journalists found that Mercer's money was not only behind Cambridge Analytica and its "heirs." It also emerged that he and his daughters had become very active in financing a number of ultra-conservative politicians and political groups at the state and national levels in the U.S., and they supported the pro-Brexit UK political figure, Nigel Farage. Mercer also was the primary funding behind Steve Bannon's *Breitbart* website, which gave voice to distinctly non-mainstream political views and voices in America and Europe. Between 2006 and 2016, Mercer is estimated to have donated nearly $35 M to the GOP [44],[53] in addition to his investments in technology and platforms in the U.S., UK, EU, and elsewhere. This is a thoroughly modern billionaire, who understands that the combination of wealth, technology, and communications platforms could afford him the influence to reshape his world much more to his liking.

Sovereign Wealth Funds (SWFs)—The extent to which the financial system is routinely involved in global influence operations tends to go relatively unnoticed for many people. The activities of venture capital, private equity, and hedge funds are

---

[52] Takambou, Mimi Mefo and Muhammad, Bello; "Shell to pay $111 million for 1970 Niger Delta oil spills"; *DW*. August 8, 2021. https://www.dw.com/en/shell-to-pay-111-million-for-1970-niger-delta-oil-spills/a-58697881.

[53] *Bowers, John; "A hedge fund house divided: Renaissance Technologies". OpenSecrets Blog. Center for Responsive Politics. June 7, 2016.*https://www.opensecrets.org/news/2016/06/a-hedge-fund-house-divided-renaissance-technologies/.

sometimes noted for their influence on people and locations, but the role of SWFs in geopolitics and influence operations is seldom discussed. SWFs are entities that manage and invest a nation's savings for the benefit of its citizens. They may invest directly in companies or sectors, or they may invest in private equity and hedge funds, or a mixture of all the above. They are often disproportionately large, relative to individual investors and even to institutional investors. Although the invested funds are generated from the nation's wealth, SWFs often act more like private MNCs, shaping their behaviors and the uses of their funds to suit their managers' and sovereign leaderships' goals and objectives on the international stage.

An excellent example of how an SWF was recently involved in a large-scale influence operation involving multiple U.S. allies involves the story of the Saudi Arabian Public Investment Fund's (PIF's) role in the Kingdom's 2019/2020 attempt to acquire the Premier League's Newcastle United football club. European football is a prestige sport in the Middle East (as in much of the world), and Saudi Arabia has long desired to elevate its international status by owning a top-ranked football club. The issue was not heavily discussed in mainstream media outside of British sport pages. However, on social media, Twitter, Parler, and Instagram, an influence operation was raging, attempting to encourage English authorities to back the PIF's acquisition of the Newcastle club. As can be seen in an excerpt from a posting from an apparent online independent investigator, the moves by the PIF on the corporate and financial side appear to have been bolstered by a Saudi troll farm:

> I'm now being trolled by Saudi twitter. That's OK comes with the territory.
>
> What is interesting, though, is at least 2 of the trolls have names which look European with English personal profiles but who mostly re(tweet) in Arabic & seem remarkably pro Saudi Govt.
>
> Well, well.
>
> > Not only that but these 2 "fake westerners" not only continually retweet/like each other's tweets about me—but a cursory scroll down their timelines show they regularly do this to each other.
> >
> > Looks like I've stumbled across a Saudi troll factory.
> >
> > Ooh I looked further down "KateStewart22" timeline
> >
> > A mix of:
> >
> > 90% pro Saudi Arabian govt tweeting
> >
> > 5% nice pictures of flowers
> >
> > & oh what's that?
> >
> > It's only Retweets of Boris Johnson & Leave.eu people.
> >
> > Sometime I feel this dark cave we've ventured into is endless...
>
> And all of a sudden I now find that in fact the Saudi bot complex is absorbing & massive and I never knew… [45][54]

---

[54] https://threadreaderapp.com/thread/1197105949472890880.html.

Although the 2019/2020 takeover attempt ended when the PIF walked away from negotiations late in the year, there is reporting as of the summer of 2021 that the PIF has not given up and in fact is keeping a close eye on arbitration currently underway between Newcastle's current owner and the Premier League over the deal [46].[55] One of the key points of concern in the arbitration is the question of whether Mohammed bin Salman (who is the Chairman of the PIF as well as Saudi Arabia's Deputy Prime Minister) would engage with the club and the league as a head of state or as the head of an investment entity.

Case study: NSO Group and Black Cube

The NSO Group is arguably the most notorious of a growing group of Israeli-founded and led private companies offering technologies and staff drawn heavily from Israel's Defense Force and Mossad personnel. In the summer of 2021, a series of investigative reports were published in the *New York Times, Washington Post*, and other international media sources, detailing how NSO's Pegasus software has been, and continues to be, used by authoritarian regimes and wealthy individuals to locate and monitor dissidents and journalists, and personal and business enemies. This was just the most recent in several years of public accusations against NSO Group, in cases ranging from mobile device hacking, to surveillance, harassment, and even claims of enabling "wet work" by state or mercenary operatives.

NSO was implicated previously in Hollywood's Harvey Weinstein sexual harassment case, as well as the murder of journalist Jamal Khashoggi by Saudi operatives. Also implicated in overlapping cases was the Israel-based Black Cube. Black Cube was at one time funded in part by London-based real estate billionaire Vincent Tchenguiz, who later sued the company (critically, Tchenguiz is the former beneficial owner of Cambridge Analytica parent entity SCL Group).

NSO is held in private equity, reportedly mostly by an American PE firm, Francisco Partners, although as of this writing, Francisco Partners does not include NSO in its public list of investments. Reporting increasingly expresses concerns about the seemingly blurred lines between these private companies and the Israeli government, intimating that tacit, if not explicit, government support for the companies is the norm.

[55] Kelly, Ciaran; "Yasir Al-Rumayyan takes new role as consortium waits on Newcastle takeover arbitration outcome"; *Chronicle Live*. June 28, 2021. https://www.chroniclelive.co.uk/sport/football/football-news/newcastle-takeover-yasir-al-rumayyan-20916754.

Derived from *New York Times* and *Fast Company* articles on NSO and Black Cube [47, 48],[56,57]

Recommendations to Mitigate Vulnerabilities

Protecting Free Speech and Media Reporting from Foreign Influences

Consolidation of the media industries coupled with laxness of foreign ownership policies and the elimination of the Fairness Act combine to create a significant vulnerability to the United States. Media moguls currently control combined interests in news media and a variety of entertainment enterprises [49].[58] It is simple for foreign interests, like China, to exert influence over news media by impacting—or threatening to impact—specific entertainment projects or enterprises.

Disney is one example of how intertwined organizations became over the past three decades. Disney includes ABC News, ESPN, movie studios, and the Shanghai Disney Resort. Multiple authors highlighted how China is able to exert influence over news reporting by threatening other Disney projects or enterprises [7, 50].[59,60]

Mitigating foreign influence across multiple media categories and multinational corporations becomes complex. A review of the proposed updates to Federal Communications Commission (FCC) industry-specific questions reveals some gaps. Current FCC policies refer to broadcasting, telecommunications and submarine cables, which does not inherently include all television, internet, cable or satellite-based streaming or distribution networks [51].[61] Media, telecommunication, or internet do not appear under the national infrastructure list. A conclusion here is that infrastructure perceptions focus on technical infrastructure rather than content.

Arguably, the most politically sensitive issue is the need to separate news media from undo corporate influence due to foreign influences over commercial ventures. FCC industry specific questions account for direct foreign influence, but not the

[56] Bergman and Mazzetti (2021). "Israeli Companies Aided Saudi Spying Despire Khashoggi Killing"; *New York Times*. July 17, 2021. https://www.nytimes.com/2021/07/17/world/middleeast/israel-saudi-khashoggi-hacking-nso.html.

[57] Pangburn, DJ; "Israeli Operatives Black Cube Reportedly Spied on Critics of NSO Group"; *Fast Company*. February 13, 2019. https://www.fastcompany.com/90305114/israeli-operatives-black-cube-reportedly-spied-on-critics-of-nso-group.

[58] Business Insider (2012). Retrieved from: https://www.businessinsider.com/these-6-corporations-control-90-of-the-media-in-america-2012-6.

[59] Seese G., and Ryan, S., (2021). Great Power Competition, 2nd Ed., Springer. Pending publication.

[60] Clark, C., (2020). Has China Compromised Every Major Mainstream Media Entity. Retrieved from: https://thefederalist.com/2020/05/04/has-china-compromised-every-major-mainstream-media-entity/.

[61] FCC, (2020). Standard Questions For Applications Referred To Executive Branch. Federal Communications Commission, DA/FCC #: DA-20-1545. FCC Record Citation: 35 FCC Rcd 14906 (18). US Government. December 30, 2020.

influences related to commercial pressures placed indirectly on corporate ventures in other business verticals.

While politically sensitive in Congress, the Fairness Act, or Fairness Doctrine, did offer some mitigation to the parochial reporting that emerged in today's news media [52].[62] Under the Fairness Doctrine, broadcasters needed to present contrasting views and devote some airtime to discuss controversial matters deemed important to public interest. This dated back to 1949 when there were three networks. The basic idea was the need to present balanced reporting. The Supreme Court upheld the doctrine in Red Lion Broadcasting Co. v. FCC, 1969 [52].[63] Commercial interests under the Reagan Administration led to the repeal of the policy in 1987 [53].[64]

The Fairness Doctrine had two basic elements: It required broadcasters to devote some of their airtime to discussing controversial matters of public interest, and to voice contrasting views regarding those matters. Re-establishing the Fairness Act can help to overcome the heavily divided political views resident in commercial news media today.

Recommendation: All delivery platforms and content considerations should be included in FCC authorities and policies to remove major gaps that allow foreign influence over freedom of speech and accurate and balanced news reporting.

Protecting Federal Elections

Federal Election Campaign Laws are clear and unambiguous that foreign money is not welcome in the American democratic process. Specifically, "§ 30,121. Contributions and donations by foreign nationals," states the following:

> (a) Prohibition. It shall be unlawful for—
>
> (1) a foreign national, directly or indirectly, to make—
>
> (A) a contribution or donation of money or other thing of value, or to make an express or implied promise to make a contribution or donation, in connection with a Federal, State, or local election;
>
> (B) a contribution or donation to a committee of a political party; or.
>
> (C) an expenditure, independent expenditure, or disbursement for an electioneering communication (within the meaning of Sect. 30,104(f)(3) of this title); or.
>
> (2) a person to solicit, accept, or receive a contribution or donation described in subparagraph (A) or (B) of paragraph (1) from a foreign national [54].[65]

---

[62] Ruane, K., (2011). Fairness Doctrine History and Constitutional Issues. Congressional Research Service 7–5700. www.crs.gov. R40009. https://sgp.fas.org/crs/misc/R40009.pdf.

[63] Ruane, (2011).

[64] Rendall, Steve (January 1, 2005). "The Fairness Doctrine: How We Lost it, and Why We Need it Back". Extra!. Retrieved October 2, 2017. Retrieved from https://fair.org/extra/the-fairness-doctrine/.

[65] Federal Campaign Election Commission (2019). Federal Election Campaign Laws. Retrieved from: https://www.fec.gov/resources/cms-content/documents/feca.pdf.

The following questions speak to the need for ultimate beneficiary analysis:

Which Federal department/agency has the responsibility to analyze FEC Form 13 s to validate all donors are indeed US Persons?

- Is any correlation for Foreign Agent Registration Act (FARA) registered entities performed?
- If the donor is a commercial entity, is Ultimate Beneficial Owner (UBO) analysis performed?
- Is Foreign Ownership Control and Influence (FOCI) analysis performed?
- If the donor is a Social Welfare Organization (SWO) or a 501c3/4 nonprofit entity, is financial segregation analysis performed to ensure foreign charitable donations to these organizations does not end up in a candidate's war chest?

If this task seems arduous, we encourage the reader to review the following, and assess how this kind of analysis could possibly be performed with the scant information that is provided:

FEC Form 13: Report of Donations Accepted for Inaugural Committee

http://docquery.fec.gov/pdf/286/201704180300150286/201704180300150286.pdf

Regrettably, completion of the FEC Form 13 appears to be the extent of the legal accountability to finance a political campaign in the United States today. Whatever is left of the McCain-Feingold act post the Supreme Court's Citizens United decision is appallingly inadequate, and places American democracy at grave risk from foreign influence.

Recommendation: The United States must revise FEC statutes to ensure that foreign contributions cannot influence U.S. elections. Legislative changes must include language that prevents American surrogates from performing pass-through contributions of foreign funding. The analytical steps referred to in the questions above should be institutionalized to ensure the sanctity of our electoral processes.

ITAR

CFR Title 22: Foreign Relations §121.1 mandates the development and maintenance of the United States Munitions List (USML). The intent is to prevent the export of weapons and munitions to foreign entities unaligned with US interests. Compliance is overseen and implemented by the Department of State's International Trafficking in Arms Regulations (ITAR) [55],[66] and at least in the opinions of the authors, ITAR is often grossly over-applied, without even so much as the appearance of an attempt to stay true to the intent of the policy: preventing the export of weapons to America's adversaries. That said, as the twenty-first century has seen information emerge as

[66] U.S. Department of State (2021). Directorate of Defense Trade Controls. Retrieved from: https://www.pmddtc.state.gov/ddtc_public?id=ddtc_kb_article_page&sys_id=4f06583fdb78d300d0a370131f961913.

a fourth domain of war (sea, air, and land being the three traditional domains), the terms "influence," "information warfare," and "information operations," currently appear nowhere in the USML [56].[67]

Recommendation: Applying ITAR to commercial influence capabilities seems like a logical and overdue application of the regulation. This process should entail mandating that capabilities including search engine optimization (SEO), public relations and media platforms with influence capabilities, firms with core competencies that include trolls, bots, social media trending, etc. be subject to inclusion on the USML. In turn, any desired export of these capabilities would be subject to ITAR waivers and approvals. This could restrict US-based commercial influence/information platforms from providing services to authoritarian governments unaligned with US interests without explicit written waivers from the US Department of State.

Conclusion

Hostile foreign influence activities increasingly leverage the private sector to exploit seams and gaps in national-level laws and policies to disrupt and corrupt political and financial processes both domestically and abroad. Commercial leverage results in influence over various media mechanisms. Many people are influenced through these media platforms, which include news and social media. These activities incite discontent and disruptive behaviors between and amongst assorted demographics and socio-economic strata to gain advantage for foreign governments, political parties, companies, and individuals. It is imperative that leadership in Washington, D.C. rise above parochial party interests to protect Americans from undo international influences, and to gain oversight of the export of these capabilities abroad. In some cases, this means passing and updating laws and amending policies described above to protect Americans and their partners and allies from obfuscated misinformation efforts by Great Power competitors and their proxies. Further, the U.S. Government and allies must regulate the exportation of capabilities to competitors and adversaries for profit.

[67] e-CFR (2021). Title 22: Foreign Relations PART 121—THE UNITED STATES MUNITIONS LIST. Retrieved from: https://www.ecfr.gov/cgi-bin/text-idx?SID=1aaef52443f7950148d7a1e4c9190206&mc=true&node=se22.1.121_11&rgn=div8.

References

1. Ripley, R. B., & Lindsay, J. M. (eds.) (1997). *US foreign policy after the Cold War*. University of Pittsburgh Press.
2. X, (George Kennan). (1947). The Sources of Soviet Conflict. Foreign Affairs. July 1947. Retrieved from https://www.foreignaffairs.com/articles/russian-federation/1947-07-01/sources-soviet-conduct.
3. Nitze, P. (1950). United States Objectives and Programs for National Security. NSC 68. Retrieved from https://history.state.gov/milestones/1945-1952/NSC68.
4. PPD-23. (1948). Report by the Policy Planning Staff. Foreign Relations Of The United States, 1948, General; The United Nations, Vol I, Part 2. Retrieved from https://history.state.gov/historicaldocuments/frus1948v01p2/d4.
5. FRUS, 1952–1954. 1984. See U.S. Department of State. Papers Relating to the Foreign Relations of the United States, 1952–1954. Washington, D.C., GPO.
6. Cull, N. J. (2009). The cold war and the United States information agency: American propaganda and public diplomacy, 1945–1989. *Naval War College Review, 62*(2), 14.
7. Seese, G., & Ryan, S. (2021). *Great power competition* (2nd ed.). Springer.
8. Conway, I., Cassedy, K., Ryan, S., & Danish, C. (2019). Reconsidering great power competition below armed conflict. The Cipher Brief. December 13, 2019. Retrieved from https://www.thecipherbrief.com/reconsidering-great-power-competition-below-armed-conflict.
9. Zeihan, P. (2020). *Disunited nations: The scramble for power in an ungoverned world*. HarperCollins.
10. International Consortium of Investigative Journalists; *Global banks defy U.S. crackdowns by serving oligarchs, criminals and terrorists*. International Consortium of Journalists. September 20, 2020. https://www.icij.org/investigations/fincen-files/global-banks-defy-u-s-crackdowns-by-serving-oligarchs-criminals-and-terrorists/.
11. Thomas. (2017). The Evolving Nature of Russia's Way of War. Military Review July-August 2017. Army University Press. Retrieved from: https://www.armyupress.army.mil/Journals/Military-Review/English-Edition-Archives/July-August-2017/Thomas-Russias-Way-of-War/.
12. Darczewska, Żochowski. (2017, June). Active measures: Russia's key export. *Point of View* Issue 64. Warsaw: Centre for Eastern Studies.
13. Report of the Select Committee on Intelligence United States Senate on Russian Active Measures Campaigns and Interference in the 2016 U.S. Election Volume 5: Counterintelligence Threats and Vulnerabilities.
14. Cadwalladr & Graham-Harrison. (2018). Cambridge Analytica: links to Moscow oil firm and St Petersburg university. March 17, 2018. Retrieved from: https://www.theguardian.com/news/2018/mar/17/cambridge-academic-trawling-facebook-had-links-to-russian-university.
15. Cadwalladr & Jukes. (2018). Arron Banks 'met Russian officials multiple times before Brexit vote.' June 9, 2018. Retrieved from: https://www.theguardian.com/politics/2018/jun/09/arron-banks-russia-brexit-meeting.
16. House of Commons Digital, Culture, Media and Sport Committee. (2019). Disinformation and 'fake news': Final Report. Retrieved from: https://publications.parliament.uk/pa/cm201719/cmselect/cmcumeds/1791/179102.htm.
17. Menn, Li, & Culliford. (2021). Parler partially reappears with support from Russian technology firm. Reuters, January 18, 2021. Retrieved from: https://finance.yahoo.com/news/parler-partially-reappears-support-russian-234401397.html.
18. Sanders. (2019). See You in Which Court? (Substack) Retrieved from: https://wildwest.substack.com/p/see-you-in-which-court.
19. Siegelman. (2018). Chart: Emerdata Limited—the new Cambridge Analytica/SCL Group? (Medium, March 25, 2018). Retrieved from: https://wsiegelman.medium.com/chart-emerdata-limited-the-new-cambridge-analytica-scl-group-63283f47670d.
20. Grossman & Ramali. (2020). Outsourcing Disinformation. (Lawfare December 13, 2020) Retrieved from: https://www.lawfareblog.com/outsourcing-disinformation.

21. Bellingcat Investigation Team. (2020). *The GRU's MH17 Disinformation Operations Part 1: The Bonanza Media Project*. Bellingcat. November 12, 2020. https://www.bellingcat.com/news/uk-and-europe/2020/11/12/the-grus-mh17-disinformation-operations-part-1-the-bonanza-media-project/.
22. Arnold & Strohecker. (2021). Angola gets breathing space from Chinese creditors says finance minister. Reuters. January 11, 2021. Retrieved from: https://www.reuters.com/article/us-angola-economy/angola-gets-breathing-space-from-chinese-creditors-says-finance-minister-idUSKBN29G18X.
23. Nyabiage, J., (2021). Lack of detail in Angola's debt deals with China could hide future risk. South China Morning Post. January 18, 2021. Retrieved from https://www.scmp.com/news/china/diplomacy/article/3118140/lack-detail-angolas-debt-deals-china-could-hide-future-risk.
24. Chen, D. (2017). China Has a New Grand Strategy and the West Should Be Ready. The Diplomat. Retrieved from: https://thediplomat.com/2017/10/china-has-a-new-grand-strategy-and-the-west-should-be-ready/.
25. Ward, A. (2018). Why Trump Cut Millions in Military Aid to Pakistan. Vox. September 4, 2018. Retrieved from https://www.vox.com/2018/9/4/17818396/pakistan-aid-military-trump-pompeo-afghanistan.
26. Haqqani, H. (2021). Islamabad will come to regret aiding the Taliban's Resurgence. Foreign Affairs. July, 22, 2021. Retrieved from: https://www.foreignaffairs.com/articles/afghanistan/2021-07-22/pakistans-pyrrhic-victory-afghanistan?utm_medium=promo_email&utm_source=special_send&utm_campaign=mktg_actives_we_all_lost_afghanistan&utm_content=20210816&utm_term=all-actives.
27. Chellaney, B. (2011). China's Strategic Subterfuge. Great Power Competition in the Indian Ocean. Stagecraft and Statecraft. https://chellaney.net/2011/06/10/618/.
28. Robinson, J. (2016). Did you catch all the ways Hollywood pandered to China this year?. *Vanity Fair*. August 5, 2016. https://www.vanityfair.com/hollywood/2016/08/did-you-catch-the-ways-hollywood-pandered-to-china-this-year.
29. Mwakideu, C. (2021). Experts warn of China's growing media influence in Africa. *Deutsche-Welle (DW)*. January 29, 2021. https://www.dw.com/en/experts-warn-of-chinas-growing-media-influence-in-africa/a-56385420.
30. Lutz, A., Business Insider. These 6 Corporations could control 90 percent of the media in America. June 14, 2012. https://www.businessinsider.com/these-6-corporations-control-90-of-the-media-in-america-2012-6.
31. Seese, G., & Ryan, S. (2020). Great Power Competition Conference 2. Counter Intuitiveness of Influence Science, Social Judgment Theory, and Strategic Messaging in Great Power Competition (Pending publication). December 2020.
32. Rudd, K. (2021). Why the Quad Alarms China. Its Success Poses a Major Threat to Beijing's Ambitions. Foreign Affairs. August 6, 2021. Retrieved from https://www.foreignaffairs.com/articles/united-states/2021-08-06/why-quad-alarms-china?utm_medium=newsletters&utm_source=twofa&utm_campaign=Why%20the%20Quad%20Alarms%20China%20&utm_content=20210813&utm_term=FA%20This%20Week%20-%20112017.
33. https://asia.nikkei.com/Spotlight/Coronavirus/US-China-fight-over-COVID-origin-escalates-with-dash-of-fake-news.
34. Brooking, E. T., & Kianpour, S. (2020). *Iranian digital influence efforts: Guerrilla broadcasting for the twenty-first century*"; the Atlantic Council. February 11, 2020. https://www.atlanticcouncil.org/in-depth-research-reports/report/iranian-digital-influence-efforts-guerrilla-broadcasting-for-the-twenty-first-century/.
35. BBC News; "Iranian blogger Sattar Beheshti dies after arrest"; *BBC News*. November 8, 2012. https://www.bbc.com/news/world-middle-east-20259533.
36. Deutsche-Welle; "Ruhollan Zan: Iran executes blogger who inspired protest"; DW.com, English version. December 12, 2020. https://www.dw.com/en/ruhollah-zam-iran-executes-blogger-who-inspired-protests/a-55914201.

37. Gray, R., & Testa, J. (2013). The Inside Story Of One Website's Defense Of Assad. A small Minnesota news outlet caused a storm when it ran a story claiming Syria's rebels carried out a chemical attack near Damascus in August. A look inside the murky world of Mint Press News. October 1, 2013, Retrieved from https://www.buzzfeednews.com/article/rosiegray/the-inside-story-of-one-websites-defense-of-assad.
38. U.S. Attorney's Office, Southern District of New York press release. "Manhattan U.S. Attorney Announces Kidnapping Conspiracy Charges Against An Iranian Intelligence Officer And Members Of An Iranian Intelligence Network"; U.S. Department of Justice, Southern District of New York; July 13, 2021. https://www.justice.gov/usao-sdny/pr/manhattan-us-attorney-announces-kidnapping-conspiracy-charges-against-iranian.
39. Wood, J. (2018). Outsourcing war: How foreigners and mercenaries power UAE's military. *Middle East Eye*; July 13, 2018. https://www.middleeasteye.net/news/outsourcing-war-how-foreigners-and-mercenaries-power-uaes-military.
40. Salisu, M. (2019). Hiring of forces and security in the Arab Gult: The case of private security companies in Qatar. *Chinese Business Review*; July-September 2019, Vol. 18, No. 3, 10–18. http://www.davidpublisher.com/Public/uploads/Contribute/5df2f989a60c4.pdf.
41. Montague, J., & Panja, T. (2019). Ahead of Qatar World Cup, a Gulf Feud Plays Out in the Shadows. *New York Times*. February 1, 2019. https://www.nytimes.com/2019/02/01/sports/world-cup-2022-qatar.html.
42. Palumbo, M., & Draper, J. (2020). How Saudis, Qataris and Emiratis took Washington. *Asia Times*. June 10, 2020. https://asiatimes.com/2020/06/how-saudis-qataris-and-emiratis-took-washington/.
43. Takambou, M. M., & Muhammad, B. (2021). Shell to pay $111 million for 1970 Niger Delta oil spills. *DW*. August 8, 2021. https://www.dw.com/en/shell-to-pay-111-million-for-1970-niger-delta-oil-spills/a-58697881.
44. Bowers, J. (2016). *A hedge fund house divided: Renaissance Technologies. OpenSecrets Blog. Center for Responsive Politics. June 7, 2016.*https://www.opensecrets.org/news/2016/06/a-hedge-fund-house-divided-renaissance-technologies/.
45. https://threadreaderapp.com/thread/1197105949472890880.html.
46. Kelly, C. (2012). Yasir Al-Rumayyan takes new role as consortium waits on Newcastle takeover arbitration outcome. *Chronicle Live*. June 28, 2021. https://www.chroniclelive.co.uk/sport/football/football-news/newcastle-takeover-yasir-al-rumayyan-20916754.
47. Bergman & Mazzetti. (2021). Israeli companies Aided Saudi spying Despire Khashoggi Killing. *New York Times*. July 17, 2021. https://www.nytimes.com/2021/07/17/world/middleeast/israel-saudi-khashoggi-hacking-nso.html.
48. Pangburn, D. J. (2019). Israeli operatives black cube reportedly spied on critics of NSO group. *Fast Company*. February 13, 2019. https://www.fastcompany.com/90305114/israeli-operatives-black-cube-reportedly-spied-on-critics-of-nso-group.
49. Business Insider. (2012). Retrieved from: https://www.businessinsider.com/these-6-corporations-control-90-of-the-media-in-america-2012-6.
50. Clark, C. (2020). Has China Compromised Every Major Mainstream Media Entity. Retrieved from: https://thefederalist.com/2020/05/04/has-china-compromised-every-major-mainstream-media-entity/.
51. FCC. (2020). Standard Questions For Applications Referred To Executive Branch. Federal Communications Commission, DA/FCC #: DA-20–1545. FCC Record Citation: 35 FCC Rcd 14906 (18). US Government. December 30, 2020.
52. Ruane, K. (2011). Fairness doctrine history and constitutional issues. Congressional Research Service. 7-5700. www.crs.gov. R40009. https://sgp.fas.org/crs/misc/R40009.pdf.
53. Rendall, S. (2005). The Fairness Doctrine: How We Lost it, and Why We Need it Back". Extra!. Retrieved October 2, 2017. Retrieved from https://fair.org/extra/the-fairness-doctrine/.
54. Federal Campaign Election Commission. (2019). Federal Election Campaign Laws. Retrieved from: https://www.fec.gov/resources/cms-content/documents/feca.pdf.
55. U.S. Department of State. (2021). Directorate of Defense Trade Controls. Retrieved from: https://www.pmddtc.state.gov/ddtc_public?id=ddtc_kb_article_page&sys_id=4f06583fdb78d300d0a370131f961913.

56. e-CFR. (2021). Title 22: Foreign Relations PART 121—THE UNITED STATES MUNITIONS LIST. Retrieved from: https://www.ecfr.gov/cgi-bin/text-idx?SID=1aaef52443f7950148d7a1e4c9190206&mc=true&node=se22.1.121_11&rgn=div8.

Dr. Sean Ryan is an Assistant Professor. His teaching includes management, leadership and strategy. He retired after 30 years as a Army Colonel having served in multiple capacities around the globe. He earned an MBA, Masters in Strategic Studies, and a doctorate in Management focused on Leadership and Organizational Change.

Ian Conway has more than twenty years of experience in national security issues, and now works as an independent analyst supporting legal investigations, corporate investigations, enhanced due diligence, competitive intelligence, and risk assessments. His investigative work has spanned six continents and a wide variety of industry sectors. He specializes in topics including complex influence operations; industrial espionage; sanctions and embargos violations; asset recovery; and compromises of intellectual property.

Kathleen Cassedy is an independent contractor and open source specialist with two decades of experience in the National Security sector. She spent the last five years identifying, researching, and analyzing and writing on modern Russian and Chinese political and economic warfare tactics; the role of foreign influence operations in gray zone problem sets; the rise of commercial influence operations in regional power competition; and the global geopolitical influence of multi-national entities, non-state actors, and super-empowered individuals.

The Future of Cyber-Enabled Influence Operations: Emergent Technologies, Disinformation, and the Destruction of Democracy

Joe Littell

Abstract Nation-states have been embracing online influence campaigns through disinformation at breakneck speeds. Countries such as China and Russia have completely revamped their military doctrine to information-first platforms [1, 2] (Mattis, Peter. (2018). China's Three Warfares in Perspective. *War on the Rocks. Special Series: Ministry of Truth.* https://warontherocks.com/2018/01/chinas-three-warfares-perspective/, Cunningham, C. (2020). A Russian Federation Information Warfare Primer. *Then Henry M. Jackson School of International Studies. Washington University.* https://jsis.washington.edu/news/a-russian-federation-information-warfare-primer/.) to compete with the United States and the West. The Chinese principle of "Three Warfares" and Russian Hybrid Warfare have been used and tested across the spectrum of operations ranging from competition to active conflict. With the COVID19 pandemic limiting most means of face-to-face interpersonal communication, many other nations have transitioned to online tools to influence audiences both domestically and abroad [3] (Strick, B. (2020). COVID-19 Disinformation: Attempted Influence in Disguise. *Australian Strategic Policy Institute. International Cyber Policy Center.* https://www.aspi.org.au/report/covid-19-disinformation.) to create favorable environments for their geopolitical goals and national objectives. This chapter focuses on the landscape that allows nations like China and Russia to attack democratic institutions and discourse within the United States, the strategies and tactics employed in these campaigns, and the emergent technologies that will enable these nations to gain an advantage with key populations within their spheres of influence or to create a disadvantage to their competitors within their spheres of influence. Advancements in machine learning through generative adversarial networks [4] (Creswell, A; White, T; Dumoulin, V; Arulkumaran, K; Sengupta, B; Bharath, A. (2017) Generative Adversarial Networks: An Overview. IEE-SPM. April 2017. https://arxiv.org/pdf/1710.07035.pdf.) that create deepfakes [5] (Whittaker, L; Letheren, K; Mulcahy, R. (2021). The Rise of Deepfakes: A Conceptual

J. Littell (✉)
Army Cyber Institute at the West Point, United States Military Academy, West Point, NY 10996, USA
e-mail: joseph.littell@westpoint.edu

A. Farhadi et al. (eds.), *The Great Power Competition Volume 3*,
https://doi.org/10.1007/978-3-031-04586-8_10

Framework and Research Agenda for Marketing. https://journals.sagepub.com/doi/abs/10.1177/1839334921999479.) and attention-based transformers [6] (https://arxiv.org/abs/1810.04805.) (Devlin et al., 2018) that create realistic speech patterns and interaction will continue to plague online discussion and information spread, attempting to cause further partisan divisions and decline of U.S. stature on the world stage and democracy as a whole.

Keywords Disinformation · Cyberspace · Cyber-enabled influence · Influence operation

Executive Summary

Online influence campaigns have become a relatively cheap and impactful way for nations to drive narratives both within their borders and internationally. Since the discovery of Russia and the Internet Research Agency's interference in the 2016 U.S. presidential election and the 2016 U.K. European Union Referendum, seventy other nations have increased their funding for similar campaigns [7].[1] Malign Influence Operations use inauthentic users to push narratives and manipulate social media recommendation algorithms to target key demographics with troves of publicly available data, amplifying already reticent societal fissures and standing as a major danger to democratic institutions throughout the world. Democracy requires four essential functions, as described by Hoover Institute Fellow Larry Diamond. The first requirement is a political system for choosing and replacing the government through free and fair elections. The second requirement is the active participation of people, as citizens, in politics and civic life. The third is the protection of the human rights of all citizens. Finally, the fourth requirement is the rule of law, in which laws and procedures apply equally to all citizens [8].[2] Emergent technologies built from advances in machine learning, such as deepfakes [5],[3] will compound current challenges by adding additional layers of believability to these inauthentic users [9][4] and their malign influence campaigns, making it more difficult for audiences to discern fact from fiction.

Russia and China have utilized these techniques heavily across the last two decades and expanded on them greatly in the COVID era. Other smaller nations have flocked

[1] Bradshaw, S; Bailey, H; Howard, P. (2020). 2020 Global Inventory of Organized Social Media Manipulation. *Computational Propaganda Research Project. Oxford Internet Institute. Oxford University.* https://demtech.oii.ox.ac.uk/wp-content/uploads/sites/127/2021/02/CyberTroop-Report20-Draft9.pdf.

[2] https://diamond-democracy.stanford.edu/speaking/lectures/what-democracy.

[3] Whittaker, L; Letheren, K; Mulcahy, R. (2021). The Rise of Deepfakes: A Conceptual Framework and Research Agenda for Marketing. https://journals.sagepub.com/doi/abs/10.1177/1839334921999479.

[4] Bastos, M., & Mercea, D. (2018). The public accountability of social platforms: lessons from a study on bots and trolls in the Brexit campaign. *Philosophical transactions. Series A, Mathematical, physical, and engineering sciences, 376*(2128), 20,180,003. https://doi.org/10.1098/rsta.2018.0003.

to methods to bolster their spheres of influence as well as confound international audiences due to the low cost of entry as well as almost relatively nonexistent regulatory barriers to participate at scale. Numerous prevalent examples exist during the last year. Russian attempts at interfering with the 2020 U.S. Presidential election through planting false stories and pushing various false narratives of a stolen election are well known [10].[5] China has used online influence campaigns to crush popular support for democracy in Hong Kong [11],[6] limiting knowledge and creating confusion around allegations of ethnic cleansing and genocide against the Muslim Uyghurs population in Xinjiang province [12],[7] and revitalize its global image in the wake of the COVID19 pandemic [13].[8] Azerbaijan has effectively used disinformation during its ongoing military excursions into Armenia [14].[9] Myanmar's military coup and consolidation of power removed democratically elected officials while killing ethnic minorities in the north and west of the country [15].[10]

As confusion around these events continues, it undermines a basic understanding of the situation and erodes the ability of governments to act collectively. Groups like Bellingcat use collaborative efforts to teach tactics to investigate and rapidly disseminate their findings to a global audience. The thankless efforts of subject matter experts and analysts will unfortunately not be enough to prevent the fomenting of dissidence against democratic ideals. We need active development of machine learning algorithms to handle the flood of coordinated influence campaigns, education on the effects these campaigns have on our population, and expertise and human input to make any sort of headway. Coordination efforts would require a central agency or government department to bring together the vast and various elements of academia, business, and government efforts already underway so that they may share best practices and efforts amongst each other.

---

[5] National Intelligence Council. (2021). Foreign Threats to the 2020 US Federal Elections. *Intelligence Community Assessment.* https://www.dni.gov/files/ODNI/documents/assessments/ICA-declass-16MAR21.pdf.

[6] Wood, D; McMinn, S; Feng,E. (2019). China Used Twitter to Disrupt Hong Kong Protests, but Efforts Began Years Earlier. *National Public Radio.* https://www.npr.org/2019/09/17/758146019/china-used-twitter-to-disrupt-hong-kong-protests-but-efforts-began-years-earlier.

[7] Uyghur Human Rights Project. (2020). The Happiest Muslims in the World: Disinformation, Propaganda, and the Uyghur Crisis. *Uyghur Human Rights Project.* https://docs.uhrp.org/pdf/Disinformation_Propganda_and_the_Uyghur_Crisis.pdf.

[8] Bernard, R; Bowsher, G; Sullivan, R; Gibson-Fall, F. (2021). Disinformation and Epidemics: Anticipating the Next Phase of Biowarfare. *Health Security. Volume 19, Number 1.* https://www.liebertpub.com/doi/pdf/10.1089/hs.2020.0038.

[9] Giles, C; Bhat, U. (2020). Nagoro-Karabakh: The Armenian-Azeri 'Information Wars'. *BBC Reality Checks and Anti-disinformation Unit.* https://www.bbc.com/news/world-europe-54614392.

[10] Beech, H. (2021). 'Now We Are United': Myanmar's Ethnic Divisions Soften after Coup. *New York Times.* https://www.nytimes.com/2021/04/30/world/asia/myanmar-ethnic-minority-coup.html.

Introduction

A once-in-a-century global pandemic swept across the world, shuttering economies [16],[11] killing millions, and infecting millions more [17][12] with long-term health complications that are, as of yet, not fully understood [18].[13] Entire countries have shut down normal day-to-day habits to quell the spread of a novel coronavirus with well-documented origins from China. Mass transit, eating in a restaurant, watching movies in a theatre, and most typical extended family get-togethers came to a halt to stop the spread of a dangerous respiratory virus [19].[14] Phrases such as "social distancing," "mask-up," and "stop the spread" became a part of common vernacular. As most nations scrambled to keep their citizens safe and control the situation to the best of their ability, China and Russia saw the chaos unfolding as an opportunity to spread disinformation surrounding COVID19.

Russia began its disinformation campaigns to sow disarray among democratic nations both near and far. Initially fueled by underplaying the threat of COVID19 [20],[15] they quickly began going after COVID19 restrictions and masking requirements, exacerbating a burgeoning anti-lockdown movement [21][16] across the United States and parts of Europe [22].[17] Their goal: have Americans and Europeans question their governments' attempts to protect the population as some form of removal of human rights. As the monumental push to create, test, and distribute a vaccine to protect the population from the virus came with unparalleled cooperation between

[11] McKinsey and Company. (2021). Covid-19: Implications for Business. *McKinsey and Company. Executive Briefing.* https://www.mckinsey.com/business-functions/risk-and-resilience/our-insights/covid-19-implications-for-business#.

[12] Center for System Sciences and Engineering (CSSE). (2021). COVID-19 Dashboard. *Johns Hopkins University. ARCGIS.* https://www.arcgis.com/apps/dashboards/bda7594740fd40299423467b48e9ecf6.

[13] Blomberg, B; Mohn, K; Brokstad, KA et al. (2021). Long COIVD in a prospective cohort of home-isolated patients. *Nature Medicine.* https://www.nature.com/articles/s41591-021-01433-3.

[14] Centers for Disease Control and Prevention. (2020). CDC Calls on Americans to wear mask to prevent COIVD-19 Spread. *Centers for Disease Control and Prevention.* https://www.cdc.gov/media/releases/2020/p0714-americans-to-wear-masks.html.

[15] Loucaides, D; Perrone, A. (2021). Germany's COVID sceptics fueled by Russian media and far-Right Conspiracies. *openDemocracy.* https://www.opendemocracy.net/en/germanys-covid-sceptics-fuelled-by-russian-media-and-far-right-conspiracies/.

[16] Benson, T. (2020). Trolls and bots are flooding social media with disinformation encouraging states to end quarantine. *Business Insider. Tech.* https://www.businessinsider.com/trolls-bots-flooding-social-media-with-anti-quarantine-disinformation-2020-4.

[17] Emmot, R. (2020). Russia deploying coronavirus disinformation to sow panic in West, EU document says. *Reuters.* https://www.reuters.com/article/us-health-coronavirus-disinformation/russia-deploying-coronavirus-disinformation-to-sow-panic-in-west-eu-document-says-idUSKBN21518F.

government, academia, and industry in a record time, Russia moved to prop up long-standing anti-vaccination conspiracy theories to slow the adoption of the COVID19 vaccine [23].[18]

China, by contrast, looked less to pull the rest of the world down but rather to elevate its own standing. Downtrodden by the news reports about Chinese failures to initially contain the virus as the source of the pandemic, China began a disinformation campaign in February 2020 with multiple, spontaneous origins to lessen their fault [24].[19] China bolstered the campaign through additional influence operations, sending medical supplies and doctors to Europe [25][20] and later vaccines to South America [26].[21] As the number of U.S. and E.U. cases and deaths increased, China used its social media sphere to antagonize the perceived failures of democracy to handle the virus in comparison to their "superior system of government" [24][22] When these attempts failed, China decided to claim instead that COVID19 was actually a bioweapon from the United States, part of a conspiracy to destroy China [27].[23]

While the two competing powers had different goals, their use of social media and disinformation stood closely aligned. Inauthentic accounts, astroturfing, information laundering, and algorithmic manipulation were mainstays used to push messages further and faster. A new era of propaganda and disinformation is coming of age, ready to divide citizens and discredit democracy. Without properly understanding the problem and how it will be used more readily by nation-states in addition to the traditional powers, it will be near impossible to stop.

---

[18] Miller, H. (2021). Russia's Anti-Vaccine Propaganda is Tatamount to a Declaration of War. *Center for Medical Economic and Innovation*. https://medecon.org/russias-anti-vaccine-propaganda-is-tantamount-to-a-declaration-of-war/.

[19] Chen, E. (2021). Chinese COVID-19 Misinformation A Year Later. *The Jamestown Foundation Global Research and Analysis*. https://jamestown.org/program/chinese-covid-19-misinformation-a-year-later/.

[20] Reuters Staff. (2020). China sends medical supplies, experts to help Italy battle coronavirus. *Reuters Healthcare and Pharma*. https://www.reuters.com/article/us-health-coronavirus-italy-respirators/china-sends-medical-supplies-experts-to-help-italy-battle-coronavirus-idUSKBN2101IM.

[21] Mallapaty, S. (2021). China's COVID vaccines are going global—but questions remain. *Nature Magazine* https://www.nature.com/articles/d41586-021-01146-0.

[22] Chen, E. (2021). Chinese COVID-19 Misinformation A Year Later. *The Jamestown Foundation Global Research and Analysis*. https://jamestown.org/program/chinese-covid-19-misinformation-a-year-later/.

[23] Kinetz, E. (2021). Anatomy of a conspiracy: With COVID, China took leading role. *Associated Press*. https://apnews.com/article/pandemics-beijing-only-on-ap-epidemics-media-122b73e134b780919cc1808f3f6f16e8.

Background

Propaganda and disinformation are not new methods for changing the behavior of a population, nor are they relegated to nonmilitary roles and impacts. In "Munitions of the Mind," author Philip Taylor states that the use of military poems and hymns to build morale and strike fear into adversaries was commonplace within ancient Babylonia by 1300 BCE [28].[24] The Assyrians, espec ially, used these methods heavily during their campaigns against the nation-states of the Kassites, Ur, and Urak to build their armies' confidence while making their enemies fear their great feats on the battlefield [28].[25] A natural progression of the adage, "history is written by the victors," continued until the invention of the Gutenberg printing press sparked a revolution in the ability to spread information.

Although printing methods preceded Gutenberg's landmark invention, 1445 CE stands as a crossroads of propaganda [29].[26] No longer were nations or the church responsible for the spread of information and knowledge; individuals with access to printing presses could furnish their own views en masse to the greater population. Text—and literacy—broke free from the original Latin as scholars, scientists, and religious figures could now produce works in their native tongues [30].[27] It is no wonder that the Renaissance and Reformation, two monumental social upheavals of modern history, took place in the wake of the printing press. Control of information has been a critical part of nations' stories—the only difference now is the medium through which they spread.

Pushback against the Catholic church throughout Europe was led, in part, by propaganda from Protestant believers. The seismic power shift was primarily due to the change in technologies that opened up access to information and who could shape information into shared knowledge. The advent of radios and motion pictures led to their immediate use as media for propaganda. The Nazi Party utilized radio heavily to consolidate power following 1933 [31].[28] Both Axis and Allied powers used newsreels at the start of motion pictures to spread propaganda during World War II [32][29] and used Hollywood to spread American influence thereafter. It, therefore, was the logical outcome that a new mode of communication—the internet, and especially social media—shepherded in a new age of propaganda and disinformation.

[24] Taylor, P. (2003). Munitions of the mind: A history of propaganda. *Manchester University Press.* https://www.jstor.org/stable/j.ctt155jd69.

[25] Taylor, P. (2003). Munitions of the mind: A history of propaganda. *Manchester University Press.* https://www.jstor.org/stable/j.ctt155jd69.

[26] Edwards, M. (1994). Printing, Propaganda, and Martin Luther. *Berkeley: University of California Press.* https://publishing.cdlib.org/ucpressebooks/view?docId=ft3q2nb278;chunk.id=0;doc.view=print.

[27] https://publishing.cdlib.org/ucpressebooks/view?docId=ft3q2nb278;chunk.id=0;doc.view=print.

[28] https://www.ushmm.org/collections/bibliography/nazi-propaganda-1.

[29] Jowett, G; O'Donnell, V. (2012). Propaganda and Persuasion, 5th Edition. *Sage Publications.*

When applied to the United States and Western liberal democracies as a whole, propaganda and disinformation can attack the foundational elements and requirements needed for self-governance directly. If a population is led to believe that elections are stolen, that, in turn, leads to a breakdown in participation in the democratic process. If that process is skewed to one individual group or belief structure, it will lead to the dissolution of rights or protections of subsets of the citizenry under the rule of law. This process of breaking down the foundation of democracy is akin to removing the leg of a table: while the structure may remain mostly intact, it cannot hold weight, and its contents fall, slipping into disarray and chaos.

Inauthentic Users and Astroturfing

In a May 2021 presentation to the U.S. Agency for Global Media and the Aspen Institute, Peter Pomerantsev spoke of how, ideally, democracies operate under the notion of one vote, one voice; but with online disinformation campaigns, adversarial groups and countries can drown out an individual's voice with a flood of activity from inauthentic users [33].[30] This process of building a false consensus is commonly referred to as "astroturfing," a play on a grassroots initiative in that some outside force is manufacturing the consensus to make it seem as if it is authentic. An example of astroturfing that was uncovered was around a movement to discredit feminists through the Twitter hashtag #endfathersday [34].[31] Several inauthentic accounts claiming to be women, particularly women of color, began posting vitriolic speech towards men under the hashtag. Members of the platform quickly began calling out inauthentic users under the hashtag #YourSlipIsShowing.

Inauthentic users are those who use the anonymity of the internet to create fake personas to push ideas, disinformation, and state propaganda. These false users can drive the astroturfing initiative to drown out authentic views of a situation, pushing forth a narrative that is most beneficial to the malign influence campaigns' goals. Inauthentic users fall under two broad categories, automated false accounts or "bots," and human-run false accounts or "sock puppets" [7].[32]

Bots are defined as accounts that are created and their actions automated to resemble those actions of a genuine user. These bots can be quickly created and

[30] https://watch.eventive.org/m4df2021/play/6074567c86f143003e0cbd99/6074697b69154f0030 86bb13.

[31] Broderick, R. (2014). Activists Are Outing Hundreds Of Twitter Users Believed To Be 4chan Trolls Posing As Feminist. *BuzzFeed News*. https://www.buzzfeednews.com/article/ryanhatesthis/your-slip-is-showing-4chan-trolls-operation-lollipop.

[32] Bradshaw, S; Bailey, H; Howard, P. (2020). 2020 Global Inventory of Organized Social Media Manipulation. *Computational Propaganda Research Project. Oxford Internet Institute. Oxford University.* https://demtech.oii.ox.ac.uk/wp-content/uploads/sites/127/2021/02/CyberTroop-Report20-Draft9.pdf.

employed into an information space for both spreading disinformation and manipulating engagement [35],[33] and algorithmic confounding [36].[34] Due to their reliance on automation, much of their interactions becomes easier to track due to temporal aspects, posting at regular intervals, social subgraphs (who they are interacting with), and similarity of content, as it is easier to repeat content than create new, unique content for so many additional users [37].[35] That said, advances in technology, which will be addressed later in the chapter, are making these trackable features more difficult.

Bots then contrast with sock puppets accounts, which are operated by real people under a fake persona. This allows the accounts to have a more authentic and personalized feel. Although tracking by metadata and I.P. addresses are still possible, the variety of the content produced, and ability to converse with anyone interacting with the influence campaign more directly, sock puppet accounts can be more difficult to ascertain.

Although bots, sock puppets, and so-called "troll farms" came into national media attention and prominence following the 2016 U.S. presidential election and British European Union Referendum, their use has a long history. The earliest documented attempts were on Russian Politicheskie web forums in the late 1990s and early 2000s [38].[36] Journalist Anna Polyanskaya described the shift from liberal democratic themes to strict authoritarian ideals in her expose in Vesnik in 2003. Polyanskaya observed a notable shift as Russian President Vladimir Putin began to consolidate his political power, citing various inconsistencies in the overall tone of posts, which leaned pro-Putin, compared to polling across RuNet, which leaned liberal and was constrained by single votes per I.P. address [38].[37]

Later in the 2000s, a Russian arm of the hacker collective known as Anonymous released emails tied to the pro-Kremlin youth organization, Nashi [39].[38] These emails described web "brigades" of paid internet users who flooded various elements of the internet to counter claims and harass dissidents. These users were paid upwards of $3 per post depending on the amount of feedback [39].[39] By the 2010s, Russia

[33] Ferrara E. (2020) Bots, Elections, and Social Media: A Brief Overview. In: Shu K., Wang S., Lee D., Liu H. (eds) Disinformation, Misinformation, and Fake News in Social Media. *Lecture Notes in Social Networks. Springer, Cham.* https://doi.org/10.1007/978-3-030-42699-6_6.

[34] Bakir, V. and McStay, A. (2018). Fake News and the Economy of Emotions. *Digital Journalism, 6:2, 154-175,* https://doi.org/10.1080/21670811.2017.1345645.

[35] Anderson, J; Rainie, L; Vogels, e > (2021). Expers Say the 'New Normal' in 2025 Will Be Far More Tech-Driven, Presenting More Big Challenges. Pew Research Center. *Emerging Technologies. Future of the Internet.* https://www.pewresearch.org/internet/2021/02/18/experts-say-the-new-normal-in-2025-will-be-far-more-tech-driven-presenting-more-big-challenges/.

[36] Polysanskaya, A; Krivov, A; Lomko, I. (2003). The Virtual Eye of Big Brother. *Vestnik Online.* http://www.vestnik.com/issues/2003/0430/win/polyanskaya_krivov_lomko.htm.

[37] Polysanskaya, A; Krivov, A; Lomko, I. (2003). The Virtual Eye of Big Brother. *Vestnik Online.* http://www.vestnik.com/issues/2003/0430/win/polyanskaya_krivov_lomko.htm.

[38] Elder, M. (2012). Polishing Putin: hacked emails suggest dirty tricks by Russian youth group. *The Guardian.* https://www.theguardian.com/world/2012/feb/07/putin-hacked-emails-russian-nashi.

[39] Elder, M. (2012). Polishing Putin: hacked emails suggest dirty tricks by Russian youth group. *The Guardian.* https://www.theguardian.com/world/2012/feb/07/putin-hacked-emails-russian-nashi.

moved beyond its internal manipulation and began heavily utilizing online brigades against both regional neighbors and internationally against the U.S. and West as a whole. By 2014, Russia was fomenting large-scale operations in support of hybrid warfare through the support of far-right and ethnocentric movements in the U.S. and Europe.

Although Russia was the first suspected peddler in online manipulation, they were not the only country entering into this burgeoning field. China began utilizing online manipulation through paid internet commentators as early as October 2004, when the CCP Propaganda Department of Changsha began hiring users to manipulate local online forums [40].[40] Quickly, the national CCP Ministry of Education began censoring and astroturfing, or creating fake grassroots movements, across Chinese university bulletin board systems and forums [41].[41]

Networks of inauthentic users working together utilize various tools available to them to spread disinformation and propaganda to susceptible users in order to muddy the waters [42],[42] gain an information advantage [43],[43] or enact a real world response [44].[44] Typically, they coordinate their functions to use social media platforms recommendation algorithms to their advantage [45],[45] to move fringe ideas and disinformation from their inauthentic network to more authentic ones to add legitimacy to the disinformation campaign [46],[46] thus, creating more misinformation as it spreads. These networks can do this with nightmarish precision due to the vast mountains of information available for sale [47],[47] stolen from email servers or

[40] Bandurski, D. (2008). China's Guerrilla War for the Web. *China Media Project. China Internet Research Conference.* https://chinamediaproject.org/2008/07/07/feer-chinas-guerrilla-war-for-the-web/.

[41] King, G; Pan, J; Roberts, M. (2017) > How the Chinese Government Fabricates Social Media Post for Strategic Distraction, not Engaged Argument. *American Political Science Review, vol. 111, pg. 484–501.* https://gking.harvard.edu/files/gking/files/50c.pdf.

[42] Lock, I; Ludolph, R. (2019). Organizational propaganda on the Internet: A systematic review. *Public Relations Inquiry. vol 9, issue 1, pg 103–127.* https://journals.sagepub.com/doi/full/10.1177/2046147X19870844.

[43] Kenny, C; Bergman, M. (2019). Understanding and Combating Russian and Chinese Influence Operations. *Center for American Progress, Foreign Policy and Security.* https://www.americanprogress.org/issues/security/reports/2019/02/28/466669/understanding-combating-russian-chinese-influence-operations/.

[44] Select Committee on Intelligence. United States Senate. (2018). Russian Active Measures Campaigns and Interference in the 20,216 U.S. Election Volume 2: Russia's Use of Social Media with Additional Views. *116*th *Congress. Senate. 1*st *Session. Report 116-XX.* https://www.intelligence.senate.gov/sites/default/files/documents/Report_Volume2.pdf.

[45] Meserole, C. (2018). How misinformation spreads on social media—and what to do about it. *Brookings Institute.* https://www.brookings.edu/blog/order-from-chaos/2018/05/09/how-misinformation-spreads-on-social-media-and-what-to-do-about-it/.

[46] Arjomand, N. (2019). Information Laundering and Globalized Media—Part I: The Problem. *Center for International Media Assistance.* https://www.cima.ned.org/blog/information-laundering-and-globalized-media-part-i-the-problem/.

[47] Sherman, J. (2021). Federal Privacy Rules Must Get "Data Broker" Definitions Right. Lawfare. *Consumer Data.* https://www.lawfareblog.com/federal-privacy-rules-must-get-data-broker-definitions-right.

scraped from publicly available sources [48][48] from the very same platforms they are manipulating. By building this false notion of consensus through astroturfing, real voices are drowned out, directly hindering the ability of citizens to play a part in active participation within civil society [49].[49] Without the ability to have a shared foundation between citizens, the ability to engage in the collective action of self-governance faces a significant threat of being destroyed.

Algorithmic Confounding

Search engines [50],[50] e-commerce platforms [51],[51] and social media networks [52][52] all utilize machine learning algorithms to recommend items to their users. These items range from a new book or movie to the news they read. These algorithms are split into three main types: content filtering, collaborative learning, and knowledge-based systems [53].[53] Each algorithm works slightly differently and can be used individually or in tandem with the other base algorithms to tailor the end user's experience. Content filtering, simply put, is using the choices of a single user and finding similarly categorized items to recommend, like offering a keyboard with a purchase of a computer mouse. Collaborative learning, also called collaborative filtering, takes the preference of one user and compares it to other similar users. The users with a large overlap in preference may be driven to other items frequented by like-minded users. For example, if one user enjoys a particular rock band's music, they may be recommended a different band's music because another user also enjoys the first band and regularly listens to the recommended band as well. Finally, knowledge-based systems require rules written by subject matter experts with vast amounts of specific domain knowledge in the product being recommended. Usually, knowledge-based systems are hybrid algorithms that work with content modeling, collaborative learning, or both.

---

[48] The National Law Review. (2019). Data Scraping Survives! (At Least for Now) Key Takeaways from 8th Circuit Ruling on the HIQ vs. LinkedIn Case. *The National Law Review.* https://www.natlawreview.com/article/data-scraping-survives-least-now-key-takeaways-9th-circuit-ruling-hiq-vs-linkedin.

[49] Diamond, L. (2004). What is Democracy. *Lecture to Hilla University for Humanistic Studies.* https://diamond-democracy.stanford.edu/speaking/lectures/what-democracy.

[50] Google Developers. (2021). Introduction to Recommendation Systems. *Google.* https://developers.google.com/machine-learning/recommendation.

[51] Hardesty, L. (2019). The history of Amazon's recommendation algorithm. *Amazon Science.* https://www.amazon.science/the-history-of-amazons-recommendation-algorithm.

[52] Ilic, A. Kabiljo, M. (2015) Recommending items to more than a billion people. *Facebook Engineering, Core Data, ML Applications.* https://engineering.fb.com/2015/06/02/core-data/recommending-items-to-more-than-a-billion-people/.

[53] Oracle. (2020). Introduction to Recommendation Engines. *Oracle Cloud Infrastructure Data Science.* https://www.oracle.com/a/ocom/docs/introduction-to-recommendation-engines-wp.pdf.

These recommendations alone have a profound effect on marketing and the end user's decision-making process. Researchers from Princeton University's departments of computer science and sociology showed that recommendation engines increased homogeneity through the selection bias and reinforced feedback from retraining the algorithm, commonly referred to as live systems, on those initial choices, making them more skewed in their distribution of items recommended [54].[54] When applied to controversial or political topics, it begins to skew the populace from a shared basis of understanding. This can be leveraged to make greater divides between sets of people despite not being as far apart as they might seem. Without a shared perspective and culture, consensus building becomes much more difficult, if not impossible. Without consensus, a representative democracy will begin to burst at the seams.

In Brazil, recommendations from YouTube helped to radicalize numerous individuals after updates to its recommendation engine. New York Times' reporters Max Fisher and Amanda Taub traced how Brazilian far-right politicians and conspiracy theorists came to prominence by embracing the video platform, elevating their exposure through the use of music and video games while exposing political beliefs. This allowed once-fringe lawmakers like Jair Bolsonaro to become president [55][55] and others to win in dramatic landslides. YouTube's algorithm drives upward of 70 percent of viewership, keeping users watching, and driving billions of dollars in ad revenue to parent company Alphabet [56][56] YouTube also acts as a producer of the content, since their ad revenue sharing pays the influencers to create and publish content. As such, it adds liquidity to some of these operations, funding those who otherwise might not be able to push the disinformation.

Engagement on most of the major platforms is driven by some form of user voting system. Users can like, upvote, share, and retweet content to make it more visible in the recommendation system. This can and is easily manipulated to twist public opinion on controversial issues. In 2014, the terrorist group known as the Islamic State of Iraq and Syria, or ISIS as it is more commonly referred, began taking over popular hashtags, a grouping mechanism for similar topics, to espouse their violent ideology and recruit new members [57].[57] Similarly, on Reddit, a group of users and moderators manipulated Reddit's use of stickied posts, or posts that are the first seen on a smaller "subreddit," to create an inorganic flow of engagement. This caused posts on r/The_Donald, a subreddit for then U.S. presidential candidate Donald Trump, to surge to the top of r/all, the most popular posts throughout the platform

[54] Chaney, A; Stewart, B; Engelhardt, B. (2018). How Algorithmic Confounding in Recommendation Systems Increases Homogeneity and Decreases Utility. *Twelfth ACM Conference on Recommender Systems (RecSys'18)*. https://scholar.princeton.edu/bstewart/publications/how-algorithmic-confounding-recommendation-systems-increases-homogeneity-and.

[55] Fisher, M; Taub, A. (2019). How YouTube Radicalized Brazil. *New York Times*. https://www.nytimes.com/2019/08/11/world/americas/youtube-brazil.html.

[56] Solsman, J. (2018). YouTube's AI is the puppet master over most of what you watch. *CNET*. https://www.cnet.com/news/youtube-ces-2018-neal-mohan/.

[57] Girginova, K. (2017). Hojacking Heads and Hashtags. *Global Journal, vol 10, Issue 56*. https://globalejournal.org/global-e/august-2017/hijacking-heads-hashtags.

starting in March 2016. This continued throughout the summer and the election until November 24 of that year when the CEO of Reddit, Steve Huffman, intervened [58][58] and stopped the manipulation.

Coupling the user manipulation of engagement with the algorithmic confounding present in most recommendation engines allows for disinformation peddlers to move their message from the white noise of social media to virality. While fringe materials may not be naturally seen by the majority of the populace, abusing these systems can vastly increase the number of people who view false and misleading materials. A study by Massachusetts Institute of Technology's Media Lab, authors Soroush Vosoughi, Deb Roy, and Sinan Aral found that false news stories spread further and faster than true stories [59][59]; reaching upwards of 100,000 users at the top 1% of stories, where traditional news at the same level rarely peaked 1,000 users. False stories also spread at a rate 6 times that of their truthful counterparts.

Information Laundering

One of the biggest hurdles our adversaries face is making their disinformation reaches a critical mass. While astroturfing and algorithmic confounding can be used to get more individuals to see the content, information laundering will spread the message more quickly due to the added legitimacy [60].[60] Information laundering is the act of a legitimate source, whether a political commentator, social media influencer, or politicians themselves, taking disinformation and spreading it across their network. By attaching their name or brand to a set of disinformation, this acts to legitimize it even if the user in question never directly states they believe in it. This methodology was shown in the 2021 report from the Center for Countering Digital Hate called the Disinformation Dozen, where 12 individuals were responsible for up to 63% of anti-vaccination content on social media [61].[61]

The use of intermediaries in many cases invokes a parasocial relationship between the trusted influencer, and the target audience [62].[62] The audience trusts the influencer's opinion as they would any other real world interpersonal relationship, causing

[58] Huffman, S. (2016). TIFU by editing some comments and creating an unnecessary controversy. *Reddit Announcements*. https://www.reddit.com/r/announcements/comments/5frg1n/tifu_by_editing_some_comments_and_creating_an/.

[59] Vosoughi, S; Roy, D; Aral, S. (2018). The spread of true and false news online. *Science vol. 359. Issue 6380 pp. 1146–1151*. https://science.sciencemag.org/content/359/6380/1146.

[60] Meleshevich, K. Schafer, B. Online Information Laundering: The Role of Social Media. *Alliance for Securing Democracy. The German Marshall Fund of the United States*. https://securingdemocracy.gmfus.org/online-information-laundering-the-role-of-social-media/.

[61] Center for Countering Digital Hate. (2021). This Disinformation Dozen. *Center for Countering Digital Hate*. https://www.counterhate.com/disinformationdozen.

[62] Parasocial interaction. https://www.oxfordreference.com/view/10.1093/oi/authority.20110803100305809.

strangers to have a much stronger hold on citizen's lives than there would be in traditional outreach and marketing. With the increase of social media usage over time particularly through the 2020 global SARS COVID-19 pandemic [63],[63] more and more people relied on social media to get their much-needed interpersonal relationships. This reliance led to a huge growth in disinformation spread and acceptance of conspiracy theories, particularly surrounding anti-government and anti-democracy movements [64].[64]

Information laundering acts as a pipeline, moving disinformation from an untrusted source to a trusted one for greater dissemination and belief. Bret Schafer and Kirill Meleshevich of the Alliance for Securing Democracy describe information laundering as a three-step process [60].[65] The first step is placement. This is done through the inauthentic accounts, whether bots or sockpuppets, putting the initial disinformation into the information environment. Typically, bots will be used for acting in tandem to spread the disinformation through networks, while simultaneously using some form of algorithmic confounding to elevate it to multiple users. By using multiple accounts in tandem, the disinformation is made legitimate since it becomes more difficult to attribute it to a single source, and therefore debunk.

The next step is layering. In order to gain authenticity, and the potential removal of content, the disinformation must travel through various layers, usually some form of intermediary to spread. In many cases the intermediaries and unwitting participants who spread misinformation because it fits a social narrative that they agree with. While algorithmic confounding can help with this, it is usually not the only pathway. For example, during the lead up in the 2016 U.S. Presidential election, Russian hackers broke into the Democratic National Convention's chairman Jon Podesta's personal and professional emails, and passed selected information to Wikileaks, a seemingly neutral third party [65].[66] Wikileaks' founder, Julian Assange, was heavily tied to Russian intelligence agencies and often acted as a disinformation conduit for them, going as far to obscure the true source of the leaked emails, insinuating it was a former DNC staffer, Seth Rich, who was tragically murdered in July of 2016 [66].[67]

The final step of information laundering is integration. This step requires legitimate news outlets to share the disinformation to a greater scale. In 2020 alone,

---

[63] Molla, R. (2021). Posting less, posting more, and tired of it all: How the pandemic has changed social media. *Vox Magazine. Recode.* https://www.vox.com/recode/22295131/social-media-use-pandemic-covid-19-instagram-tiktok.

[64] Arendt, D; Blaha, L. (2015). Opinions, influence, and zealotry: a computational study on stubbornness. *Computational and Mathematical Organization Theory.* 20, 184–209. https://link.springer.com/article/10.1007/s10588-015-9181-1.

[65] Meleshevich, K. Schafer, B. Online Information Laundering: The Role of Social Media. *Alliance for Securing Democracy. The German Marshall Fund of the United States.* https://securingdemocracy.gmfus.org/online-information-laundering-the-role-of-social-media/.

[66] Intelligence Community Assessment. (2017). Assessing Russian Activities and Intentions in Recent US Elections. *Officer of the Director of National Intelligence. United States of America.* ICA- 2017-01D. https://www.dni.gov/files/documents/ICA_2017_01.pdf.

[67] Mueller, R. (2019). Report On The Investigation In Russian Interference In The 2016 Presidential Election. U.S. *Department of Justice. Volume 1. CFT 600.8(c)* https://www.justice.gov/archives/sco/file/1373816/download.

multiple cases of this integration have occurred. Leading up to the 2020 U.S. Presidential election, Russian groups tried to plant incriminating documents about then Democratic Nominee Joseph Biden in the preceding months. While many news organizations refused to publish the article due to the inability to corroborate the claims, a few mainstream outlets did, adding legitimacy to the influence campaign.

This legitimization of disinformation can be as simple as reaching a social media influencer who shares it with their followers, elevating its visibility to potentially millions of users. Or it can be nuanced trafficking through multiple legitimate and illegitimate social media platforms, websites, blogs, and algorithms to end up at its target. For example, throughout 2019 and 2020, Russia ran a website called "Peace Data" to influence left leaning citizens in the English-speaking world with the use of known freelance reporters were to add legitimacy to the site [67].[68] These freelance reporters approached on Twitter and asked to write articles on various topics and were paid between $100 and $250 per article. Many of these articles, according to the freelancers, were then spun by the editorial staff to have a particular political angle.

Privacy

While most of the previous techniques may seem like these groups are releasing disinformation on the grand ocean of social media and hoping for something to stick, this couldn't be further from reality. Data brokerage was a $200 billion dollar industry in 2020 and has no signs of slowing down in growth [68].[69] Nearly all social media companies rely on data and advertising, and the sale of both, in order to remain profitable. The adage, "If you're not paying for it, you're not the customer; you're the product" [69][70] has become a staple for the current internet age. As such, a cottage industry has sprouted up around the buying and selling of data, the building of user profiles, demographics of user history, purchases, and website viewings, while predicting their potential actions [70].[71] Hundreds of entities and companies exist solely to buy and sell data for advertising, with many of which are hiding or refusing to show what data they own on U.S. citizens [71].[72]

---

[68] https://www.reuters.com/article/us-usa-election-facebook-russia/duped-by-russia-freelancers-ensnared-in-disinformation-campaign-by-promise-of-easy-money-idUSKBN25T35E.

[69] Knowledge Sourcing Intelligence. (2021).Global Data Broker Market Size, Share, Opportunities, COVID-19 Impact, And Trends By Data Type (Consumer Data, Business Data), By End-User Industry (BFSI, Retail, Automotive, Construction Others), And By Geography—Forecast From 2021. *Report KSI0601611226, Page 131.* https://www.knowledge-sourcing.com/report/global-data-broker-market.

[70] Lewis, A. (2010). User-Driven Content. *Meta Filter Community Blog.* https://www.metafilter.com/95152/Userdriven-discontent#32560467.

[71] Brathwaite, S. (2021). What Does a data broker do? *Security Made Simple.* https://www.securitymadesimple.org/cybersecurity-blog/what-does-a-data-broker-do.

[72] Privacy Rights Clearninghouse. (2021). Data Brokers. https://privacyrights.org/data-brokers.

Social media platforms themselves act more as first part data miners, as they have huge troves of data produced by their users daily. Their proprietary recommendation algorithms claim that they can target specific groups at highly accurate rates [72, 73].[73,74] For example, Facebook allows advertisers to push their content to specific demographics, in specified locations, with selected interests and behaviors, some of which would be illegal in other contexts [74].[75] Facebook has had multiple instances where they allowed people to exclude Black Americans and other minorities from housing advertisements, a violation of the Fair Housing Act [75].[76] Facebook has refused requests to show what advertisements were targeted toward service members, either for fraudulent companies like multi-level marketing schemes or political advertisements. As a result, companies can specifically target U.S. service members or veterans directly, using this highly trusted group to launder misinformation into the mainstream. There is no oversight or regulation of this targeted advertising industry. Individuals and companies are allowed to spend whatever they want with their targeting limited only by the metrics a company. For example, Russia purchased hundreds of thousands of dollars' worth of ads from Facebook leading up to the 2016 U.S. Presidential Election in order to influence its results [76].[77]

Due to limited laws governing data privacy, the onus is placed on the user to protect themselves. This is onerous because, as previously stated, the data being collected can be completely unknown to the end user. In many cases, organizations that do completely other tasks, may be collecting data on a user to sell. In 2020, Avast, a large antivirus company with upwards of 430 million active monthly users, was collecting web browser history across accounts and sessions and selling them on the market [77].[78] Sessions replays act as recreation of a user's action on a Document Object Model, essentially the framework that makes a website or web application, allowing the data collected to be more that strictly clickthrough rates, but to include hovering over certain sections, how they navigate the page itself, and what they seem to focus on [78].[79]

---

[73] Facebook for Business. (2021). Help your ads find the people who will love your business. *Facebook.* https://www.facebook.com/business/ads/ad-targeting.

[74] Twitter For Business. (2021). Twitter Ads Targeting. *Twitter.* https://business.twitter.com/en/advertising/targeting.html.

[75] Facebook for Developers. (2021). Targeting Search, Behaviors. *Facebook.* https://developers.facebook.com/docs/marketing-api/audiences/reference/targeting-search#behaviors.

[76] HUD Public Affairs. (2019). HUD charges Facebook over company's targeted advertising practices with housing discrimination. *U.S. Department of Housing and Urban Development Archives, HUD No. 19–035.* https://archives.hud.gov/news/2019/pr19-035.cfm.

[77] Shane, S. and Goel, V. (2017). Fake Russian Facebook Accounts Bought $100,100 in Political Ads. *The New York Times.* https://www.nytimes.com/2017/09/06/technology/facebook-russian-political-ads.html.

[78] Cox, J. (2020). Leaded Documents Expose the Secretive Market for your Browsing Data. *Vice Magazine.* https://www.vice.com/en/article/qjdkq7/avast-antivirus-sells-user-browsing-data-investigation.

[79] Fullstory. (2021). The Definitive guide to session replay. *Fullstory Learning Center.* https://www.fullstory.com/resources/the-definitive-guide-to-session-replay.

Furthermore, even when users turn off their locational data within a specified app, they are still being tracked through their phone through their Mobile Advertising ID [79].[80] These MAIDs, as they are commonly referred to, allow cellular phone manufacturers to tailor aids to you while not linking personally identifiable information to the person who owns and uses the phone. However, like many other issues with privacy, in practice MAIDs are not anonymous, and there is an industry specifically for unmasking MAIDs and connecting them to real people [80].[81] Being able to track individuals by their device opens several possibilities for near peers to build a pattern of life, directly influence them, or depending on their web content, compromise them.

One of the most notorious cases of this kind of targeted influence is Cambridge Analytica and their scandals involving the 2016 U.S. Presidential Elections and the British EU Referendum. Cambridge Analytica claimed to be able to microtarget users to greater political engagement and voter turnout. The Company, which was a subsidiary of a private intelligence company, used data collected for academic research to microtarget U.S. and U.K. citizens for specific political movements at scale [81].[82] They also incorporated psychographic information, alleging in their marketing materials that they could psychologically target individuals. Whistleblower Christopher Wylie also claimed ties between Cambridge Analytica and Russian government intermediaries in a Senate Judiciary hearing [82].[83] While many users understood that their data might be advertised in products, they may not have been aware that they could potentially be used by foreign countries to cause political turmoil.

[80] Ropek, L. (2021). 'Anonymous' Mobile Advertising IDs Aren't So Anonymoust, And They Are Everywhere. *Gizmondo*. https://gizmodo.com/anonymous-mobile-advertising-ids-aren-t-so-anonymous-1847292216.

[81] Cox, J. (2021). Inside the Industry That Unmasks People at Scale. *Vice Magazine*. https://www.vice.com/en/article/epnmvz/industry-unmasks-at-scale-maid-to-pii.

[82] Kang, C and Frenkel, S. (2018). Facebook Says Cambridge Analytica Harvested Data of Up to 87 Million Users. *The New York Times*. https://www.nytimes.com/2018/04/04/technology/mark-zuckerberg-testify-congress.html.

[83] Wylie, C. (2018) In the Matter of Cambridge Analytica and Other Related Issues. *Written Statement to the United States Senate Committee on the Judiciary*. https://www.judiciary.senate.gov/imo/media/doc/05-16-18%20Wylie%20Testimony.pdf.

Deplatforming and Forced Migration

Following the fallout around Cambridge Analytica and foreign influence campaigns interference in U.S. Politics, a number of social media companies began working through the problem [83–85].[84,85,86] Many used the tactic of removing the content from suspected foreign accounts, or groups infiltrated by foreign entities, which were typically bot or sock puppets. While each platform had their own methodology for identifying accounts or groups, they generally kept that methodology secret to the public. Certain platforms, like Twitter [83],[87] were open about what they removed and archived it for use by other academics and researchers, other platforms kept it a secret and closed ranks. This occurred while simultaneously removing real users who violated that platforms terms of service, typically around inciting violence and hate speech [86, 87],[88,89] or general public safety following the COIVD19 Global Pandemic [88].[90] This removal of real people, commonly referred to as deplatforming, had some short-term success in reducing disinformation and misinformation on the platform [89].[91]

Studies done by the iDrama Lab, led by Jeremy Blackburn of Binghamton University, on two deplatformed subreddits showed that user participation on the platform will decrease after a deplatforming event, reduce the amount of new users, and reduce the amount of overall content [90].[92] A similar study by researchers out of Georgia Institute of Technology on two separate subreddits confirmed the iDrama research, as well as showed that rates of hate speech declined by the users who remained

---

[84] Twitter. (2021). Information Operations. *Twitter Transparency*. https://transparency.twitter.com/en/reports/information-operations.html.

[85] Facebook. (2021). The State of Influence Operations 2017–2020. *Facebook Threat Report*. https://about.fb.com/wp-content/uploads/2021/05/IO-Threat-Report-May-20-2021.pdf.

[86] Reddit Administrative Staff. (2019). Suspected Campaign from Russia on Reddit. *Reddit Security*. https://www.reddit.com/r/redditsecurity/comments/e74nml/suspected_campaign_from_russia_on_reddit/.

[87] Twitter. (2021). Information Operations. *Twitter Transparency*. https://transparency.twitter.com/en/reports/information-operations.html.

[88] Huffman, S. (2020). Update to Our Content Policy. *Reddit Announcements*. https://www.reddit.com/r/announcements/comments/hi3oht/update_to_our_content_policy/.

[89] Twitter. (2021). Permanent suspension of @realDonaldTrump. *Twitter*. https://blog.twitter.com/en_us/topics/company/2020/suspension.

[90] Facebook. Taking Actions Against Misinformation Across Our Apps. *Facebook*. https://www.facebook.com/combating-misinfo.

[91] Dwoskin, E. and Timberg, C. (2021). Misinformation Dropped Dramatically the week after Twitter banned Trump and some allies. *The Washington Post*. https://www.washingtonpost.com/technology/2021/01/16/misinformation-trump-twitter/.

[92] Ribeiro, MH et all. (2021). Do Platform Migrations Compromise Content Moderations? Evidence from r/The Donald and r/Incels. 24th *ACM Conference on Computer-Supported Cooperative Work and Social Computing* https://arxiv.org/pdf/2010.10397.pdf.

on the platform [91].[93] For the platforms themselves, this serves as a positive as it removes some of the malevolent forces within them, and shows they have an active moderation and environment free from issues that might lose advertising.

While deplatforming has some benefits, it also has some detriments. One of the biggest is that it can have the tendency to draw attention to a group or movement that hasn't gained mainstream attention [92].[94] While Reddit banned the Qanon Conspiracy for violent rhetoric in 2018, it grew continuously through 2020, reaching its height following the U.S. Presidential Election [93],[95] potentially due in part through the media coverage undermining that deplatforming effort. Adding the taboo of removal from a major social media can push a conspiracy, particularly one centered around persecution, to critical mass [94].[96] While deplatforming will remove potential problems from a specific platform, and potentially make things better within it, it will not remove the ideas and narrative behind the problem.

If users are removed from a platform, or that platform itself is deplatformed from a host, the users within a group will typically attempt to migrate to another more open platform [95].[97] This migration will have multiple results. The first are beneficial, in that typically there isn't a 1 to 1 movement, generally losing active members along the way. The group will also likely not have as large of an audience to recruit from as they would on a large, corporate social media platform [90].[98] Typically, the benefits end there, and instead become benefits for foreign influence campaigns. Users that migrate typically become more insular and radicalized [90].[99] Depending on the political leaning of the group, and the limited moderation, this can be a dangerous combination for political violence. Parler, following mass exodus of conservative users from Twitter, grew substantially through 2020. However, following the Capitol Insurrection on January 6th, 2021, Parler lost its hosting through Amazon Web Services.

---

[93] Chandrasekharan, E et all. (2017). You Can't Stay Here: The Efficacy of Raddit's 2015 Ban Examined Through Hate Speech. *Proceedings of the ACM on Human–Computer Interaction, Vol1, article 31. Pages 1–22* https://dl.acm.org/doi/10.1145/3134666.

[94] Aliapoulious, M et al. (2021). An Early Look at the Parler Online Social Network. *Proceedings of the International AAAI Conference on Web and Social Media, 15 pages 943–951.* https://arxiv.org/pdf/2101.03820.pdf.

[95] Dickson, EJ. (2021). The Qanon Community Is in Crisis—But On Telegram, It's Also Growing. *Rolling Stone Magazine.* https://www.rollingstone.com/culture/culture-news/qanon-telegram-channels-increase-1117869/.

[96] Nickas, J; Isaac, M; Frenkel, S. (2021). Millions Flock to Telegram and Signal as Fear Grows Over Big Tech. *The New York Times.* https://www.nytimes.com/2021/01/13/technology/telegram-signal-apps-big-tech.html.

[97] Johnson NF et all. (2019). Hidden resilience and adaptive dynamics of the global online hate ecology. *Nature 573, pages 261–265.* https://www.nature.com/articles/s41586-019-1494-7.

[98] Ribeiro, MH et all. (2021). Do Platform Migrations Compromise Content Moderations? Evidence from r/The Donald and r/Incels. *24th ACM Conference on Computer-Supported Cooperative Work and Social Computing* https://arxiv.org/pdf/2010.10397.pdf.

[99] Ribeiro, MH et all. (2021). Do Platform Migrations Compromise Content Moderations? Evidence from r/The Donald and r/Incels. *24th ACM Conference on Computer-Supported Cooperative Work and Social Computing* https://arxiv.org/pdf/2010.10397.pdf.

This led Parler to turn to DDoS-Guard for these services, a company with ties to the Russian Federation, who's owners are based in Rostov-on-Don, Russia [96].[100]

These smaller offshoots, also have more questionable security practices compared to their larger corporate counterparts. Activists attempted to catalog Parler posts following the Capitol Insurrection and found a vulnerability that allowed them to pull all data from the platform at an admin level, including deleted posts, messages, and locational data [97].[101] Similarly, Gab, a platform who hosted the 2018 Tree of Life synagogue shooter [98],[102] was hacked the next month, leaking private messages and content similar to that of the Parler leak [99].[103] Both of these platforms include U.S. congress members and U.S. Senators, along with a slew of state legislators, mayors, and governors. Another study done by the iDrama lab, in cooperation with Boston University's SecLAB, showed that many users deplatformed go to these alternative platforms only to become more vitriolic, although losing much of the size of their following in doing so [100].[104]

Generative Media

New machine learning algorithms that produce realistic images, videos, audio, and text are being created every day. These algorithms are generally grouped under the name of Generative Adversarial Networks [101],[105] for the way that two algorithms work against each other, like a counterfeiter against law enforcement, to generate realistic media. The most notable example is the titular deepfakes, named as a portmanteau of deep neural networks that create fake media. These videos, as they were found, typically replaced women in adult films with the faces of celebrities [102].[106] Over the course of 2019 and 2020, these algorithms raised great concern

---

[100] Menn, J; Li, K; Culliford, E. (2021). Parler partially reappears with support from Russian technology firm. *Reuters Internet News*. https://www.reuters.com/article/us-usa-trump-parler-russia/parler-partially-reappears-with-support-from-russian-technology-firm-idINKBN29N23N.

[101] Holt, J; Brookie, G, Brooking, E. (2021). Fast Thinking: How the Capitol riot was coordinated online. *The Atlantic Council*. https://www.atlanticcouncil.org/content-series/fastthinking/fast-thinking-how-the-capitol-riot-was-coordinated-online/.

[102] Saldiva, G; Van Sant, S; Bowman, E. (2018). Suspect Charged With 29 Federal Counts in Pittsburgh Synagogue Massacre. *National Public Radio*. https://www.npr.org/2018/10/27/661347236/multiple-casualties-in-shooting-near-pittsburgh-synagogue.

[103] Greenburg, A. (2021). Far-Right Platform Gab Has Been Hacked—Including Private Data. *Wired Magazine*. https://www.wired.com/story/gab-hack-data-breach-ddosecrets/.

[104] Ali, S et all. (2021). Understanding the Effect of Deplatforming on Social Networks. WebSci' 12: 13th ACM Web Science Conference pages 187–195. https://seclab.bu.edu/people/gianluca/papers/deplatforming-websci2021.pdf.

[105] Goodfellow, I et all. (2014). Generative Adversarial Nets. Proceedings of the 27th *International Conference on Neural Information Processing Systems, Volume 2, pages 2672–2680*. https://arxiv.org/pdf/1406.2661.pdf.

[106] Cole, S. (2017). AI-Assisted Fake Porn Is Here. *Vice Magazine*. https://www.vice.com/en/article/gydydm/gal-gadot-fake-ai-porn.

among the legal circles [103],[107] government officials [104],[108] defense analysts [105],[109] academia [106],[110] and industry [107][111] for the potential for their use in disinformation campaigns.

Many of these algorithms are being open sourced and shared on programming repositories like github and bitbucket [108, 109],[112,113] open for use by the general public, and foreign adversaries. In fact, many have already been found in the wild. Foreign Intelligence Services have already used deep fakes to add credibility to sock puppet accounts meant for collecting information of U.S. service members and those of the intelligence community on LinkedIn [110].[114] Domestically, U.S. political groups have used sock puppet accounts to weave a narrative leading up to elections on Facebook [111].[115] These technologies have been so democratized that a Pennsylvanian mother sent deepfaked pictures of her daughter's cheerleading rival [112].[116]

---

107 Citron, D. (2019). The National Security Challenge of Artifical Intelligence, Manipulated Media, and "Deep Fakes." *Prepared Written Testimony and Statement for the Record. House Permanent Select Committee on Intelligence.* https://intelligence.house.gov/uploadedfiles/citron_testimony_for_house_committee_on_deep_fakes.pdf.

108 Rubio, M and Warner, M. (2019). Rubio, Warner Express Concern Over Growing Threat Posed by Deepfakes. *Joint Statement from Senators Marco Rubio and Mark Warner.* https://www.rubio.senate.gov/public/index.cfm/2019/10/rubio-warner-express-concern-over-growing-threat-posed-by-deepfakes.

109 Watts, C. (2019). The National Security Challenge of Artifical Intelligence, Manipulated Media, and "Deep Fakes." *Prepared Written Testimony and Statement for the Record. House Permanent Select Committee on Intelligence.* https://intelligence.house.gov/uploadedfiles/clint_watts_-_house_select_committee_on_intelligence_-_ai__deep_fakes_-_13_june_2019.pdf.

110 Doermann, D. (2019). The National Security Challenge of Artifical Intelligence, Manipulated Media, and "Deep Fakes." *Prepared Written Testimony and Statement for the Record. House Permanent Select Committee on Intelligence.* https://intelligence.house.gov/uploadedfiles/doermann-statement-final.pdf.

111 Clark, J. (2019). The National Security Challenge of Artifical Intelligence, Manipulated Media, and "Deep Fakes." *Prepared Written Testimony and Statement for the Record. House Permanent Select Committee on Intelligence.* https://intelligence.house.gov/uploadedfiles/clark_deepfakes_sfr.pdf.

112 Deepfakes. (2019). FaceSwap Manifesto. *Github pages.* https://github.com/deepfakes/faceswap#manifesto.

113 Radek. (2018). Myfakeapp. *Bitbucket.* https://bitbucket.org/radeksissues/myfakeapp/src/master/.

114 Boyd. M. (2021). Deepfakes and LinkedIn: malign interference campaigns. *Malwarebytes Labs, Social Engineering.* https://blog.malwarebytes.com/social-engineering/2019/11/deepfakes-and-linkedin-malign-interference-campaigns/.

115 Collins, B. (2019). Facebook says a pro-Trump media outlet used artificial intelligence to create fake people and push conspiracies. *National Broadcasting Company News.* https://www.nbcnews.com/tech/tech-news/facebook-says-pro-trump-media-outlet-used-artificial-intelligence-create-n1105951.

116 Associated Press. (2021). Cheerleader's mom accused of making 'deepfakes' of rivals. The Associated Press https://apnews.com/article/pennsylvania-doylestown-cheerleading-0953a60ab3e3452b87753e81e0e77d7f.

These algorithms have a deep amount of potential for abuse by Russia, China, and Iran in disinformation campaigns. As stated above, they are already being used in profile pictures on bot and sock puppet account profiles to add a layer of believability. nVidia Corporation's StyleGAN [113, 114][117,118], and StyleGAN2 [115, 116][119,120] all create detailed and realistic still images that are completely synthetic using a technique called style transfer, which takes characteristics from one image and applies them to another. Both trained models have been released to the public, and both have been used in created accounts.

The reach of GANs and Deepfakes goes much further than that. An algorithm created by Samsung Moscow can create realistic moving images and videos based off a single image of a person [117].[121] With more images, from different angles, the quality will increase substantially, making seamless videos over the top of a staged video. This is a substantial decrease from something like the 'face swap' app that gave way to the initial rounds of deepfakes, which required hundreds if not thousands of images of the target to create a realistic fake. While many news agencies argue the main threat is political, in practice most targets thus far have been people without the means to pushback against [118].[122] This means the targeting of government officials, particularly those vulnerable in overseas assignments [119],[123] for blackmail and extortion [120],[124] which are already commonplace in a foreign intelligence agency's arsenal [121].[125]

[117] Kerras, T; Laine, S; Aila T. (2019). A Style-Based Generator Architecture for Generative Adversarial Networks. *2019 IEEE/CVF Conference on Computer Vision and Pattern Recognition (CVPR)*. https://arxiv.org/abs/1812.04948.

[118] nVidia Labs. (2019). Style-Gan Repository. *Github*. https://github.com/NVlabs/stylegan.

[119] Kerras, T et all. (2020). Analyzing and Improving the Image Quality of StyleGAN. *2019 IEEE/CVF Conference on Computer Vision and Pattern Recognition (CVPR)*. https://arxiv.org/abs/1912.04958.

[120] nVidia Labs. (2020). Style-Gan2 Repository. *Github*. https://github.com/NVlabs/stylegan2.

[121] Zakhorav, E et all. (2019) Few-Shot Adversarial Learning of Realistic Neural Talking Head Models. *2019 IEEE/CVF International Conference on Computer Vision (ICCV)*. https://arxiv.org/pdf/1905.08233.pdf.

[122] Gosse, C. and Burkell, J. (2020). Politics and porn: how news media characterizes problems presented by deepfakes. *Critical Studies in Media Communications, Volume 30 Issue 5*. https://nca.tandfonline.com/doi/full/10.1080/15295036.2020.1832697?journalCode=rcsm20#.YPrydo5KhPa.

[123] Littell, J. (2019). Don't Believe Your Eyes (or Ears): The Weaponizations of Artificial Intelligence, Machine Learning, And Deepfakes. *War on the Rocks*. https://warontherocks.com/2019/10/dont-believe-your-eyes-or-ears-the-weaponization-of-artificial-intelligence-machine-learning-and-deepfakes/.

[124] Weiss, M. (2019). The Hero Who Betrayed His Country. *The Atlantic*. https://www.theatlantic.com/international/archive/2019/06/estonia-russia-deniss-metsavas-spy/592417/.

[125] Nemtsova, A. (2019). What to Do When the Russian Government Wants to Blackmail You. *The Atlantic*. https://www.theatlantic.com/international/archive/2019/03/russia-government-blackmail-kompromat/585850/.

Couple the visualization with synthetic audio [122],[126] and you have very realistic fakes [123][127] that can be made cheaply, quickly, and distributed through the aforementioned information laundering methodology. Foreign adversaries could cause protests similar to the 2014 Benghazi consulate attack based on fake video or audio. This could be equally worrisome for both domestic and foreign operations, where soft targets could be overwhelmed by unknowing populations upset at fake media. Russia has already expertly orchestrated real world protests using Facebook groups and advertising [124],[128] adding in perfectly crafted fake media could bring that capability to scale.

And while not Generative Adversarial Networks or deepfakes per se, generative text [125][129] stands to expand and automate many processes already in use. OpenAI's Generative Pretrained Transformers (GPT) [126][130] alone has multiple applications already in the wild. Users can create realistic social media bots that pull topics from currently tending topics and create appropriate comments within them. This allows for robust botnets and sock puppet accounts that will circumvent some of the trusted methods for hunting and removing their accounts. Other media can be created from these models as well. Meme generators based of a string of text can create visual media [127][131] that can be more pervasive and influential than simple text [128].[132] Even mainstream news agencies are utilizing model generated text to fill out, or in other cases entirely write articles on a specific subject with minimal inputs [129].[133] Given Russians preponderance to create entire news sites and paying real journalists for legitimacy, they could potentially automate much of the process for a similar effect.

[126] Tung, L. (2016). Adobe's VoCo voice project: Now you really can put words in someone else's mouth. *ZDnet.* https://www.zdnet.com/article/adobes-voco-voice-project-now-you-really-can-put-words-in-someone-elses-mouth/.

[127] Dessa Corporation. (2019). RealTalk: We Recreated Joe Rogan's Voice Using Artificial Intelligence. YouTube. https://youtu.be/DWK_iYBl8cA.

[128] Kosoff, M. (2017). How Russia Secretly Orchestrated Dozens of U.S. Protests. *Vanity Fair.* https://www.vanityfair.com/news/2017/10/how-russia-secretly-orchestrated-dozens-of-us-protests.

[129] Kawthekar, P; Rewari, R; Bhooshan, S. (2017). Evaluating Generative Models for Text Generation. *Stanford University.* https://web.stanford.edu/class/archive/cs/cs224n/cs224n.1174/reports/2737434.pdf.

[130] Brown, T et all. (2020). Language Models are Few-Shot Learners. *Advances in Neural Information Processing Systems 33.* https://arxiv.org/abs/2005.14165.

[131] Anand, A. (2020). Another 10 gems of GPT-3. *Dev.to Blog.* https://dev.to/amananandrai/another-10-gems-of-gpt-3-2639.

[132] Huntington, H. (2017). The affects and Effect of Internet Memes: Assessing Perceptions and Influence of Online User-Generated Poltical Discourse as Media. *Colorado State University Department of Journalism and Media Communication.* https://mountainscholar.org/bitstream/handle/10217/183936/Huntington_colostate_0053A_14303.pdf.

[133] Hutson, M. Robo-writers: the rise and risks of language generating AI. *Nature News Feature.* https://www.nature.com/articles/d41586-021-00530-0.

Data Poisoning and Contamination

As automation of disinformation takes hold, processes to find and remove it will also turn to automation to counter it. As such, much like cyber security, an arms race will begin between those nations who look to use online spaces for the spread of disinformation, and those who look to remove it. In order to counter the sheer volume of disinformation being created, and forensically determine authenticity of various pieces of media, algorithmic methods will need to be used, as human analysts alone will not be able to process the volume, variety, and veracity of the data. This, in turn, will lead to innovative ways to get around these algorithmic filters.

The most likely course of action is to develop adversarial machine learning to disrupt and outright disable the algorithms trying to identify various methods, media, and pathways of disinformation. Typically, this is referred to as data poisoning [130][134] or, if the model is already created and running, data contamination [131].[135] Data poisoning serves as an adversarial machine learning attack that focuses on adding data that the algorithm learns from at the time of model creation. It attempts to bias the decisions made by the algorithmic model so that selected data goes through undetected. For a clearer example, if Russia wanted to have more believable accounts, they could use deepfake images from a GAN to make profile pictures. If the platform, or another nation wanted to find these accounts, they may train a computer vision algorithm to look for them [132].[136] Given the game of cat and mouse, the Russian creating the media may have anticipated some form of intervention, and instead overlaid their images with a small marker, only visible to the computer, but not to the human who labels the data. Thus, the algorithm made to identify and mark these suspicious accounts will most likely be accounts that are identified by the placed marker, leaving other accounts, that were not meant to be found easily, to pass through the algorithmic net.

Comparatively speaking, adversarial data contamination works similarly, but instead of the algorithm being built off of the data, it's being updated by new data after having been already trained. Overall, the methodology would remain the same, but there would be a larger number of accounts that need to be identified in order to slowly shift the bias in the direction our adversaries would like. This type of attack is much more likely, as platforms and others have already begun to algorithmically filter out fake accounts and synthetic media.

---

[134] Xiao, H et all. (2015). Is Feature Selection Secure against Training Data Poisoning. *Proceedings in the 3nd International Conference on Machine Learning*. http://proceedings.mlr.press/v37/xiao15.pdf.

[135] Xiao, H. et all (2015). Support vector machines under adversarial label contamination. *Neurocomputing Volume 160, pages 53–62*. https://www.sec.in.tum.de/i20/publications/support-vector-machines-under-adversarial-label-contamination/file/main-revision.pdf.

[136] Nguyen, T et all. (2019). Deep Learning for Deepfakes Creation and Detection: A Survey. *Cornell University Arxiv*. https://arxiv.org/pdf/1909.11573.pdf.

Conclusion

More and more countries are heavily leaning into online disinformation to further their political goals both domestically and abroad. Many of which have put information at the forefront of their military and diplomatic campaigns to gain the advantage in this ever more connected world. Data is readily available for potentially building psychological profiles of groups at scale, or individually for more precise influencing. This leads to many different dangers from government and military service members to the very democratic institutions they uphold. Facts can be drowned out from the conversation to allow fringe beliefs to seem as though they are commonplace.

Russia and China have decades of online manipulation in their repertoire already being turned onto the United States, and the greater ideals of western democracies. Daron Acemoglu and James A. Robinson, authors of "The Narrow Corridor," describe the fragile balance between state and society that allow liberal democracy to avoid falling into the alternatives of lawlessness or authoritarianism [133].[137] If the state becomes too strong, it risks despots removing liberties from the people, if the society becomes too strong, it falls into a state of anarchy, where the state's ability to protect the populace from groups within fails. Long term influence campaigns can tilt the balance of power either way, allowing for the house of cards, that is our freedom, to come tumbling down.

To keep democracy from falling to tatters, we must institute better protection of our privacy, in particular our data, while our national defense pushes back against outside actors who look to break it down. Without full-fledged, proactive countering of our near peers, we will be continuously playing catchup, fighting off our back foot merely trying to keep our heads above water. Without change to our current stature, we will inevitably see greater fissures in our society, and less trust in our state. Without the ability of citizens to actively participate in their government, through free and fair elections, and build consensus across different belief structures and groups of citizenries, democracy cannot function in a state that is accepted. It will give way to other, harsher forms of government, that will see these failures as an opportunity to seize power at the expense of the rights of many of its citizens.

References

1. Mattis, P. (2018). China's Three Warfares in Perspective. *War on the Rocks. Special Series: Ministry of Truth.* https://warontherocks.com/2018/01/chinas-three-warfares-perspective/.
2. Cunningham, C. (2020). A Russian Federation Information Warfare Primer. *Then Henry M. Jackson School of International Studies. Washington University.* https://jsis.washington.edu/news/a-russian-federation-information-warfare-primer/.

[137] Acemoglu, D; Robinson, J. (2019). The Narrow Corridor. States, Societies, and the Fate of Liberty. *Penguin Random House Publishing.*

3. Strick, B. (2020). COVID-19 Disinformation: Attempted Influence in Disguise. *Australian Strategic Policy Institute. International Cyber Policy Center.* https://www.aspi.org.au/report/covid-19-disinformation.
4. Creswell, A., White, T., Dumoulin, V., Arulkumaran, K., Sengupta, B., & Bharath, A. (2017). Generative Adversarial Networks: An Overview. IEE-SPM. April 2017. https://arxiv.org/pdf/1710.07035.pdf.
5. Whittaker, L., Letheren, K., & Mulcahy, R. (2021). The Rise of Deepfakes: A conceptual framework and research agenda for marketing. https://doi.org/10.1177/1839334921999479.
6. https://arxiv.org/abs/1810.04805.
7. Bradshaw, S., Bailey, H., & Howard, P. (2020). 2020 Global Inventory of Organized Social Media Manipulation. *Computational Propaganda Research Project. Oxford Internet Institute. Oxford University.* https://demtech.oii.ox.ac.uk/wp-content/uploads/sites/127/2021/02/CyberTroop-Report20-Draft9.pdf.
8. https://diamond-democracy.stanford.edu/speaking/lectures/what-democracy.
9. Bastos, M., & Mercea, D. (2018). The public accountability of social platforms: lessons from a study on bots and trolls in the Brexit campaign. *Philosophical transactions. Series A, Mathematical, physical, and engineering sciences, 376*(2128), 20180003. https://doi.org/10.1098/rsta.2018.0003.
10. National Intelligence Council. (2021). Foreign threats to the 2020 US federal elections. *Intelligence Community Assessment.* https://www.dni.gov/files/ODNI/documents/assessments/ICA-declass-16MAR21.pdf.
11. Wood, D., McMinn, S., & Feng, E. (2019). China Used Twitter to Disrupt Hong Kong Protests, but Efforts Began Years Earlier. *National Public Radio.* https://www.npr.org/2019/09/17/758146019/china-used-twitter-to-disrupt-hong-kong-protests-but-efforts-began-years-earlier.
12. Uyghur Human Rights Project. (2020). The Happiest Muslims in the World: Disinformation, Propaganda, and the Uyghur Crisis. *Uyghur Human Rights Project.* https://docs.uhrp.org/pdf/Disinformation_Propganda_and_the_Uyghur_Crisis.pdf.
13. Bernard, R., Bowsher, G., Sullivan, R., & Gibson-Fall, F. (2021). Disinformation and Epidemics: Anticipating the Next Phase of Biowarfare. *Health Security. Volume 19, Number 1.* https://doi.org/10.1089/hs.2020.0038.
14. Giles, C., & Bhat, U. (2020). Nagoro-Karabakh: The Armenian-Azeri 'Information Wars'. *BBC Reality Checks and Anti-disinformation Unit.* https://www.bbc.com/news/world-europe-54614392.
15. Beech, H. (2021). 'Now We Are United': Myanmar's Ethnic Divisions Soften after Coup. *New York Times.* https://www.nytimes.com/2021/04/30/world/asia/myanmar-ethnic-minority-coup.html.
16. McKinsey and Company. (2021). Covid-19: Implications for Business. *McKinsey and Company. Executive Briefing.* https://www.mckinsey.com/business-functions/risk-and-resilience/our-insights/covid-19-implications-for-business.
17. Center for System Sciences and Engineering (CSSE). (2021). COVID-19 Dashboard. *Johns Hopkins University. ARCGIS.* https://www.arcgis.com/apps/dashboards/bda7594740fd40299423467b48e9ecf6.
18. Blomberg, B., Mohn, K., & Brokstad, K. A. et al. (2021). Long COIVD in a prospective cohort of home-isolated patients. *Nature Medicine.* https://www.nature.com/articles/s41591-021-01433-3.
19. Centers for Disease Control and Prevention. (2020). CDC Calls on Americans to wear mask to prevent COIVD-19 Spread. *Centers for Disease Control and Prevention.* https://www.cdc.gov/media/releases/2020/p0714-americans-to-wear-masks.html.
20. Loucaides, D., & Perrone, A. (2021). Germany's COVID sceptics fueled by Russian media and far-Right Conspiracies. *openDemocracy.* https://www.opendemocracy.net/en/germanys-covid-sceptics-fuelled-by-russian-media-and-far-right-conspiracies/.
21. Benson, T. (2020). Trolls and bots are flooding social media with disinformation encouraging states to end quarantine. *Business Insider. Tech.* https://www.businessinsider.com/trolls-bots-flooding-social-media-with-anti-quarantine-disinformation-2020-4.

22. Emmot, R. (2020). Russia deploying coronavirus disinformation to sow panic in West, EU document says. *Reuters*. https://www.reuters.com/article/us-health-coronavirus-disinformation/russia-deploying-coronavirus-disinformation-to-sow-panic-in-west-eu-document-says-idUSKBN21518F.
23. Miller, H. (2021). Russia's Anti-Vaccine Propaganda is Tatamount to a Declaration of War. *Center for Medical Economic and Innovation*. https://medecon.org/russias-anti-vaccine-propaganda-is-tantamount-to-a-declaration-of-war/.
24. Chen, E. (2021). Chinese COVID-19 Misinformation A Year Later. *The Jamestown Foundation Global Research and Analysis*. https://jamestown.org/program/chinese-covid-19-misinformation-a-year-later/.
25. Reuters Staff. (2020). China sends medical supplies, experts to help Italy battle coronavirus. *Reuters Healthcare and Pharma*. https://www.reuters.com/article/us-health-coronavirus-italy-respirators/china-sends-medical-supplies-experts-to-help-italy-battle-coronavirus-idUSKBN2101IM.
26. Mallapaty, S. (2021). China's COVID vaccines are going global—but questions remain. *Nature Magazine* https://www.nature.com/articles/d41586-021-01146-0.
27. Kinetz, E. (2021). Anatomy of a conspiracy: With COVID, China took leading role. *Associated Press*. https://apnews.com/article/pandemics-beijing-only-on-ap-epidemics-media-122b73e134b78091 9cc1808f3f6f16e8.
28. Taylor, P. (2003). Munitions of the mind: A history of propaganda. *Manchester University Press*. https://www.jstor.org/stable/j.ctt155jd69.
29. Edwards, M. (1994). Printing, Propaganda, and Martin Luther. *Berkeley: University of California Press*. https://publishing.cdlib.org/ucpressebooks/view?docId=ft3q2nb278;chunk.id=0;doc.view=print.
30. https://publishing.cdlib.org/ucpressebooks/view?docId=ft3q2nb278;chunk.id=0;doc.view=print.
31. https://www.ushmm.org/collections/bibliography/nazi-propaganda-1.
32. Jowett, G., & O'Donnell, V. (2012). *Propaganda and Persuasion*, 5th Edn. Sage Publications.
33. https://watch.eventive.org/m4df2021/play/6074567c86f143003e0cbd99/6074697b69154f003086bb13.
34. Broderick, R. (2014). Activists Are Outing Hundreds Of Twitter Users Believed To Be 4chan Trolls Posing As Feminist. *BuzzFeed News*. https://www.buzzfeednews.com/article/ryanhatesthis/your-slip-is-showing-4chan-trolls-operation-lollipop.
35. Ferrara E. (2020). Bots, Elections, and Social Media: A Brief Overview. In K. Shu, S. Wang, D. Lee, & H. Liu (eds) *Disinformation, misinformation, and fake news in Social Media*. Lecture Notes in Social Networks. Springer, Cham. https://doi.org/10.1007/978-3-030-42699-6_6.
36. Bakir, V., & McStay, A. (2018). Fake news and the economy of emotions. *Digital Journalism, 6*(2), 154–175. https://doi.org/10.1080/21670811.2017.1345645
37. Anderson, J., Rainie, L., Vogels, e. (2021). Expers Say the 'New Normal' in 2025 Will Be Far More Tech-Driven, Presenting More Big Challenges. Pew Research Center. *Emerging Technologies. Future of the Internet*. https://www.pewresearch.org/internet/2021/02/18/experts-say-the-new-normal-in-2025-will-be-far-more-tech-driven-presenting-more-big-challenges/.
38. Polysanskaya, A., Krivov, A., &Lomko, I. (2003). The Virtual Eye of Big Brother. *Vestnik Online*. http://www.vestnik.com/issues/2003/0430/win/polyanskaya_krivov_lomko.htm.
39. Elder, M. (2012). Polishing Putin: hacked emails suggest dirty tricks by Russian youth group. *The Guardian*. https://www.theguardian.com/world/2012/feb/07/putin-hacked-emails-russian-nashi.
40. Bandurski, D. (2008). China's Guerrilla War for the Web. *China Media Project. China Internet Research Conference*. https://chinamediaproject.org/2008/07/07/feer-chinas-guerrilla-war-for-the-web/.
41. King, G., Pan, J., & Roberts, M. (2017). How the Chinese Government Fabricates Social Media Post for Strategic Distraction, not Engaged Argument. *American Political Science Review, vol. 111, pg. 484–501*. https://gking.harvard.edu/files/gking/files/50c.pdf.

42. Lock, I., & Ludolph, R. (2019). Organizational propaganda on the Internet: A systematic review. *Public Relations Inquiry. vol 9, issue 1, pg 103–127.* https://doi.org/10.1177/2046147X19870844.
43. Kenny, C., & Bergman, M. (2019). Understanding and Combating Russian and Chinese Influence Operations. *Center for American Progress, Foreign Policy and Security.* https://www.americanprogress.org/issues/security/reports/2019/02/28/466669/understanding-combating-russian-chinese-influence-operations/.
44. Select Committee on Intelligence. United States Senate. (2018). Russian Active Measures Campaigns and Interference in the 20216 U.S. Election Volume 2: Russia's Use of Social Media with Additional Views. *116*th *Congress. Senate. 1*st *Session. Report 116-XX.* https://www.intelligence.senate.gov/sites/default/files/documents/Report_Volume2.pdf.
45. Meserole, C. (2018). How misinformation spreads on social media—and what to do about it. *Brookings Institute.* https://www.brookings.edu/blog/order-from-chaos/2018/05/09/how-misinformation-spreads-on-social-media-and-what-to-do-about-it/.
46. Arjomand, N. (2019). Information Laundering and Globalized Media—Part I: The Problem. *Center for International Media Assistance.* https://www.cima.ned.org/blog/information-laundering-and-globalized-media-part-i-the-problem/.
47. Sherman, J. (2021). Federal Privacy Rules Must Get "Data Broker" Definitions Right. Lawfare. *Consumer Data.* https://www.lawfareblog.com/federal-privacy-rules-must-get-data-broker-definitions-right.
48. The National Law Review. (2019). Data Scraping Survives! (At Least for Now) Key Takeaways from 8th Circuit Ruling on the HIQ vs. LinkedIn Case. *The National Law Review.* https://www.natlawreview.com/article/data-scraping-survives-least-now-key-takeaways-9th-circuit-ruling-hiq-vs-linkedin.
49. Diamond, L. (2004). What is Democracy. *Lecture to Hilla University for Humanistic Studies.* https://diamond-democracy.stanford.edu/speaking/lectures/what-democracy.
50. Google Developers. (2021). Introduction to Recommendation Systems. *Google.* https://developers.google.com/machine-learning/recommendation.
51. Hardesty, L. (2019). The history of Amazon's recommendation algorithm. *Amazon Science.* https://www.amazon.science/the-history-of-amazons-recommendation-algorithm.
52. Ilic, A. & Kabiljo, M. (2015) Recommending items to more than a billion people. *Facebook Engineering, Core Data, ML Applications.* https://engineering.fb.com/2015/06/02/core-data/recommending-items-to-more-than-a-billion-people/.
53. Oracle. (2020). Introduction to Recommendation Engines. *Oracle Cloud Infrastructure Data Science.* https://www.oracle.com/a/ocom/docs/introduction-to-recommendation-engines-wp.pdf.
54. Chaney, A., Stewart, B., & Engelhardt, B. (2018). How Algorithmic Confounding in Recommendation Systems Increases Homogeneity and Decreases Utility. *Twelfth ACM Conference on Recommender Systems (RecSys '18).* https://scholar.princeton.edu/bstewart/publications/how-algorithmic-confounding-recommendation-systems-increases-homogeneity-and.
55. Fisher, M., & Taub, A. (2019). How YouTube Radicalized Brazil. *New York Times.* https://www.nytimes.com/2019/08/11/world/americas/youtube-brazil.html.
56. Solsman, J. (2018). YouTube's AI is the puppet master over most of what you watch. *CNET.* https://www.cnet.com/news/youtube-ces-2018-neal-mohan/.
57. Girginova, K. (2017). Hojacking Heads and Hashtags. *Global Journal, vol 10, Issue 56.* https://globalejournal.org/global-e/august-2017/hijacking-heads-hashtags.
58. Huffman, S. (2016). TIFU by editing some comments and creating an unnecessary controversy. *Reddit Announcements.* https://www.reddit.com/r/announcements/comments/5frg1n/tifu_by_editing_some_comments_and_creating_an/.
59. Vosoughi, S., Roy, D., & Aral, S. (2018). The spread of true and false news online. *Science, 359*(6380), 1146–1151. https://science.sciencemag.org/content/359/6380/1146.
60. Meleshevich, K., & Schafer, B., Online Information Laundering: The Role of Social Media. *Alliance for Securing Democracy. The German Marshall Fund of the United States.* https://securingdemocracy.gmfus.org/online-information-laundering-the-role-of-social-media/.

61. Center for Countering Digital Hate. (2021). This Disinformation Dozen. *Center for Countering Digital Hate.* https://www.counterhate.com/disinformationdozen.
62. Parasocial interaction. https://doi.org/10.1093/oi/authority.20110803100305809.
63. Molla, R. (2021). Posting less, posting more, and tired of it all: How the pandemic has changed social media. *Vox Magazine. Recode.* https://www.vox.com/recode/22295131/social-media-use-pandemic-covid-19-instagram-tiktok.
64. Arendt, D., & Blaha, L. (2015). Opinions, influence, and zealotry: a computational study on stubbornness. *Computational and Mathematical Organization Theory, 20*, 184–209. https://doi.org/10.1007/s10588-015-9181-1.
65. Intelligence Community Assessment. (2017). Assessing Russian Activities and Intentions in Recent US Elections. *Officer of the Director of National Intelligence. United States of America.* ICA- 2017–01D. https://www.dni.gov/files/documents/ICA_2017_01.pdf.
66. Mueller, R. (2019). Report On The Investigation In Russian Interference In The 2016 Presidential Election. U.S. *Department of Justice. Volume 1. CFT 600.8(c)* https://www.justice.gov/archives/sco/file/1373816/download.
67. https://www.reuters.com/article/us-usa-election-facebook-russia/duped-by-russia-freelancers-ensnared-in-disinformation-campaign-by-promise-of-easy-money-idUSKBN25T35E.
68. Knowledge Sourcing Intelligence. (2021).Global Data Broker Market Size, Share, Opportunities, COVID-19 Impact, And Trends By Data Type (Consumer Data, Business Data), By End-User Industry (BFSI, Retail, Automotive, Construction Others), And By Geography—Forecast From 2021. *Report KSI0601611226, Page 131.* https://www.knowledge-sourcing.com/report/global-data-broker-market.
69. Lewis, A. (2010). User-Driven Content. *Meta Filter Community Blog.* https://www.metafilter.com/95152/Userdriven-discontent#32560467.
70. Brathwaite, S. (2021). What Does a data broker do? *Security Made Simple.* https://www.securitymadesimple.org/cybersecurity-blog/what-does-a-data-broker-do.
71. Privacy Rights Clearninghouse. (2021). Data Brokers. https://privacyrights.org/data-brokers.
72. Facebook for Business. (2021). Help your ads find the people who will love your business. *Facebook.* https://www.facebook.com/business/ads/ad-targeting.
73. Twitter For Business. (2021). Twitter Ads Targeting. *Twitter.* https://business.twitter.com/en/advertising/targeting.html.
74. Facebook for Developers. (2021). Targeting Search, Behaviors. *Facebook.* https://developers.facebook.com/docs/marketing-api/audiences/reference/targeting-search#behaviors.
75. HUD Public Affairs. (2019). HUD charges Facebook over company's targeted advertising practices with housing discrimination. *U.S. Department of Housing and Urban Development Archives, HUD No. 19–035.* https://archives.hud.gov/news/2019/pr19-035.cfm.
76. Shane, S., & Goel, V. (2017). Fake Russian Facebook Accounts Bought $100,100 in Political Ads. *The New York Times.* https://www.nytimes.com/2017/09/06/technology/facebook-russian-political-ads.html.
77. Cox, J. (2020). Leaded documents expose the secretive market for your browsing data. *Vice Magazine.* https://www.vice.com/en/article/qjdkq7/avast-antivirus-sells-user-browsing-data-investigation.
78. Fullstory. (2021). The Definitive guide to session replay. *Fullstory Learning Center.* https://www.fullstory.com/resources/the-definitive-guide-to-session-replay.
79. Ropek, L. (2021). 'Anonymous' Mobile Advertising IDs Aren't So Anonymoust, And They Are Everywhere. *Gizmondo.* https://gizmodo.com/anonymous-mobile-advertising-ids-aren-t-so-anonymous-1847292216.
80. Cox, J. (2021). Inside the Industry That Unmasks People at Scale. *Vice Magazine.* https://www.vice.com/en/article/epnmvz/industry-unmasks-at-scale-maid-to-pii.
81. Kang, C., & Frenkel, S. (2018). Facebook Says Cambridge Analytica Harvested Data of Up to 87 Million Users. *The New York Times.* https://www.nytimes.com/2018/04/04/technology/mark-zuckerberg-testify-congress.html.
82. Wylie, C. (2018) In the Matter of Cambridge Analytica and Other Related Issues. *Written Statement to the United States Senate Committee on the Judiciary.* https://www.judiciary.senate.gov/imo/media/doc/05-16-18%20Wylie%20Testimony.pdf.

83. Twitter. (2021). Information Operations. *Twitter Transparency.* https://transparency.twitter.com/en/reports/information-operations.html.
84. Facebook. (2021). The State of Influence Operations 2017–2020. *Facebook Threat Report.* https://about.fb.com/wp-content/uploads/2021/05/IO-Threat-Report-May-20-2021.pdf.
85. Reddit Administrative Staff. (2019). Suspected Campaign from Russia on Reddit. *Reddit Security.* https://www.reddit.com/r/redditsecurity/comments/e74nml/suspected_campaign_from_russia_on_reddit/.
86. Huffman, S. (2020). Update to Our Content Policy. *Reddit Announcements.* https://www.reddit.com/r/announcements/comments/hi3oht/update_to_our_content_policy/.
87. Twitter. (2021). Permanent suspension of @realDonaldTrump. *Twitter.* https://blog.twitter.com/en_us/topics/company/2020/suspension.
88. Facebook. Taking Actions Against Misinformation Across Our Apps. *Facebook.* https://www.facebook.com/combating-misinfo.
89. Dwoskin, E., & Timberg, C. (2021). Misinformation Dropped Dramatically the week after Twitter banned Trump and some allies. *The Washington Post.* https://www.washingtonpost.com/technology/2021/01/16/misinformation-trump-twitter/.
90. Ribeiro, M. H. et al. (2021). Do Platform Migrations Compromise Content Moderations? Evidence from r/The Donald and r/Incels. *24*th *ACM Conference on Computer-Supported Cooperative Work and Social Computing* https://arxiv.org/pdf/2010.10397.pdf.
91. Chandrasekharan, E. et al. (2017). You Can't Stay Here: The Efficacy of Raddit's 2015 Ban Examined Through Hate Speech. *Proceedings of the ACM on Human-Computer Interaction,* Vol. 1, Article 31, pp. 1–22. https://doi.org/10.1145/3134666.
92. Aliapoulious, M. et al. (2021). An early look at the Parler online social network. *Proceedings of the International AAAI Conference on Web and Social Media*, Vol. 15, pp. 943–951. https://arxiv.org/pdf/2101.03820.pdf.
93. Dickson, E. J. (2021). The Qanon Community Is in Crisis—But On Telegram, It's Also Growing. *Rolling Stone Magazine.* https://www.rollingstone.com/culture/culture-news/qanon-telegram-channels-increase-1117869/.
94. Nickas, J., Isaac, M., & Frenkel, S. (2021). Millions Flock to Telegram and Signal as Fear Grows Over Big Tech. *The New York Times.* https://www.nytimes.com/2021/01/13/technology/telegram-signal-apps-big-tech.html.
95. Johnson, N. F. et al. (2019). Hidden resilience and adaptive dynamics of the global online hate ecology. *Nature, 573,* 261–265. https://www.nature.com/articles/s41586-019-1494-7.
96. Menn, J., Li, K., & Culliford, E. (2021). Parler partially reappears with support from Russian technology firm. *Reuters Internet News.* https://www.reuters.com/article/us-usa-trump-parler-russia/parler-partially-reappears-with-support-from-russian-technology-firm-idINKBN29N23N.
97. Holt, J., Brookie, G., & Brooking, E. (2021). Fast Thinking: How the Capitol riot was coordinated online. *The Atlantic Council.* https://www.atlanticcouncil.org/content-series/fastthinking/fast-thinking-how-the-capitol-riot-was-coordinated-online/.
98. Saldiva, G., Van Sant, S., & Bowman, E. (2018). Suspect Charged With 29 Federal Counts in Pittsburgh Synagogue Massacre. *National Public Radio.* https://www.npr.org/2018/10/27/661347236/multiple-casualties-in-shooting-near-pittsburgh-synagogue.
99. Greenburg, A. (2021). Far-right platform gab has been hacked—including private data. *Wired Magazine.* https://www.wired.com/story/gab-hack-data-breach-ddosecrets/.
100. Ali, S. et al. (2021). Understanding the Effect of Deplatforming on Social Networks. WebSci '12: 13th ACM Web Science Conference, pp. 187–195. https://seclab.bu.edu/people/gianluca/papers/deplatforming-websci2021.pdf.
101. Goodfellow, I. et al. (2014). Generative Adversarial Nets. Proceedings of the 27th *International Conference on Neural Information Processing Systems*, Vol. 2, pp. 2672–2680. https://arxiv.org/pdf/1406.2661.pdf.
102. Cole, S. (2017). AI-assisted fake porn is here. *Vice Magazine.* https://www.vice.com/en/article/gydydm/gal-gadot-fake-ai-porn.

103. Citron, D. (2019). The National Security Challenge of Artifical Intelligence, Manipulated Media, and "Deep Fakes." *Prepared Written Testimony and Statement for the Record. House Permanent Select Committee on Intelligence.* https://intelligence.house.gov/uploadedfiles/citron_testimony_for_house_committee_on_deep_fakes.pdf.
104. Rubio, M., & Warner, M. (2019). Rubio, warner express concern over growing threat posed by Deepfakes. *Joint Statement from Senators Marco Rubio and Mark Warner.* https://www.rubio.senate.gov/public/index.cfm/2019/10/rubio-warner-express-concern-over-growing-threat-posed-by-deepfakes.
105. Watts, C. (2019). The national security challenge of Artifical INTELLIGENCE, MANIPULATED MEDIA, and "Deep Fakes." *Prepared Written Testimony and Statement for the Record. House Permanent Select Committee on Intelligence.* https://intelligence.house.gov/uploadedfiles/clint_watts_-_house_select_committee_on_intelligence_-_ai__deep_fakes_-_13_june_2019.pdf.
106. Doermann, D. (2019). The national security challenge of Artifical intelligence, manipulated media, and "Deep Fakes." *Prepared Written Testimony and Statement for the Record. House Permanent Select Committee on Intelligence.* https://intelligence.house.gov/uploadedfiles/doermann-statement-final.pdf.
107. Clark, J. (2019). The national security challenge of Artifical intelligence, manipulated media, and "Deep Fakes." *Prepared Written Testimony and Statement for the Record. House Permanent Select Committee on Intelligence.* https://intelligence.house.gov/uploadedfiles/clark_deepfakes_sfr.pdf.
108. Deepfakes. (2019). FaceSwap Manifesto. *Github pages.* https://github.com/deepfakes/faceswap#manifesto.
109. Radek. (2018). Myfakeapp. *Bitbucket.* https://bitbucket.org/radeksissues/myfakeapp/src/master/.
110. Boyd. M. (2021). Deepfakes and LinkedIn: malign interference campaigns. *Malwarebytes Labs, Social Engineering.* https://blog.malwarebytes.com/social-engineering/2019/11/deepfakes-and-linkedin-malign-interference-campaigns/.
111. Collins, B. (2019). Facebook says a pro-Trump media outlet used artificial intelligence to create fake people and push conspiracies. *National Broadcasting Company News.* https://www.nbcnews.com/tech/tech-news/facebook-says-pro-trump-media-outlet-used-artificial-intelligence-create-n1105951.
112. Associated Press. (2021). Cheerleader's mom accused of making 'deepfakes' of rivals. The Associated Press https://apnews.com/article/pennsylvania-doylestown-cheerleading-0953a60ab3e3452b87753e81e0e77d7f.
113. Kerras, T., Laine, S., & Aila T. (2019). A style-based generator architecture for generative adversarial networks. *2019 IEEE/CVF Conference on Computer Vision and Pattern Recognition (CVPR).* https://arxiv.org/abs/1812.04948.
114. nVidia Labs. (2019). Style-Gan Repository. *Github.* https://github.com/NVlabs/stylegan.
115. Kerras, T. et al. (2020). Analyzing and improving the image quality of StyleGAN. *2019 IEEE/CVF Conference on Computer Vision and Pattern Recognition (CVPR).* https://arxiv.org/abs/1912.04958.
116. nVidia Labs. (2020). Style-Gan2 Repository. *Github.* https://github.com/NVlabs/stylegan2.
117. Zakhorav, E. et al. (2019) Few-shot adversarial learning of realistic neural talking head models. *2019 IEEE/CVF International Conference on Computer Vision (ICCV).* https://arxiv.org/pdf/1905.08233.pdf.
118. Gosse, C., & Burkell, J. (2020). Politics and porn: How news media characterizes problems presented by deepfakes. *Critical Studies in Media Communications,* 30(5). https://doi.org/10.1080/15295036.2020.1832697?journalCode=rcsm20#.YPrydo5KhPa.
119. Littell, J. (2019). Don't Believe Your Eyes (or Ears): The Weaponizations of artificial intelligence, machine learning, and Deepfakes. *War on the Rocks.* https://warontherocks.com/2019/10/dont-believe-your-eyes-or-ears-the-weaponization-of-artificial-intelligence-machine-learning-and-deepfakes/.

120. Weiss, M. (2019). The hero who betrayed his country. *The Atlantic*. https://www.theatlantic.com/international/archive/2019/06/estonia-russia-deniss-metsavas-spy/592417/.
121. Nemtsova, A. (2019). What to Do When the Russian Government Wants to Blackmail You. *The Atlantic*. https://www.theatlantic.com/international/archive/2019/03/russia-government-blackmail-kompromat/585850/.
122. Tung, L. (2016). Adobe's VoCo voice project: Now you really can put words in someone else's mouth. *ZDnet*. https://www.zdnet.com/article/adobes-voco-voice-project-now-you-really-can-put-words-in-someone-elses-mouth/.
123. Dessa Corporation. (2019). RealTalk: We Recreated Joe Rogan's Voice Using Artificial Intelligence. YouTube. https://youtu.be/DWK_iYBl8cA.
124. Kosoff, M. (2017). How Russia secretly orchestrated dozens of U.S. protests. *Vanity Fair*. https://www.vanityfair.com/news/2017/10/how-russia-secretly-orchestrated-dozens-of-us-protests.
125. Kawthekar, P., Rewari, R., & Bhooshan, S. (2017). Evaluating generative models for text generation. *Stanford University*. https://web.stanford.edu/class/archive/cs/cs224n/cs224n.1174/reports/2737434.pdf.
126. Brown, T. et al. (2020). Language models are few-shot learners. *Advances in Neural Information Processing Systems, 33*. https://arxiv.org/abs/2005.14165.
127. Anand, A. (2020). Another 10 gems of GPT-3. *Dev.to Blog*. https://dev.to/amananandrai/another-10-gems-of-gpt-3-2639.
128. Huntington, H. (2017). The affects and effect of internet memes: assessing perceptions and influence of online user-Generated Poltical discourse as media. *Colorado State University Department of Journalism and Media Communication*. https://mountainscholar.org/bitstream/handle/10217/183936/Huntington_colostate_0053A_14303.pdf.
129. Hutson, M. Robo-writers: the rise and risks of language generating AI. *Nature News Feature*. https://www.nature.com/articles/d41586-021-00530-0.
130. Xiao, H et al. (2015). Is feature selection secure against training data poisoning. *Proceedings in the 3nd International Conference on Machine Learning*. http://proceedings.mlr.press/v37/xiao15.pdf.
131. Xiao, H. et al. (2015). Support vector machines under adversarial label contamination. *Neurocomputing, 160*, 53–62. https://www.sec.in.tum.de/i20/publications/support-vector-machines-under-adversarial-label-contamination/file/main-revision.pdf.
132. Nguyen, T. et al. (2019). Deep learning for Deepfakes creation and detection: a survey. *Cornell University Arxiv*. https://arxiv.org/pdf/1909.11573.pdf.
133. Acemoglu, D., & Robinson, J. (2019). The narrow corridor. states, societies, and the fate of liberty. Penguin Random House Publishing.

Joe Littell enlisted in the Army in 2003 as an infantryman and attained the rank of Sergeant before commissioning in 2010. Upon commission, Major Littell has served as a Platoon Leader, Company Executive Officer, and Battalion Logistics Officer while assigned to the 83rd Chemical Battalion. As a 1LT, MAJ Littell applied for, assessed, and completed the Psychological Operations Qualification Course and served within the ARSOF community as a Tactical Detachment Commander and Company Commander with 9th PSYOP Battalion (Airborne). MAJ Littell currently serves as a research scientist at the Army Cyber Institute at West Point on the Information Warfare team working on computational propaganda, narrative warfare, radicalization, and microtargeting through publicly and commercially available data. He holds a BS in Computer Science from the University of South Florida and a MS in Data Science from Duke University.

The Cyber Pandemic that Could Redefine the Great Power Competition: Preparing the Defense Industrial Base

Adib Farhadi, Ian Galloway, and Ayman Bekdash

Abstract The Fourth Industrial Revolution (4IR) stands poised to transform the geopolitics and geoeconomics of the Great Power Competition (GPC) as digital and cyber worlds permeate societies, governments, and nation-states. The United States must reconceptualize its public–private approach to cybersecurity, starting with the Defense Industrial Base (DIB). As policymakers continue to grapple with the COVID-19 pandemic, the World Economic Forum (WEF) warns that "we should prepare for a COVID-like global cyber pandemic that will spread faster and further than a biological virus, with an equal or greater economic impact" (Davis & Pipkaite in What the COVID-19 pandemic teaches us about cybersecurity—And how to prepare for the inevitable global cyberattack. World Economic Forum, 2020, [3]). Despite this warning, the new cybersecurity regulations being rolled out under what is commonly known as Cybersecurity Maturity Model Certification (CMMC) are inadequate to protect the DIB from future cyber threats. Underpinned by the conceptual framework of stakeholder capitalism, this chapter posits that cybersecurity regulations must be reimagined to foster greater process and systems agility, transparency, and trust between the government and the private sector.

Keywords Great power competition · Cyber security · Defense industrial base

Introduction

Data and technology form the most critical foundations of today's digital era. Near-complete digitization of nearly every critical process and system means that cyber-disruptions can set off a chain of events that culminates in the failure of an entire system. The global community no longer enjoys the luxury of moving back to an analog era—we are too dependent on new technology, too dependent on data, and

A. Farhadi (✉)
University of South Florida, Tampa, FL 33620, USA
e-mail: farhadi@usf.edu

I. Galloway · A. Bekdash
DGC International, McLean, VA 22102, USA

A. Farhadi et al. (eds.), *The Great Power Competition Volume 3*,
https://doi.org/10.1007/978-3-031-04586-8_11

too dependent on the Internet of Things (IoT) driving the Fourth Industrial Revolution. Companies, specifically those in the Defense Industrial Base (DIB), that do not take their cybersecurity posture seriously will find themselves frozen out of trillions in government funding as well as struggle to compete with state-backed competitors in emerging revisionist nations. The result will profoundly impact geopolitics, geoeconomics, and the Great Power Competition (GPC).

Many businesses are racing into a global game of corporate and nation-state competition with outdated cyber-priorities, operating with partial cybersecurity solutions, traditional information security controls, and a slow-moving cyber-posture. Looking forward, entrepreneurs and other business leaders that seek to work with the Department of Defense (DOD) must ensure that cybersecurity is a core pillar of business operations. Businesses must change their approach to ensure that cybersecurity is not merely window dressing but a truly effective tool that improves the competitiveness of the DIB in the GPC. The threat landscape only grows larger, and most businesses remain unprepared for the multitude of threats.

Rapidly evolving cyber threats necessitate a timely shift in the U.S. national security paradigm—from a competitive state-oriented agenda to a cooperative public–private strategy that can better accommodate business and human security needs. A cyber-pandemic principally threatens U.S. national security in three key areas: (1) economic espionage (and tech transfer) of American corporations (business security), (2) the vulnerability and exploitation of personal data (human security), and (3) offensive cyberattacks on critical infrastructure (state security and business security, depending on whether the infrastructure is privately or publicly owned or both).

State-sponsored and state-aided cyberattacks definitively change the balance of power in favor of adversarial attackers over U.S. business' cyber-defenses. Chinese and Russian cyber-attacks were "targeting U.S. and military corporate networks as early as 2003….in a series of intrusions known as Titan Rain" [16]. Computer network exploitation endures as part of the Chinese and Russian intelligence toolkit, specifically targeting military technology developed by the United States at a high cost. This ability to "leap-frog" competitor businesses or militaries is the single biggest challenge facing the United States and key to understanding the future of the modern Great Power Competition.

Critically, any future solution to the cybersecurity challenge needs to be grounded in a framework that allows U.S. companies to defeat strategic state-capitalist actors by focusing on long-term, over-the-horizon threats. This chapter argues that significant policy changes will be needed to position the United States better to deter cyberattacks of a foreign state and non-state actors, including (1) greater cooperation with existing industry groups, (2) increased emphasis in identifying and investing in the non-traditional DIB, (3) changes to the CMMC Third-Party Assessor Organization, (C3PAO) auditing mechanism to make it more robust, (4) implementation of automated tools such as a cyber hygiene score, (5) public/private partnerships and forums for security and incident management, and (6) a cybersecurity infrastructure bill. Such reforms will result in better outcomes for both the federal government and businesses.

Underpinned by stakeholder capitalism, new fiduciary responsibilities must elevate cyber-governance requirements in both public and private institutions. A cybersecurity infrastructure bill, like the one passed for physical infrastructures such as roads and bridges, is not only necessary but critical to ensuring parity concerning cybersecurity posture for all DIB companies. Significant investments in the cyber-infrastructure of businesses, including hardware, software, and human talent, should be made in exchange for stronger fiduciary cyber-responsibilities and reporting requirements for public and private companies. A transformed corporate culture and private sector that emphasizes cyber-hygiene ensure that threats are taken seriously and build trust between consumers, corporations, and government. Public–private collaboration that invests in business cybersecurity first and foremost will position the United States uniquely within the modern GPC. As Robert Metzger noted: "Government and industry leaders must accept that the best of present defenses may only drive adversaries to aggression directed where defenses are weak or absent" [18]. "National Security" spending and defense budgets garner near-universal bi-partisan support, which makes the DIB a logical starting point for PPPs as a model for cyber-collaboration deployed within a framework of stakeholder capitalism.

Cybersecurity Maturity Model Certification (CMMC)

Cybersecurity breaches, including unauthorized access to networks, applications, and data, have caught the attention of governments worldwide and, as a result, are working on rolling out rules and standards intended to protect controlled unclassified information in public and classified procurements. Such rules and standards are being developed under the auspices of what is commonly known as CMMC, a unifying new certification model to ensure that cybersecurity contractors adequately protect sensitive information. This government-led approach will be insufficient and is too narrowly focused on meeting the needs of the DIB because the tools and certification processes will inhibit continuous system improvements, hinder agile development, and reduce transparency and trust between stakeholders.

Given the DIB's significant dependence on digital devices and information, inadequate cyber-posture is a significant threat to U.S. hegemony. Indeed, U.S. hegemony derives specifically through the interconnectedness of global markets and institutions that are most at risk from a future cyber-pandemic. After notable government breaches in 2015, most memorably, of the Office of Personnel Management and the Internal Revenue Service, which resulted in 21.5 million records stolen, the government issued mandates requiring cybersecurity contractors to protect CUI residing in non-Federal information systems and organizations [18]. Deadlines for CMMC rollout continue to be delayed or missed, with many businesses still non-compliant with the mandate [15]. This includes many companies that form the underlying foundation of the industrial innovation base, with foreign actors stealing large amounts of sensitive data, trade secrets, and intellectual property every day. Technology and data loss contribute to the erosion of the DIB and harm U.S. government interests with the GPC.

A unified framework is needed to combat adversarial attacks that would result in the eventual dethroning of the United States as a major contender in the GPC. To its credit, with CMMC 2.0, a revised and updated version of CMMC, the U.S. government is attempting to consolidate multiple cybersecurity and resilience frameworks proposed over the last several years to combat rising cyber insecurity. A recent study identified more than 25 research activities across 36 industries that attempted to clarify the disparate frameworks [23]. Furthermore, a simple search in Google Scholar brings up more than 10,000 results for the "cybersecurity maturity model" and around 12,000 hits for "cyber resilience maturity assessment" [23].

CMMC attempts to address the significant question of how to protect the business from cyberattacks; however, it does not achieve its aims (even the latest 2.0 version), because the framework fails to adequately promote transparency between the private and public sectors; it does not effectively incentivize continuous improvement in cyber hygiene practices; and because it will create further regulation that does not improve the U.S. security posture, particularly when compared to the speed and innovation of cyberattacks from adversarial state and non-state actors. This static framework creates significant issues for the DIB, especially when many small and medium-sized businesses that comprise it do not currently comply with the most basic Federal Acquisition Regulations (DFARS & FAR) and National Institutes of Standards and Technology (NIST) requirements that already exist.

Even if the DIB can retroactively catch up and meet these requirements, and even if the current CMMC model is successful in getting them there, the industry will inevitably be caught off-guard once again if premeditated action and foresight are not built back into the model. For example, CMMC control SC.3.177 states that companies must use Federal Information Processing Standards (FIPS) Encryption to secure their data. However, traditional forms of encryption will be rendered useless by Quantum Computing. In the current model, businesses are expected to meet a level of cyber hygiene. Once they achieve that level, there is no incentive for them to seriously re-evaluate their protections until they are required to recertify, giving executives an excuse for complacency and leaving their information technology systems vulnerable to attack.

As a result, CMMC is ill-designed to counter foreign-state actors executing synchronized strategies to close innovation gaps. The DIB must assume that foreign state-sponsored tools perform magnitudes better than even the best open-source tools available to U.S. corporations. Foreign state and private actors involved in cybercrime and espionage target the DIB "to close [the] capability gaps" between them and the United States [10]. The results of such attacks from state actors are astonishing, totaling billions of dollars a year in net losses.

Understanding the Traditional DIB and Why It Matters in a Cyber-Pandemic

Despite a nearly universal agreement about the importance of shielding critical services and assets from digital harm, governments have thus far had difficulty in accurately assessing where the most significant vulnerabilities lie [8]. While the Federal Government identifies key cybersecurity contracts to enforce CMMC before others, the USG has yet to identify critical businesses and industries to support hardening immediately comprehensively. To achieve stakeholder collaboration and elevate cybersecurity within a national security policy, we must understand how CMMC impacts the DIB and how the DIB of approximately 300,000 companies is viewed in government. Cybersecurity defines its industrial base into two broad categories, as noted in the figure below:

DoD's Characterization of the Defense Industrial Base

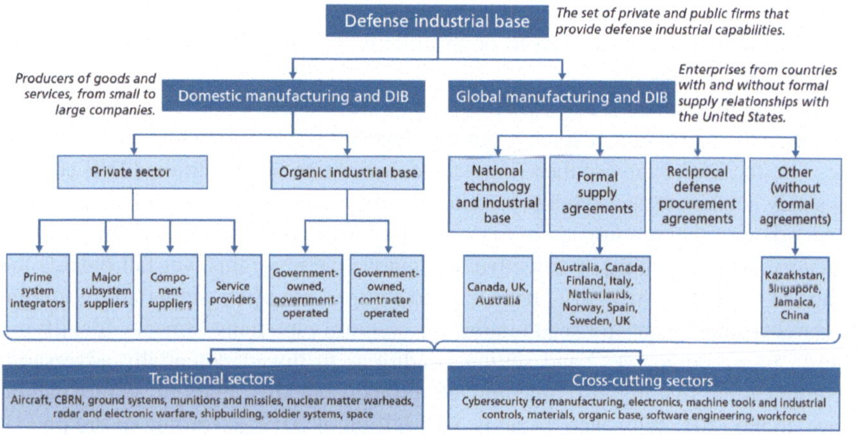

SOURCE: Derived from U.S. Department of Defense, 2018.
NOTE: CBRN = chemical, biological, radiological, and nuclear.

Aerospace and defense sector leaders such Lockheed Martin and Raytheon remain dominant players but make up only a tiny percentage of the overall DIB. Supply chain sensitivities have traditionally limited the fidelity of comprehensive industry data by company size. Small companies that participate indirectly in cybersecurity supply chains are particularly difficult to quantify [7]. The composition of corporate supply chains is considered proprietary by many prime contractors. Data from the North American Industry Classification System (NAICS) codes shows that over 99% of U.S. firms have a revenue of <$100 million. For this chapter, they are considered small or medium-sized companies. Many of these companies sell commercial products and services, obfuscating which companies are classified as part of the "aerospace and defense" sector.

Imbalanced applications of cyber policy also occur in the government sector outside of the defense and intelligence communities [2]. DIB sector leaders have well-established operational cyber defense programs, while less well-resourced small and mid-sized companies lag behind these benchmarks. The frequency of outsourcing and subcontracting in modern manufacturing further challenges cyber policy application. Airbus has 1676 publicly disclosed "tier one" suppliers and has over 12,000 "tier two and below" suppliers. General Motors has 856 and over 18,000 of each, respectively [14]. The "lower tier" businesses are significantly more likely to lack critical cybersecurity infrastructure, including a security information and event management (SIEM) system and the in-house cyber expertise with which to operate it.

Notwithstanding the cybersecurity improvements needed for the "traditional" DIB and sector leaders, investments and programs that target "non-traditional" DIB and lower-tier suppliers are imperative. As noted in Farhadi and Galloway, the operating space of the future GPC will likely no longer be waged with conventional weapon systems built by corporations in the "Aerospace and Defense" sector, urgently warranting the identification and integration of non-traditional and commercial technology companies into the DIB (2022, In Press). The scope of non-traditional DIB could also be expanded to include innovative commercial technology companies, including those identified through initiatives of the Defense Innovation Unit (DIU) in Silicon Valley.

CMMC practitioners and cybersecurity policymakers could coordinate efforts with the DIU and venture capital groups such as In-Q-Tel to bring innovative companies into the DIB. Many of these innovative, lower-tier, non-traditional DIB companies may need increased cyber protection immediately despite not holding a government contract that requires them to do so. Startups at universities also fall into this category, many working on effective and innovative programs, yet fall outside the scope of CMMC. University programs and startups are often supported by foreign researchers and hungry for seed funding and venture capital investments. Identifying these "future companies" and setting conditions to thwart potentially adversarial venture capital funding also preemptively protects the DIB. Foreign venture capital flows to the DIB sector were roughly $6.5 billion in 2019, according to Preqin [4]. While discussing adversarial capital deployed into the DIB and non-traditional DIB is critical, it is better addressed outside this article. The non-traditional DIB, including thousands of small and medium-sized companies, is critical suppliers of products and services to the traditional DIB. The lack of focus on critical "non-traditional DIB," including innovative early-stage companies, poses a threat to U.S. competitiveness within the future GPC.

CMMC Gaps and Challenges to Private Industry

"The DIB is the target of increasingly frequent and complex cyberattacks by adversaries and non-state actors. Dynamically enhancing DIB cybersecurity to meet these evolving threats and safeguarding the information that supports and enables our warfighters, is a top priority for the Department. CMMC is a key component of the Department's expansive DIB cybersecurity effort.

The CMMC program includes cyber protection standards for companies in the defense industrial base (DIB). By incorporating cybersecurity standards into acquisition programs, CMMC provides the Department assurance that contractors and subcontractors are meeting cybersecurity's cybersecurity requirements" [24].

The CMMC programs are admirable but grounded in a framework that may stymie collaboration rather than enhance it. Cyber-hygiene best practices should constantly evolve to meet new threats never remain static. However, CMMC, as it is currently structured, may lead companies to "wait and see" how their proposed security solutions will be interpreted by internal or external cyber-teams and assessed by C3PAOs (independent assessors that are certified by the government). In addition, CMMC does not resolve the many significant cybersecurity gaps that plague most American businesses, including the use of the agile methodology in IT systems and the lack of cost-effective security information and event management systems.

Security Information and Event Management Systems (SIEMs)

Security Information and Event Management Systems (SIEMs) are tools that allow a company to monitor all log data that flows in and out of its network. While large defense and technology companies leverage SIEM solutions, many small and medium-sized companies lack a comprehensive SIEM solution. These businesses may monitor segments of their networks, like sign-in logs or firewall notifications, but few centralize their log data.

CMMC requires elements of a SIEM solution to be implemented but does not directly account for the ancillary experts in cybersecurity and process owners required to manage one effectively. According to a recent 2020 RAND study, which reviewed the costs associated with CMMC: "Most small DIB firms may not be able to afford the cyber-defenses that the CMMC could mandate, and many medium-sized DIB firms may face the same challenges, especially if held to the highest compliance levels of the CMMC" [7].

Defense contractors should already be enforcing some SIEM and cybersecurity regimes; however, according to the National Defense Magazine, most are not enforcing cyber standards defined by the CMMC [12]. Specifically, of the 300,000 companies that comprise the Pentagon's supply chain, "about 290,000 of those have no cybersecurity requirements whatsoever" [12]. Typically, these companies are subcontractors to large businesses performing work in the lower tiers of a supply chain and cannot afford to meet the department's increasingly demanding cybersecurity requirements [13]. The costs of a SIEM and in-house experts in cybersecurity can be prohibitively expensive for small and medium-sized businesses. Based on 2021 data, the average salary of a cybersecurity specialist is $83,516, and the average cost of a comprehensive SIEM solution is over $600,000 per year [6]. These costs do not include CMMC certification and other big-ticket items, such as cybersecurity insurance. The high costs may: (1) deter many companies from true implementation, or (2) cause a number of innovative companies to forego working with the cybersecurity altogether,or (3) create a quasi-monopoly of secure large businesses, reducing innovation and competition within the DIB.

The Agile Methodology, Patching and Addressing Vulnerabilities

The DIB and CMMC practitioners should re-evaluate the benefits and drawbacks of the agile methodology in the context of a cyber-pandemic. The agile methodology relies on trust and collaboration to develop, deploy, and iterate solutions quickly and incrementally. CMMC policy itself would be well-served by implementing this methodology to govern its protocols. However, agile methodology also creates vulnerabilities in cybersecurity for businesses and technology companies that CMMC is not built to address. Many companies deploy products and systems on a model of "field it fast, fix it later," resulting in unfixed vulnerabilities in their software [8]. In 2013, hackers accessed a Microsoft database that contained descriptions of critical and unfixed vulnerabilities in its software, including the Windows operating system [17]. The unfixed vulnerabilities allowed hackers to develop tools (or weapons) to target the American business part of the WannaCry and NotPetya attacks in 2017 [21]. To achieve a more robust level of security and reduce threats to businesses in the digital age, governments will need to step in and hold digital service providers and the manufacturers of ICT technology accountable for ensuring their products maintain adequate safety standards [8]. Ultimately, the paradigm of "field it fast, fix it later," which continues to hold sway in the technology industry, plays a determining role in this accountability. Traditionally, slow corrective actions for patches also plague many large businesses. Languid patch implementation following a vulnerability disclosure has led to multiple hacks and cyber-defense failures. While CMMC has controls that attempt to address this problem, specifically RM 2.143, CM 5.074, and MA 2.111, they may not solve the problem systematically for reasons discussed later in this chapter.

The United States maintains a National Vulnerability Database that incorporates data from hundreds of organizations and countries. Current measurements indicate that the NVD's data feed over 30,000 unique organizations using API and web tools over two weeks. The NVD sees requests originating from nearly all countries, with industrialized countries making up the vast majority of NVD's users. Within the United States, the NVD observes users from nearly all sectors of critical infrastructure. The U.S. Department of Homeland Security's Cyber and Infrastructure Security Agency discloses vulnerabilities reported to it to the public by way of the U. S. Computer and Emergency Readiness Team within 45 days of the initial reporting, regardless of the existence or availability of patches or workarounds from affected vendors [8].

China has a similar vulnerability database system, but it operates twice as fast as its American counterpart, averaging just 13 days from vulnerability identification to public disclosure [8]. China proactively scours the web and other sources of information, looking for vulnerability information. In contrast, the United States waits for reports from vendors to be processed through the Common Vulnerabilities and Exposures database [26]. A lack of resourcing for robust open-source databases often results in companies building their databases or outsourcing to a SIEM at considerable cost. CMMC assessors and implementing companies would be well served by more accessible access to cyber-threat information all too often siloed in various public and private forums and subject to impassable hurdles. In addition, the voluntary program for sharing information on threats to cybersecurity poses challenges, given that not all DIB firms can access the service because it requires a cybersecurity Common Access Card (CAC). Many DIB firms may lack informal ties to the Intelligence Community, making them privy to important cyber-threat information (Gonzales). Open-source solutions are not a panacea and pose security challenges. Open-source code that is publicly editable is susceptible to actors with malicious intent. A company that uses open-source software would have to prove to an auditor that updates and patches are tightly controlled and do not contain code written by anonymous contributors. Given the risks and vulnerabilities, open-source databases and solutions still present an effective, low-cost solution to small businesses if supported through public–private networks.

CMMC and Resilient Supply Chains

The current digital ecosystem is underpinned by many components, subcomponents, standards bodies, equipment providers, chip manufacturers, mobile device providers, users and employees, governments, and other third parties. The digital ecosystem is subsequently undermined when companies do not control their hardware, software, and assembly processes. Computer chips and other IoT components are frequently not made in the United States, nor are many engineers who design and program them. CMMC does not explicitly address supply-chain risk management (SCRM) related to verifiable trusted hardware. A requirement in CMMC RM 4.148 states: "Develop and update as required, a plan for managing supply chain risks associated with the IT supply chain"; however, there are no direct requirements stated as to what an organization needs to mitigate against specifically. Great power can erode if state actors quickly close technological and scientific gaps without paying a high cost to achieve those gains, which allows the exploiter to "leapfrog" and advance.

CMMC Third Party Assessor Organization (C3PAO)

C3PAO stands for CMMC Third-Party Assessor Organization. C3PAOs, as currently envisioned, is intended to audit DIB companies' cybersecurity posture to ensure compliance with new cybersecurity regulations. Many questions of assessor organizations' role in the CMMC certification process remain unanswered when posed by industry groups, including: "[H]ow the [cybersecurity] and its auditors will handle the immediate influx of contractors requiring certifications; the specific criteria for determining the certification level necessary to perform a contract; how the department and its accreditation body will ensure consistency of third-party audits; and how it will address the impact on the commercial item and small business contractors, which ordinarily do not obtain significant cost recovery under reimbursable contracts with the government" [1]. Cybersecurity officials admitted that they "are not set up or resourced to do these certifications and audits of 300,000 companies. As it is, our $750 billion budget [the total public cybersecurity budget] does not cover all that we need to do. Government outsourcing of this critical certification process needs to be re-evaluated. So we needed to look outside" [12].

Utilizing private for-profit contractors to certify DIB companies creates a perverse incentive within the audit process. C3PAO's will theoretically have to meet a high-standards and legally attest to the veracity of their audits or face prosecution; however, the profit motives remain strong in a shareholder-driven economy. C3PAO's could ensure companies "pass" audits to lock in future revenue [audits]. Furthermore, CMMC audit cost is "determined by assessment model scope (level of certification sought), organizational scope and size, and complexity... with small organizations...less costly than...large manufacturers, where multiple assessors may be required, and analysis could span several weeks.

Because of this, there can be no standard pricing for CMMC assessments" [24]. This could lead to C3PAO being less likely to support small businesses to make better margins from large corporations with more complex system requirements. Small and medium-sized companies, especially those with little expertise in cybersecurity, also face the risk of being easily manipulated by "certified" consultants to buy products or services they do not need. Conflicts of interest abound when C3PAO's offer certification support, including technical expertise and IT, support for their customers to meet CMMC requirements. If the same assessor companies provide technical consulting services, they also provide auditing services. Such assessor companies will ensure certification to protect their "product." The profit motive in this instance creates a perverse incentive for the auditors to protect their new revenue stream. The lack of a unified standard to enforce opens the door for C3PAOs to bend the rules to ensure their paying contractor meets the requirements, even if the technology and security solutions do not meet standards or are substandard compared to another similar product.

Conceptual Changes to the CMMC Framework

Resilience and vulnerability disclosure requirements, a cybersecurity hygiene score, mandatory incident disclosure requirements, increased federal funding to re-tool the cyber posture of businesses, and revisions to the C3PAO audit model will better position the DIB to meet the challenges of a future cyber-pandemic. An expansion of public–private partnerships will be a critical tool in supporting the DIB and help build trust between stakeholders to improve the U.S. cybersecurity posture within GPC.

Resilience and Vulnerability Disclosure

The path towards effective cybersecurity for the DIB is a "new vulnerability disclosure process (and operational requirements); a duty to warn of imminent danger, such as in the case of an emerging attack; and a duty to assist in the case of cyber-emergencies" [9]. The DIB has come to rely on ICT companies who can implement a "new communications and warning system for urgent patches, adding 'emergency' to their repertoire of categories (emergency, critical, important, moderate, and low)" [8]. A recent search at the U.S. Consumer Product Safety Commission (CPSC) noted zero IT recalls for potential data loss hazards. Greater industry collaboration with the CPSC can drive "accountability by eliminating or significantly reducing after-market repairs (patch Tuesday) to a market that drives accountability through product recalls. Vendors could be required to deliver well-engineered products and services and present buyers with a list of the underlying components, libraries, and dependencies—a 'software bill of materials'—to drive transparency and accountability" [8]. Industry groups and regulatory bodies could revise their existing vulnerability disclosure models to adapt to the future cyber domain.

Targeted investments in key cybersecurity and technology companies will have exponential benefits to the DIB. Large technology companies have an outsized impact on the vulnerabilities of the DIB because most U.S. businesses utilize their services. Well-known hacker Jeff Moss notes that "maybe 20 companies around the world are in a position to do something to increase security and resilience for all of us" [25, p. 44]. A special auditing body focusing on the most critical cybersecurity infrastructure companies could be created as a subset of the newly proposed C3PAO framework. Organizational conflicts of interest (OCI) that emerge between critical cybersecurity companies and C3PAOs add further utility to a special assessor auditing body.

DIB Cybersecurity-Hygiene Score

The Common Vulnerability Scoring System (CVSS) developed by NIST/NVD offers the DIB a practical model that can be expanded and applied to CMMC in the form of a cyber-hygiene score. Owned and operated by a U.S.-based non-profit, CVSS uses an open framework for communicating the characteristics and severity of software vulnerabilities to the industry [20]. A transparent cyber-hygiene score specifically tailored for CMMC requirements could help differentiate companies that institute best practices and create systems visibility among lower-tier suppliers even if they are not required to be CMMC-certified [19]. DIB companies could input, as part of their required disclosures on the System for Award Management (SAM.GOV—a website database profile required to do business with the U.S. Government), the systems and software they utilize based on a set of government questions (e.g., What cloud storage system do you use? Do you utilize a SIEM, if so, which one?). Tailored methodologies would integrate with the user-provided data. They could account for how DIB companies leverage their internal cyber-operations in conjunction with technology solutions to ensure that the "managed," "reviewed," and "optimized" aspects of CMMC factor into the score. Assigning process owners and written explanations to how each technology is tied back to an operational action would influence the cyber hygiene system.

A clear, understandable cyber-hygiene score could provide additional data points to small businesses when selecting and deploying information technology systems. Understanding the CMMC-specific cybersecurity attributes of Microsoft's Azure from Amazon's Web Services, or what differentiates Enveil versus Vera for data packet security, will help drive better decision-making across the DIB. A scoring system that ranks companies qualitatively (process and systems) and quantitatively adds more value. As automation increases, scores could help predict what parts of a company's cybersecurity posture are most at risk (like how a credit score helps individuals identify complex financial data).

A public–private working group of cybersecurity experts who collaborate on behalf of the U.S. Government to rate, assess, and evaluate the top technology companies such as Amazon, Microsoft, Symantec, Okta, and Google is a logical first step. Data aggregation compiled from multiple large businesses could help inform the cybersecurity score algorithmically. The working group would regularly update the top software cyber score automatically recalculate scoring for companies who use their software. Focusing on large business' software first alleviates the burden on many small businesses and allows the government to provide product or software security alerts similar to a credit score. Furthermore, building these scores into the SAM database can inform C3PAO's who are re-certifying companies.

A transparent cyber-hygiene score would empower stakeholders, including public and private sectors, and hold companies accountable for their potential to introduce risk into the DIB. Under such a system, companies would gravitate toward measures that reduce their exposure to cyber-risks, such as their reliance on foreign entities or adversary-controlled supply chains. To enforce such a system, an industry association, non-profit, or university might be more attuned to key predictors of risk than a more generalist adjudicating entity. Companies will appreciate the clarity of knowing what they need to improve to increase their score (or keep their current score). The proposed risk assessing agency can develop open scoring algorithms by processing data related to previous and ongoing problems in cybersecurity. Such incidents could train algorithms to assess risks and forecast damages more accurately. Over time, these algorithms could harness positive track records...as well as negative experiences...failures to improve risk detection [19].

Through a cyber-hygiene score system, businesses and customers (such as cybersecurity) would collaborate to ensure the system is fair, open-source, and inclusive by involving a combination of voices from cybersecurity experts and data scientists. By combining various sources of information, transparency will enhance competition and innovation while also minimizing the likelihood of regulatory capture by larger firms looking to influence cybersecurity rulemaking. Looking at aggregate scores over time will also allow the government to see who is actively monitoring the environment and making improvements to stay secure. As the cybersecurity score becomes more refined, insurance premium costs for cyber theft, damage, etc., would be lowered as risks become quantified, measurable, and transparent. This approach is not novel. Many systems utilize risk-based algorithms to "score" companies like NAVEX for regulatory and legal compliance. Further, an impartial scoring system, more widely employed, may incentivize greater cooperative efforts across vital sectors in the spirit of great power cooperation.

Improving C3PAOs

Universities, academia, and select cybersecurity non-profits can add substantial value as C3PAOs are authorized to provide CMMC certification. Universities offer a ready group of technical experts and have a significant population of U.S.-citizen masters and doctoral students who can perform such audits. Leveraging academia would also expand the pool of professionals with a security clearance if the United States were to engage in a cyberwar with our adversaries.

Mandatory Disclosure and Public–Private Collaboration

Reporting cybersecurity incidents and vulnerabilities is an essential element of effective cybersecurity. Cybersecurity relies upon businesses to self-disclose breaches but does not mandate them. Except for some "contractors cleared to do classified work, cybersecurity neither mandates nor facilitates the use of automated systems of event monitoring" [18]. An automated SIEM is widely accepted as a key means by which sophisticated companies respond and recover from attacks. Too little of the U.S.

DIB utilizes a SIEM, primarily due to a lack of financial resources or technical expertise. An open-source public–private SIEM would greatly improve information transfer between government and industry, thereby expediting responses to threats and attacks.

Implementing technology is not the only consideration when achieving CMMC compliance. Day-to-day operational cybersecurity burdens require knowledgeable staff members and process owners to manage and maintain these technologies. To ameliorate the cost burdens, subsidies could be provided to entities such as the NDIA and AFCEA to assist with the rollout, training, and implementation of complex cybersecurity systems.

All-Small Mentor Protégé Programs (currently designed for large businesses to assist smaller businesses to grow in the federal marketplace) can be expanded to include CMMC and cybersecurity requirements. Mentor companies could receive tax credits, performance incentive fees, and credits towards meeting their set-aside subcontracting and spending targets, a stated goal of the U.S. government.

A cybersecurity infrastructure bill passed separately from the omnibus cybersecurity spending bill would cover all of the proposed systems and frameworks mentioned above and emphasize the importance of cybersecurity outside the DIB. Tax credits and small business administration loans ultimately forgiven like the COVID Paycheck Protection Program loans could be provided to companies who implement a SIEM and hire or retrain staff in cybersecurity systems and CMMC.

Conclusion

To out-compete state-sponsored competitors, the U.S. government and the DIB must align resources within a framework of stakeholder capitalism to mutually reinforce defensive cyber-measures with U.S. government oversight. The alignment must focus on radical openness and transparency while decreasing the profit motives so entrenched in shareholder economies. Building trust with and between public and private sectors in defense of business stakeholders is the only sustainable long-term policy because cybersecurity relies on collaboration and shoring up of the weakest links in an interconnected system. Individual citizens, businesses, and communities can no longer disconnect from the IoT to protect themselves from those who are untrustworthy. Conversely, the wider the circle of trust with and between "cyber stakeholders," including foreign corporations, citizens, and suppliers, will expand the traditional DIB and enhance our collective security relative to our adversaries. With that noted, given the prohibitive costs for a SIEM or DCP2 for most small and medium-sized companies, the only way to ensure that cybersecurity can rely on even the weakest links in the DIB is if the U.S. government provides significant financial and technical support.

Securing data and ensuring the methodology for doing so should be the most critical facet of CMMC 2.0. With the proper framework grounded in stakeholder capitalism and mutually beneficial enforcement protocols, creators, administrators, and users of the cyber-domain as a technological construct can supervise and repair it. The U.S. government and the DIB can develop open-source, agile groups that respond quickly to adversaries. Dialogue can be improved between private companies and government by introducing public–private partnerships that encourage information sharing and cooperation rather than discouraging it. "Voluntary event sharing [is still limited as]... many companies refrain from sharing sensitive data about their information systems and their security experience" [18].

Achieving cyber-resilient critical infrastructure for businesses poses significant challenges. Society relies on infrastructure and services extending beyond a specific organizational entity; yet, existing cybersecurity maturity models typically aim to assess a single organization [11]. Shaked et al. note that the United States must reimagine critical infrastructure and services and explore cyber-resilience as a system property [22]. This can only be achieved by developing a system that encourages cooperation, develops clear and consistent frameworks for enforcement, and emphasizes transparency concerning the systems that underpin the digital ecosystem.

Without specific initiatives such as the cybersecurity infrastructure bill, C3PAO reforms, and a cyber-hygiene score, many small and even medium and large-sized companies may become ineligible for cybersecurity contracts, thus reducing the number of potential vendors that can support the DIB. Those companies that meet the criteria at the expense of bottom-line earnings will do so reluctantly, without looking to improve appreciably; meaning that competitors who look past profit will beat us with great urgency. Absent the necessary shifts in cybersecurity policy towards PPPs, more cooperative oversight, and the prioritization of human and business security, the United States risks losing not only its competitive edge in the modern GPC but its most vital national business resources.

Special thanks to Joseph C Dorsey for his support and invaluable contributions to this article.

References

1. Burnette, R., Cassidy, S., & Clark, S. (2020). *Pentagon updating cybersecurity guidance, national defense*, 24 January, viewed 6 February 2020, https://www.nationaldefensemagazine.org/articles/2020/1/23/small-businesses-concerned-about-new-cybersecurity-certification.
2. Conti, G., & Fanelli, R. (2019). How could they not: Thinking like a state cyber threat actor. *The Cyber Defense Review,* 15 November, viewed 3 March 2020. https://cyberdefensereview.army.mil/Portals/6/Documents/CDR%20Journal%20Articles/Fall%202019/CDR%20V4N2-Fall%202019_CONTI-FANELLI.pdf?ver=2019-11-15-104103-203.
3. Davis, N., & Pipkaite, A. (2020). *What the COVID-19 pandemic teaches us about cybersecurity—And how to prepare for the inevitable global cyberattack.* World Economic Forum, 1 June, viewed 6 January 2021. https://www.weforum.org/agenda/2020/06/covid-19-pandemic-teaches-us-about-cybersecurity-cyberattack-cyber-pandemic-risk-virus/.
4. Fannin, R. (2020). *How the US-China trade war has starved some Silicon Valley start-ups,* CNBC, February, viewed 6 March 2022, https://www.cnbc.com/2020/01/31/chinese-venture-capitalists-draw-back-silicon-valley-investments.html.

5. Farhadi, A., & Galloway, I. (2022, In Press). Building trust and advancing U.S. Geoeconomic strength through public–private partnership stakeholder Capitalism. In *The great power competition volume 2: Contagion effect: Strategic competition in the COVID-19 Era*. Springer Press.
6. Foresight. (2021). *True cost of SIEM (Security Information and Event Management).* https://www.glassdoor.com/Salaries/cyber-security-specialist-salary-SRCH_KO0,25.htm, https://foresite.com/true-cost-siem-security-information-event-management/.
7. Gonzales, D., Harting, S., Adgie, M.K., Brackup, J., Polley, L., & Stanley, K. (2020). *A defense industrial base cyber protection program for unclassified defense networks,* Santa Monica, CA: RAND Corporation, viewed 3 January 2022, https://www.rand.org/pubs/research_reports/RR4227.html.
8. Hathaway, M. (2019). *Patching our digital future is unsustainable and dangerous*. Centre for International Governance Innovation, viewed 16 September 2021, https://www.cigionline.org/publications/patching-our-digital-future-unsustainable-and-dangerous/.
9. Hathaway, M., & Savage, J. (2012). Duties for Internet Service Providers. *Paper presented at Cyber Dialogue*. Canada Centre for Global Security Studies, Munk School of Global Affairs, University of Toronto, March.
10. Hallman, W. (2019). Cybersecurity: Front and center for industry. *National Defense Magazine Online,* 19 June, viewed 3 January 2021, https://www.nationaldefensemagazine.org/articles/2019/6/19/ndia-perspective-cybersecurity---front-and-center-for-industry.
11. Kott, A., & Linkov, I. (Eds.) (2019). *Cyber resilience of systems and networks*. Springer, Cham. https://doi.org/10.1007/978-3-319-77492-3
12. Lee, C. (2020). *Vital signs 2020: Small businesses concerned about new cybersecurity certification'*. National Defense, 23 January, viewed 6 March 2021, https://www.nationaldefensemagazine.org/articles/2020/1/23/small-businesses-concerned-about-new-cybersecurity-certification.
13. Lopez, T. (2019). DOD focuses on minimizing cyber threats to department, contractors. *DoD News*, December, viewed 9 July 2021, https://www.defense.gov/explore/story/Article/2034756/dod-will-help-small-companies-meet-cybersecurityrequirements/.
14. Lund, S., Manyika, J., Woetzel, J., Barriball, E., Krishnan, M., Alicke, K., Birshan, M., George, K., Smit, S., Swan, D., & Hutzler, K. (2020). *Risk, resilience, and rebalancing in global value chains*, McKinsey Global Institute, 6 August, viewed 27 April 2021, <https://www.mckinsey.com/business-functions/operations/our-insights/risk-resilienceand-rebalancing-in-global-value-chains.
15. Lynch, I. (2020). *Department of defense controlled unclassified information compliance: The impact on small business contractors*. Capitol Technology University, ProQuest Dissertations Publishing, 28000085.
16. Mattis, P., & Brazil, M. (2019). *Chinese communist espionage: An intelligence primer*. Naval Institute Press.
17. Menn, J. (2017). Exclusive: Microsoft responded quietly after detecting secret database hack in 2013. *Reuters,* 17 October, viewed 15 May 2020. www.reuters.com/article/us-microsoft-cyber-insight/exclusive-microsoft-responded-quietly-after-detecting-secret-database-hack-in-2013-idUSKBN1CM0D0.
18. Metzger, R. (2020). Cyber safety in the era of cyber warfare. *SciTech Lawyer, 16*(3), 8 February, viewed 15 March 2021. https://www.americanbar.org/groups/science_technology/publications/scitech_lawyer/2020/spring/cyber-safety-the-era-cyber-warfare/.
19. Norris, W., Balmain Rodgers, J., Blazek, C., Hewage, T., & Kobza, B. (2020). A market-oriented approach to supply chain security. *Security Challenges, 16*(4), 65–81. Institute for Regional Security. https://regionalsecurity.org.au/research-and-thought-leadership/security-challenges/.
20. National Vulnerability Database. (2021). *Vulnerability metrics*, viewed 3 March 2022, https://nvd.nist.gov/vuln-metrics/cvss.
21. Patel, A. (2017). Petya: "I Want To Believe". *F-Secure (blog)*, 29 June, viewed 8 December 2017, https://labsblog.f-secure.com/2017/06/29/petya-i-want-to-believe/.

22. Shaked, A., Tabansky, L., & Reich, Y. (2021). Incorporating systems thinking into a cyber resilience maturity model. *IEEE Engineering Management Review, 49*(2), 110–115, 1, Second quarter, June 2021, viewed 20 August 2021. https://doi.org/10.1109/EMR.2020.3046533.
23. Sharkov, G. (2020). Assessing the maturity of national cybersecurity and resilience. *Connections: The Quarterly Journal, 19*(4), pp. 5–24, viewed 5 January 2021, <https://www.researchgate.net/publication/348446458_Assessing_the_Maturity_of_National_Cybersecurity_and_Resilience>.
24. The Cybersecurity Maturity Model Certification Accreditation Body. (2021). CyberAB, viewed 8 November 2021, https://cmmcab.org/faq/.
25. Vos, A. (2020). Whose cyber is it anyways? A private perspective on a public good. *Atlantisch Perspectief, 44*(4), 41–45, viewed January 2021, https://www.jstor.org/stable/48600571.
26. Waterman, S. (2017). China's vulnerability disclosure system twice as fast as U.S. version. *CyberScoop*, 23 October, viewed 3 January 2021, www.cyberscoop.com/china-vulnerability-reporting-nvd-recorded-future/.

Dr. Adib Farhadi is an Assistant Professor, Faculty Director of Executive Education, and lead editor of the Great Power Competition book and conference series at the University of South Florida. His research is at the intersection of religion, politics, economics, and conflict, with a particular focus on the Central and South Asia (CASA) Region. He is the author of *Countering Violent Extremism by Winning Hearts and Minds* and a frequent presenter on the topics of Strategic Competition, Countering Violent Extremism (CVE), and Strategic Negotiations & Communication. Formerly, Dr. Farhadi served in senior positions for Afghanistan and has extensively advised the United States government and various other international organizations. Dr. Farhadi earned his Ph.d. in Economics from the University of Canberra, M.A. from New York University, and B.S. from East Carolina University.

Ian Galloway is a Principle and Management Executive in the Aerospace, Defense, and Energy sector with a keen interest in—and significant experience with—public affairs, international business, strategic consulting, and risk management.

At DGC International (DGCI), he specializes in developing highly innovative solutions that meet client needs in demanding, adverse, and contingency environments. Ian focuses on expanding current services; capturing new business in existing core mission areas; risk management; strategic communication; and corporate development. Recently, he helped to document and refine key supply chain, logistics and program man-agement processes and systems in support of government, military and international commercial markets. As a key member of DGCI's closely knit and highly collaborative Executive management Team, Mr. Galloway has contributed significantly to the achievements of DGCI.

He travels frequently to maintain his ties to international culture and politics—and his skills to promote the efforts of the Memorial Day Flowers Foundation and Zamani Foundation at home and abroad.

Ayman Bekdash serves as a board member of DGCI Corporation, a Virginia-based business that supports the U.S. Government. He was a significant contributor to the setup and expansion of DGCI operations in South America, Africa, the Middle East, and Central Asia. Prior to joining DGCI, he worked at Cardno Emerging Markets, an international development firm. Ayman proudly supports several charitable initiatives for military members and their families, as well as local and international charitable organizations supporting education. He has also published several articles and white papers on a variety of topics, both in the United States and abroad.

Mr. Bekdash holds a Bachelor of Science and two Bachelor of Arts degrees from the University of Rochester. He is also a former Fulbright fellow.

Global Perspectives: Cyber Alliances and Partnerships in Great Power Competition

Alexander Crowther

Abstract Recently Great Power Competition (GPC) has re-entered the category of concerns for the United States, our allies, and partners. One field of competition that involves the US together with our friends is in cyber operations. The barrier to entry is very low and no one has yet figured out how to deter cyber operations, so many players conduct a wide variety of cyber operations against us all. This chapter discusses GPC, how cyber operations fit into GPC, what the US and friends are doing about it, and what we can expect in the future.

Keywords Great power competition · Hybrid warfare · Cyber operation

Great Power Competition

Polities have always competed, whether tribes, cities, or states, because interests inevitably overlap. Although most of that competition is non-violent, violence occurs, not for the sake of violence itself, but because, as Clausewitz points out, war is politics by other means. That means that war is merely one technique to achieve political objectives because "war is thus an act of force to compel our enemy to do our will [1]." Opponents will face each other in competition, seeking to gain their political objectives, "overcoming" their enemy as Clausewitz says:

> If you want to overcome your enemy you must match your effort against his power of resistance, which can be expressed as the product of two inseparable factors, viz. *the total means at his disposal* and *the strength of his will.* [1]

This competition continues to this day. Although the United States enjoyed a unipolar moment after the Cold War, decades of counterterrorism campaigns around the world, in particular Iraq and Afghanistan have expended American lives and treasure, with a negative effect on what Clausewitz refers to above as the "means at his disposal" and "will" of the US.

A. Crowther (✉)
Florida International University, Miami, FL 33199, USA
e-mail: gcrowthe@fiu.edu

A. Farhadi et al. (eds.), *The Great Power Competition Volume 3*,
https://doi.org/10.1007/978-3-031-04586-8_12

Currently there are three major camps around the world: the US with our allies and partners, the autocratic revisionist states who seek to modify the current system for their own benefit together with their friends, and those who seek to stay out of the way and minimize damage to their countries, political systems and economies.

The current iteration of Great Power Competition has its' roots in the wake of the Cold War. China began their economic expansion and recovery from the Century of Humiliation during the rule of Deng Xiaoping between 1978 and 1989. Deng Xiaoping Thought led to Capitalism with Chinese Characteristics, where the Chinese Communist Party monopolized political power while allowing the Chinese people to adopt a more capitalist outlook and enrich themselves. The Soviet Union collapsed in the early 1990s and began to recover when Vladimir Putin took over in 2000. Although Putin manifested dislike for the US-led global system, he told the world that he would no longer support the world order at the Munich Security Conference in 2007. Their friends North Korea and Iran have faced off against the rest of the world since 1949 and 1979, respectively. They are joined by lesser lights such as Cuba and Venezuela. These states learned during the First Gulf War that direct, kinetic confrontation against the US, allies and partners was a good way to get hurt, so they began analytic efforts to figure out how to frustrate the US-led coalition and achieve their own political goals. This resulted in approaches such as Unrestricted Warfare [2] the Three Warfares [3], hybrid warfare [4], gray zone operations [5], and New Generation Warfare [6]. Unrestricted warfare was the Chinese analysis of how to confront US power in the wake of the First Gulf War which they followed with an approach to competing with the US called the "Three Warfares", consisting of psychological, media and law. They use the media to run information operations and the law (sometimes called "lawfare").

Hybrid Warfare, the Gray Zone and Political Warfare

In 2007, the American Dr. Frank Hoffman developed the concept of Hybrid Warfare. He was trying to codify changes that he saw with competition around the world [4]. Hoffman wrote that

> Hybrid threats incorporate a full range of different modes of warfare including conventional capabilities, irregular tactics and formations, terrorist acts including indiscriminate violence and coercion, and criminal disorder. Hybrid Wars can be conducted by both states and a variety of non-state actors. These multi-modal activities can be conducted by separate unites, or even by the same unit, but are generally operations and tactically directed and coordinated within the main battlespace to achieve synergistic effects in the physical and psychological dimensions of conflict. The effects can be gained at all levels of war. [4]

The Russians conceived of hybrid warfare started with Frank Hoffman's writing. They then looked at this through the lens of what they thought that the US was doing to them, in particular the "Color Revolutions". They then added significant amounts of information operations to the mix for two reasons. First, Russians have used information as a weapon for centuries. Second, they believe that the USSR collapsed due to a concerted information campaign against themselves and their allies and partners. The mix of these three concepts resulted in what could be called "Hybrid Warfare with Russian Characteristics".

At roughly the same time that Hoffman was conceptualizing hybrid warfare, security thinkers conceptualized operations in the "Gray Zone". Mazarr [5] credits Deputy Secretary of Defense Robert Work with mentioning the concept at an April 2015 U.S. Army War College conference. Mazarr goes on to identify those who are operating in the Gray Zone as "revisionist or dissatisfied powers appear to be in the market for options to transform the status quo". The Gray Zone can be conceptualized as the area below the threshold for "use of force" or "armed attack" as mentioned in the Charter of the United Nations (United Nations,1945. Available at https://www.un.org/en/about-us/un-charter). If an offending state crosses that threshold (generally accepted as damaging or destroying property or injuring or killing people) then the target state can use all elements of national power including armed force to defend themselves, as displayed in Fig. 1.

Hybrid Warfare and the Gray Zone join a variety of time-tested concepts that describe the current system. The best overarching concept that describes this approach to competition is Political Warfare.

Political Warfare is concisely described as "a form of strategy that leverages all of the diplomatic, informational, military, and economic capabilities at a nation's disposal to achieve its strategic objectives" [7]. Both the Soviet Union and the United States used political warfare as an organizing principle during the Cold War. The Soviets used political warfare from the very beginning of their regime. Dickey et al. [8] quote Stephen Blank as saying that

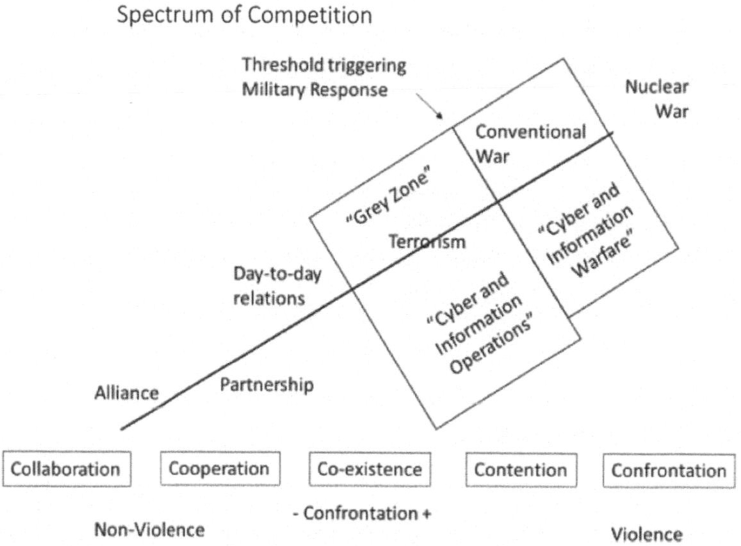

Fig. 1 The spectrum of competition and the grey zone

> The legendary tactical flexibility of the Soviet regime derives from their conceptualization of conflict as being waged on all fronts or across the board—whence the internal structure of the protagonists becomes the center of gravity. The Bolshevik vision of politics as another form of warfare endowed its practitioners with the maximum feasible number of instruments with which to wage their struggle even in the face of superior enemy military power.

The US diplomat George Kennan (who developed the concept of "containment" and described the approach in the "Long Telegram" of February 22, 1946. Kennan, 1946. Available at https://teachingamericanhistory.org/document/the-long-telegram/) codified how he thought the US should conceptualize political warfare.

Political warfare is the logical application of Clausewitz's doctrine in time of peace. In broadest definition, political warfare is the employment of "all the means at a nation's command, short of war, to achieve its national objectives. Such operations are both overt and covert. They range from such overt actions as political alliances, economic measures (as ERP [the Economic Recovery Plan, better known as the Marshall Plan]), and "white" propaganda to such covert operations as clandestine support of "friendly" foreign elements, "black" psychological warfare and even encouragement of underground resistance in hostile states" (Kennan, May 4, 1948. Available at https://www.law.upenn.edu/live/files/9964-kennan-memo-political-war farepdf).

The Russians and Chinese are using political warfare against the United States, NATO and the EU. Many individuals and organizations recognize this; however they lack a framework for response.

Both hybrid warfare and the gray zone are elements of political warfare. Political warfare describes the use of all elements of national power to achieve goals without entering into kinetic combat. Hybrid warfare is an organizational approach to political warfare that combines military and non-military as well as criminal approaches. The gray zone is where these operations occur on the spectrum of competition, above peaceful means of competition and below the threshold of armed attack which would allow the victim to respond with military force.

Those who did not benefit from the American moment bided their time while the United States was distracted. China grew economically, Russia recovered from their post-Cold War interregnum, and Iran and North Korea sought economic stability and to obtain nuclear weapons to avoid the fate of their fellow "Axis of Evil" member Iraq [9]. At the same time, different groups who saw themselves as too weak to directly confront the US sought to use terror as a weapon against what they saw as the "Global hegemon". This competition continues to this day.

Global Cyber Competition

The common thread that unites all approaches to counter the US, allies and partners is that they focus on both information and cyber operations. Each of those types of operations are easy to keep below the threshold of armed attack while also providing a significant effect. The US allies and partners all face a significant threat from cyber and information actors, in particular China, the Democratic People's Republic of North Korea (the DPRK or North Korea), Iran, and Russia. They have very few allies but a number of partners, which include Cuba, Nicaragua, Syria, Venezuela, and Yemen. They are united in their use of authoritarian governments and a dislike for the US approach to the global system. Because they are not united by any particular ideology or approach (as Communist countries were) they do not pose a major threat to the global system but their information and cyber operations provide a significant challenge since they target US and partners weaknesses.

China's grand strategic objective is regime survival. Their international goal is to replace the US as the global hegemon. Their supporting objectives include maintaining internal order, sustaining high economic growth, pacifying their periphery, and cementing their international status [10]. They use several techniques. Their main effort is to control internal information using the Great Firewall (which keeps out what they consider to be bad information) and the Great Cannon (which is a major cyber operation to shut down web sites that they find offensive). They practice the extensive pillaging of information from other states to integrate the stolen technology into their economy and skip over the normal intervening stages. They particularly target countries that have intellectual property such as the US as well as regional powers (e.g. India and Japan). Their last major approach is the subordination of both Chinese and foreign companies to Chinese objectives, as seen recently by the cancellation of the Ant Initial Public Offering (IPO) [11].

China also has a very different vision of what constitutes "Information" than does the US. As can be seen in Fig. 2, The Chinese sees the "Information Domain" as consisting of a variety of areas united.

One particular part of their approach concerns the network domain. The Chinese understand that their system prevents them from making decisions as fast as the US, particularly during a conflict. Since they cannot speed themselves up, they have decided that they only need to slow down US decision-making. Since so much of the US advantage is in communications systems, the Chinese seek to penetrate those networks and position themselves to manipulate or even deny those systems.

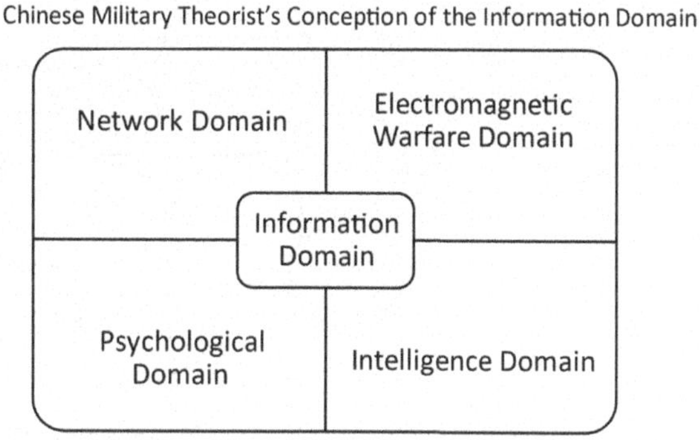

Fig. 2 The Chinese approach to the information domain (based on Sheldon & McReynolds [12])

The Russian grand strategic is regime survival. Their external objective is to disrupt the current global system that they believe is designed to keep Russia down. Their supporting objectives include being treated as an equal (especially by the US), keeping the money flowing to the oligarchs and the siloviki (former security personnel who essentially run the Russian government), fracturing NATO and EU, fracturing western democracies, and regaining their buffer zone in Eastern Europe or renewing their hegemonic control over what they call their "Near Abroad". As mentioned earlier, Vladimir Putin messaged his dislike of the international system as early as the 2007 Munich Security Conference [13].

They use cyber operations in support of both political warfare and conventional warfare [8]. Their operations have a significant information component, a cultural remnant of Soviet operations. They "recruit" (i.e. coerce) cyber criminals to work for the Russian Federation. They have a variety of competing intelligence organizations that do not cooperate with each other. It is an opaque system,Putin does not give explicit orders and everyone seeks Putin's approval. The end result is that some operations are rogue, some planned and ordered, some just happen. For instance, on May 10, 2021, the FBI confirmed that the Darkside ransomware was responsible for the compromise of the Colonial Pipeline networks [14] Was this a criminal operation, a cyber operation designed to curry favor with their Russian masters, or an operation ordered by the Russian Federation?

The Russians also have an integrated view of information as well as their own approach that they call "Information Confrontation" as displayed in Fig. 3.

Iran's grand strategic objective is the same as Russia and the other non-Chinese actors in this group: regime survival. They are Islamic, revolutionary and Iranian/Persian. Their supporting goals are to expand regional relations and bolstering

Fig. 3 The Russian perspective on information confrontation (Adapted from Giles [15])

the neighborhood policy, promote economic diplomacy to achieve economic development, increase global interaction, and counter US existential threats and hostile policies. They are much more limited than either Russia or China, especially regarding their economic and educational systems but are surprisingly sophisticated. Their main cyber operations are reconnaissance in both military and infrastructure command and control (C2) of antagonists.

North Korea is the sole Communist-Confucian hereditary dynasty and the world's most cultish, isolated, and nationalistic nation. They seek nuclear weapons, avoid international sanctions, and use cyber crime to generate resources. Due to extreme international sanctions regimes there are very limited in both their educational and economic capabilities so they literally give kindergarteners aptitude tests and start recruiting cyber talent that early, set on cyber track, and groomed very carefully because normal North Koreans do not have access to the internet.

The Chinese founded the Shanghai Cooperation Organization (SCO) on 15 June 2001 in Shanghai, China. China and Russia are the main partners, other founding members include Kazakhstan, Kyrgyzstan, Tajikistan, and Uzbekistan. India and Pakistan joined in 2017. Examining their stance on cyber issues provides some insight into how this group has provided a different framework to compete with the US vision for cyberspace. On 9 January 2015, the SCO published a leet to the UN General Assembly that addressed an"International code of conduct for information security" [16]. In it they used a variety of phrases that, when examined closely, lay out their conceptual approach to counter every US move in cyber space. They call for everyone to "comply with the UN Charter" and have "respect for the sovereignty…of all States (and) respect for the social systems of all countries". This means that, from their perspective, the US has to respect their desire to superimpose whatever political system on their people that they choose, including authoritarian systems. The document goes on to say that states should not use ICTs "to carry out activities which run counter to the task of maintaining international peace and security (or to) interfere in the internal affairs of other States or with the aim of undermining their political, economic and social stability." This means that the US should be prohibited from conducting any operations in cyberspace that they believe constitute influence operations, such as revelations of human rights abuses in Xinjiang or the suppression of democracy in Hong Kong in contravention of the 1996 Anglo-Chinese treaty that required a separate political system in Hong Kong for 50 years. The document also talks about "ensur(ing) ICT supply chain security, in order to prevent other States from exploiting their dominant position in ICT." This means that they can do whatever they feel necessary (including intellectual property theft) in order to build their ICT supply chain so that they can avoid purchasing US- and allied-built cyber systems.

The document also says that they "recognize individual rights, which may be subject to certain restrictions, but these shall only be such as are provided by law and are necessary for respect of the rights or reputations of others and the protection of national security or of public order". This is a nod to the UN Declaration of Human Rights, which China and Russia have signed in spite of the fact that they favor authoritarian systems that emphasize state control instead of citizen rights under a "rule by law" system where the legal code supports the state against the population versus the "rule of law" system championed by the US and exemplified in almost all allies and many partners. "This means that they respect human rights on paper, but they can use whatever coercive techniques that they desire against their people as long as it supports "the protection of national security or of public order" which is a very broad category indeed.

Finally, the SCO says that "all States must play the same role in international governance of the Internet" which is their code for supporting the multilateral stakeholder, which they feel that they influence to the detriment of the multistakeholder approach favored by the US.

All of these states and the SCO cite 'sovereignty' to control their populations, use cyber operations to disrupt the US, allies and partners, they tightly control internal information, they use information to gain international support for their goals, and they assist each other whenever possible. China and Russia seek to diminish US influence and make the world multipolar. China and the DPRK play US, Republic of Korea (South Korea or ROK) and Japan against each other while China, the DPRK and (to a lesser extent) Russia seek to keep the DPRK viable. Their major arena of competition is in cyberspace.

So why does the US need international cooperation in cyberspace? Why do they need to help allies and partners develop their cyber capabilities? The UN lists 193 states as well as the Vatican and Palestine as well as Taiwan (United Nations 2021. Available at https://www.un.org/en/about-us/member-states). As we see above, the list of miscreants is actually quite short. The United States, therefore, has relations with some 180 other states as well as other organizations. The way the internet was designed causes traffic to spread across the globe in small packets. There is no way to predict where the traffic will go unless you have developed a specific routing capability. Additionally, the current global system is designed to foster international cooperation in order to avoid another world war. The internet was designed and built by the US however, as it has grown and permeated all societies, the US has given way to international internet governance. This governance is typically either multilateral (states involved make decisions) or multistakeholder (all stakeholders, not just states, make decisions). Because the internet is so interwoven in the US, individuals, companies, and the government are extremely vulnerable. Because the US would like to continue to shape internet governance, they need to cooperate with others to gain a majority of votes in internet governance bodies. These two reasons combine with the aggressive nature of adversary and criminal cyber operations to motivate the US to cooperate internationally through both allies and partners.

The top tier cyber-oriented partnership that the US is a member of is the Five Eyes community, consisting of Australia, Canada, New Zealand, the UK and the US.

The most successful military alliance in history (so-called because they have managed what was hitherto impossible—to keep peace in Europe for over 70 years) is a 30-member organization made up of Canada, the US and 28 European states. This is a major platform for US security cooperation (NATO, 2021. Available at www.NATO.int).

The US also has 17 Major Non NATO Allies (MNNAs). Afghanistan, Argentina, Australia, Bahrain, Brazil, Egypt, Israel, Japan, Jordan, Korea, Kuwait, Morocco,

New Zealand, Pakistan, the Philippines, Thailand, and Tunisia. The State Department also states that "Taiwan shall be treated as an MNNA, without formal designation as such" (Department of State, 2021. Available at https://www.state.gov/major-non-nato-ally-status/).

There are also other organizations that provide the US a platform for cooperation. Some were designed and codified by treaty, such as the Organization of American States, while others are more ad-hoc organizations such as "The Quad" a recent creation that brings Australia, India, Japan and the US together to deal with China.

The US also cooperates with other organizations that the US not a member of, such as the European Union. Most members of the EU are also NATO members with the exception of Austria, Cyprus, Finland, Ireland, Malta, and Sweden (European Union, 2021. Available at https://european-union.europa.eu/principles-countries-history/country-profiles_en). The US also has bilateral or multilateral relations with several of those states. For instance, Secretary of Defense Jim Mattis signed a trilateral defense cooperation agreement with both Finland and Sweden in May 2018 [17].

Additionally, the US cooperates with a variety of other states around the world. Some are closer than others while not every relationship covers the entire spectrum of cooperation but focuses on individual topics of mutual interest. Examples include Chile, Colombia, India, Indonesia, and Mexico. It is of course impossible to list all relations that the US has with the 196 states.

Cyber Operations

As previously discussed, adversaries and criminals aggressively conduct cyber operations against the US, allies and partners. China conducts influence operations as well as stealing all information that they can get their hands on. They also routinely penetrate US and allied networks to understand how the networks function as well as putting themselves into position to be able to modify or shut down those networks in case of conflict.

Russia conducts influence operations such as the Internet Research Agency in 2016. They also conduct cyber operations against US allies and partners as part of their political warfare, such as the 2007 national denial of service operations against Estonia, operations in support of conventional military operations in Georgia in 2008, and operations against Ukraine in 2014–2015.

Russia and Iran both explore US networks, in particular industrial control system (ICT) networks to familiarize themselves with those networks and prepare to shut them down in case of conflict. North Korea performs mainly criminal operations in order to provide resources to their sanctions-starved economy. Periodically they will also conduct politically-driven cyber operations such as the Sony Hack of 2015.

Each type of operation performed by China, Russia, Iran and North Korea further motivate the US to cooperate with Allies and partners. What form does that cooperation take today, and what form should it take in the future?

US Cyber Cooperation

US cyber cooperation takes many forms but only has two major categories: offensive and defensive cyber operations. Actions include (but are not limited to) collaboration, cooperation, internet governance, and building partner capacity.

Cooperation in Offensive Cyber Operations

Offensive cyber operations (OCO) are the most controversial are of cooperation. The US collaborates with a few allies and cooperates with others. Collaboration is usually bilateral but can take place in a coalition atmosphere. Most OCO remains highly classified so beyond the remit of this paper. The best-known case of collaboration in a coalition environment is the Coalition against the Islamic State or DAESH. The global coalition against DAESH was formed in 2014 and consist of 83 different actors [18]. Because DAESH focused so much on internet-based influence operations, cyberspace was one major campaign area for the coalition. In an interview with National Public Radio. General Paul Nakasone, then-commander of Joint Task Force Ares and now Commander of United States Cyber Command, talked about Operations Glowing Symphony, the US portion of the cyber operations against DAESH (NPR. How The U.S. Hacked ISIS, September 26, 2019. Available at https://www.npr.org/2019/09/26/763545811/how-the-u-s-hacked-isis). Although the US did do unilateral operations against DAESH, they were in a coalition setting, much like US unilateral actions against Libya as part of NATO operations in 2011. When the coalition conceptualized effects that required OCO, they would talk to coalition members who have OCO capability. Those members would then communicate with each other and determine who had both the capability (derived from their force structure) and authorities (which are derived from domestic legislation). Whoever had both capability and authority would take the lead for the operation and others would conduct supporting operations. General Nakasone commented that Cyber Command and the NSA had used the coalition against DAESH as the model for creating the Russia Task Force to deal with Russian election tampering in the US (NPR. Task Force Takes On Russian Election Interference, August 14, 2019. Available at https://www.npr.org/2019/08/14/751048230/new-nsa-task-force-takes-on-russian-election-interference).

NATO has been a conundrum for US cyber operators for years. NATO allies are among the world's most connected societies while NATO itself has a modern command and control system that spans the globe, from communications to the two major strategic commands, Allied Command Operations in Mons, Belgium, and Allied Command Transformation in Norfolk Virginia to intermediate/operational commands. This network must be defended by NATO assets as national assets will be defending their own networks.

In addition to NATO networks, the alliance must deal with the wildly divergent cyber capabilities amongst the 30 allies. These capabilities run the gamut from the most sophisticated systems in the world to less well-developed systems in the newer and smaller allies. Although the very well-known article 5 of the NATO treaty requires mutual defense [19], the less well-known Article 3 requires each ally to invest in their own defense [19], which was included to prevent security free riding by allies. Article 3 does not mean that NATO cannot assist allies. Indeed, NATO has a cyber building partner capacity (BPC) program that seeks to complement national BPC efforts performed by individual allies.

NATO also has cyber partnerships that the US is a party to in their role as NATO allies. The Departments of Defense and State both work to ensure that these partnerships complement each other rather than duplicating efforts.

In the 2010 timeframe, NATO refused to conceptualize the use of OCO, emphasizing the defensive nature of the alliance. NATO thinking on cyber has steadily matured and now the alliance states that "We will enhance the effective integration of sovereign cyber effects, provided voluntarily by Allies, into collective defence and Alliance operations and missions, in the framework of strong political oversight." (NATO 2021). This means that they admit the utility of OCO in support of defensive operations, much like a conventional counterattack to clear alliance terrain might be necessary to defend the territory of a NATO ally. However, NATO itself will not perform OCO. Instead, NATO as an alliance will agree that OCO are required to support a defensive operation during the planning process. The NATO planners will then identify what effect in cyberspace that they desire. Then several the allies will decamp to the Sensitive Compartmented Information Facility (SCIF) next door and, as during counter Daesh operations, discuss who has the capability and the authority to perform operations required to achieve the effects desired by NATO planners. Once the willing NATO allies agree on who will conduct which OCO, they will return and inform NATO planners of what operations they will perform in support of NATO operations. Although this is a cumbersome process, it is a significant improvement over the days when NATO did even admit the utility of OCO.

The other type of OCO collaboration that the US could do would be to conduct these operations in support of an ally or partner that really needs the assistance. Although there are no open-source reports of the US conducting these types of operations, there are certainly occasions where partners would ask for assistance, like Ukraine in 2014/5. Any operation of this nature would require a request from an ally or partner, a recommendation from DOD to support it, and Presidential approval to conduct the operations. This structure is designed to ensure that these types of operations are extremely rare.

Cooperation in Defensive Cyber Operations

The vast majority of cyber collaboration and cooperation that the US does is defensive in nature. In general, the US allies and partners do not conduct many OCO. There are several reasons for this. First, the US, allies, and partners are among the most connected societies and economies in the world. Second, they generate large amounts of intellectual property that actors such as China seek to steal. Third, these societies tend to be richer than others, so criminals disproportionately target them. Fourth, freer societies tend to have a harder time with security operations than autocracies due to the non-coercive nature of our societies. Last, autocracies like Russia and China feel existentially threatened by our freedoms in general and democracy in particular. All these reasons make our societies targets to states, criminals, and other actors.

The US performs a variety of defensive cyber collaboration and cooperation with friendly states, other organizations (like the EU) and corporations. Each of the major US government organizations that perform cyber operations participates in their own way. The State Department is responsible for cyber diplomacy, including developing agreements with other states and organizations as well as coordinating ally and partner stances in regional and global digital fora such as the Group of Government Experts or the International Telecommunications Union. The Department of Defense works with other defense organizations around the world to collaborate, coordinate, and build partner capacity. The Department of Justice, particularly the Federal Bureau of Investigation, work with global, regional, and individual partners to assist efforts having to do with cybercrime. The Department of Homeland Security works with other non-defense and non-intelligence bodies such as Ministries of the Interior or Security as well as working with corporations that have a presence in the US. The intelligence community works with other intelligence bodies. Their efforts are guided by the Office of the Director of National Intelligence and cyber efforts are led by the NSA.

There are two major types of cooperation in defensive cyber operations. First is the actual collaboration in operations. Second is the coordination and discussing coordination. The third is building partner capacity (BPC).

There are several good examples of collaboration, either in a group or bilaterally. NATO serves a this function amongst allies and sometimes with partners. The US also collaborates bilaterally with several advanced allies such as Australia and Japan. This usually takes the either the form of organizations within the US collaborating on specific actions and programs, such as Indo-Pacific Command (Indo-Pacom) and the forementioned states or the form of bilateral collaboration between Washington DC and other capitols.

The United States coordinates with more states than it collaborates with. The State Department coordinates with fellow travelers when preparing for global for a like the International Telecommunications Union (ITU) or the Group of Government Experts (GGE). Defense coordinates with friendly Ministries of Defense, the FBI coordinates with global (e.g., International Criminal Police Organization or INTERPOL), regional police (e.g., the European Union Agency for Law Enforcement Cooperation or EUROPOL), or country-level law enforcement agencies. As an example, the FBI deployed to assist Ukraine in 2014 and 2015 when they were targeted by Russian cyber operations.

Discussing coordination is one of the main functions of several international organizations such as the ITU. The State Department works at the global and regional levels to discuss concepts and opportunities for future coordination. Once other organizations or states agree to coordinate with the U.S., they work with their counterparts in government.

BPC is the final action that the US does with allies and partners. Although there is a definitive list of who the US considers allies, the definition of 'partner' is less clear. Some like Sweden and Finland are very close partners while others are less close, and cooperation is on a case-by-case basis. BPC is a delicate mission because there is usually a power imbalance between those who are helping and those who are accepting assistance.

Performing a BPC mission requires quite a bit of homework on the part of the team that is assisting allies and partners. Before the team ever meets with the host nation (HN), they need to closely examine the cyber capacity, legislation, policies, and organizations of that HN. Document research includes looking at the national security strategy, the national cyber strategy, and the national telecommunications strategy at a minimum. The team should also be prepared to look both horizontally and vertically. Horizontal orientation involves looking at the whole-of-government and whole-of society approaches that states adopt. Whole of government refers to who are government players in cyber operations. Each country has a different approach. Some are very simple where one government agency leads the national effort. Others are much more complicated. For example, in the US DOD guard their own networks and does cyber operations on behalf of the state, DHS helps both the private and public sectors, DOJ handles cyber crime and the State Department handles cyber diplomacy. Whole-of-society is how the country approaches cybersecurity as a whole, not just the governmental efforts. The vertical approach looks at how a society handles cyber education from kindergarten through seniors, including grade school, high school, college, graduate schools, worker training, leader training and senior leader training.

Once a team has done their homework, they should meet with their "clients" or whoever they are trying to assist in building cyber capacity. Initial meetings must be handled very carefully due to the power imbalance mentioned earlier. The team should emphasize that they are not there to tell the HN what to do, but to help the HN think through their cyber situation and make decisions to improve their approach. Particularly, the team should make it very clear that the HN should not adopt the US cyber security or education systems, and that they are not trying to sell the HN anything. The US model is unique and has limitations imposed by national experience. The posse comitatus law from 1877 prevents the US military from operating within the US, for instance while most countries want to use their militaries (who are usually better resources) to assist with cybersecurity within the country. Another pair of foibles of the US system is the presence of six national cyber centers rather than one national center and a backup and the presence of 18 different organizations in the Intelligence Community. Other countries should not seek to mimic any of these approaches as they cannot sustain them.

The first several meetings with the HN should be strictly information gathering. Meetings should ideally be private and open and honest. Because each country has a unique approach, the team needs to understand what is different about this particular country. In one country, the author (who was helping to design a national cyber center) discovered that the Transportation Ministry owned telecommunications, because when the country initially ran telegraph lines the railroads had already cut byways through the mountains, so the state gave the railroads responsibility for the telegraphs, which they still maintain. In another case, the author discovered that the national Computer Emergency Response Team (CERT) is actually part of their national "Academic and Research Network" which is a national body for prevention of and protection from cyber threats to the security of public information systems…(whose)…main task is handling cyber security incidents to preserve cyber security in their country, except for those that fall under the state intelligence system or the military. In another case, the national intelligence organization had the authority to pay their cyber workforce more than the rest of the government, which had a significant impact on the ability of the rest of the government to recruit and retain cyber talent. Knowledge of the HN system early on is vital to the task of helping them to develop cyber capacity.

After discovering who is who, the team must figure out is the national system for cyber education. If high schools and universities do not produce enough properly trained people to do the mission, then the organizations themselves will have to both train and educate their workforce, a significant expense in people, money, and time.

Finally, the team must figure out what circumstances affect their partners in addition to the education system. In one country that the author worked with, the private sector offered higher wages and less strenuous work circumstances while transnational companies from other countries opened businesses in their capitol city and paid higher wages still. Other transnational companies paid even higher wages and moved locals to other countries to work. All of these provided significant pressure on public sector cyber organizations.

Once the team has figured out the local system, they must work with their local partners to work through their situation and develop recommendations for their leadership to approve (if internal) or champion (if external). This is an iterative system that will require several trips to the HN. Changes internal to the specific agency the team is assisting are the easiest. Recommendations should run the gamut from human resources to organization to operations. Usually, the hardest changes are those that require additional resources as they have budgetary implications, which require leadership to make difficult decisions.

External recommendations are much more difficult. Because cyber is a team sport, it is almost always insufficient for one agency to make enough changes to resolve the situation. Particular problems include clarification of roles and missions of involved agencies (which can involve both the executive and legislative branches), homogenization of hiring and pay authorities across organization (legislative), recommended changes to the education system (legislative and Ministry of Education), cooperation between public sector organizations (public organizations with cyber missions), cooperation between the public and private sectors, and the need for a larger budget to pay (legislative). Senior leaders from the organization that the team is assisting must become involved and be willing to spend political capital in order to influence other organizations as well as both the legislative and executive branches.

Recommendations

Russia or China don't really need to worry about allies or partners as they lack both and seek to dominate any relationship that they participate in. the United States, on the other hand, seeks allies and partners to achieve strength through common standards and purposes. Because new states become allies or partners regularly, there is a wide disparity in capabilities, particularly cyber. It is incumbent upon the US to build partner capacity, but within the rubric of those alliances and partnerships, a delicate task. The Unites States should continue to collaborate and coordinate with allies and partners and increase efforts to build partner cyber capacity. This would include providing more resources to the mission and spreading the mission among all government agencies that perform cyber missions on behalf of the US government, to wit Defense, Homeland Security, Justice and State.

This mission is possible but requires diplomacy to help figure out who to help and then a team that not only understands cyber operations and structure but is also able to talk to people from other countries with empathy and nuance. This endeavor will fail if technical experts head to other countries and just tell them what to do. Assistance teams should be very carefully vetted before being allowed to interact with allies and partners in an advisory capacity.

Conclusion

Cyber is a team sport and networks are only as strong as the weakest member, therefore it is incumbent on states that have more mature cyber structures such as the US, UK, Germany, and France are assisting their allies and partners to develop their own cyber capabilities. This is a difficult mission but an extremely important one that, properly executed, can reap significant benefits for the US, allies and partners.

References

1. von Clausewitz, C. (1984). On war (Peter Paret, Transl. & Ed.).

2. Liang, Q., & Xiangsui, W. (1999). *Unrestricted warfare*. Beijing: PLA Literature and Arts Publishing House.
3. Halper, S. (2013, May). *China: The three warfares. Office of net assessment*. Office of the Secretary of Defense. Washington, DC.
4. Hoffman, F. G. (2007). *Conflict in the 21st century: The rise of hybrid wars*. Potomac Press. https://www.potomacinstitute.org/images/stories/publications/potomac_hybridwar_0108.pdf.
5. Mazarr, M. J. (2015, Decembr) *Mastering the gray zone: Understanding a changing era of conflict*. Strategic Studies Institute. https://ssi.armywarcollege.edu/mastering-the-gray-zone-understanding-a-changing-era-of-conflict/.
6. Karber, P., & Thibeault, J. (2016, May 20). *Russia's new-generation warfare*. https://www.ausa.org/articles/russia%E2%80%99s-new-generation-warfare.
7. Douglas, A. (2018, January 12). Livermore. In *A primer on understanding the threat—Russian, Iranian, and Chinese political warfare*. Small Wars Journal. https://smallwarsjournal.com/jrnl/art/primer-understanding-threat-russian-iranian-and-chinese-political-warfare.
8. Dickey, J. V., Everett, T. B., Galvach, Z. M., Mesko, M. J., & Soltis, A. V. (2015, June). *Russian political warfare: Origin, evolution, and application*. Naval Postgraduate School. https://calhoun.nps.edu/handle/10945/45838?show=full.
9. US President George W. Bush declared during his 2002 State of the Union Address of Iran, Iraq and North Korea that "States like these, and their terrorist allies, constitute an axis of evil, arming to threaten the peace of the world". See the transcript of the State of the Union Address, January 29, 2002. https://georgewbush-whitehouse.archives.gov/stateoftheunion/2002/.
10. Blackwill, R. D. (2016, May 26). *China's strategy for Asia: Maximize power, replace America*. https://nationalinterest.org/feature/chinas-strategy-asia-maximize-power-replace-america-16359.
11. The Economist (2021, April 8). *China's rulers want more control of big tech*. https://www.economist.com/business/2021/04/08/chinas-rulers-want-more-control-of-big-tech.
12. Sheldon, R., & McReynolds, J. (2016). Civil-military integration and cybersecurity. Figure 8.2 in China and Cybersecurity Jon R. Lindsay and Tai Ming Cheung.
13. Putin, V. (2007, February 10). *Speech and the following discussion at the Munich Conference on Security Policy*. http://en.kremlin.ru/events/president/transcripts/24034.
14. FBI Statement on Network Disruption at Colonial Pipeline (2021, May 10). https://www.fbi.gov/news/pressrel/press-releases/fbi-statement-on-network-disruption-at-colonial-pipeline.
15. Giles, K. (2016). *Handbook of Russian Information Warfare*. NATO Defense College. https://www.ndc.nato.int/download/downloads.php?icode=506.
16. Shanghai Cooperation Organization. (2015, January 9). *International code of conduct for information security*. https://eucyberdirect.eu/content_knowledge_hu/2015-sco-international-code-of-conduct-for-state-behaviour-in-information-security/.
17. Defense.gov. *Secretary Mattis Hosts Finnish, Swedish, Ministers for Trilateral Meeting*.
18. Global Coalition against Daesh. https://theglobalcoalition.org/en/.
19. The North Atlantic Treaty. Washington D.C. (1949, April 4). https://www.nato.int/cps/en/natolive/official_texts_17120.htm.

Dr. Alexander Crowther is a visiting research professor with the Jack D. Gordon Institute for Public Policy and the Department of Criminology and Criminal Justice, both within the Steven J. Green School of International and Public Affairs at FIU. Dr. Crowther is also an integral part of FIU's Emerging Preeminent Program Cybersecurity@FIU. Prior to coming to FIU, Dr. Crowther worked on cyber and European issues at the National Defense University (NDU) in Washington, D.C. and continues to lecture on cybersecurity at the Organization of American States (OAS) as well cyber law/crime/terror at NDU's College of Information and Cyberspace (CIC). He has also worked with cyber issues at NATO and does Building Partner Capacity for the Institute for Security Governance.

Cyber Leadership in the Era of the Great Power Competition

Garrett Potts

Abstract In an effort to consider what the U.S. can do to mitigate threats to the relatively new battlefield of cyberspace, this chapter calls readers' attention to *how* critical infrastructure and software breaches often happen. Next, the chapter engages with questions regarding *who* can help us to learn from the breaches further. An answer is provided in light of the ideal type of the 'Cyber Leader'—defined herein as someone who demonstrates craft-expertise in the practice of cybersecurity to promote the internal goods of privacy and security. A partial account of *what* virtues cyber leaders require is also sketched before chapter conclusions are drawn.

Keywords Great power competition · Cyber leadership · Technical wisdom · Practical wisdom · Cyber security

Introduction

While previous centuries led us to mostly imagine warfare and disagreements in the form of engaged, hand-to-hand human conflict, the 21st Century beckons us to consider a relatively new dimension of the battlefield, which appears to be where malicious actors are making their greatest advancements right now—cyberspace. In recent years, we have seen a surge in the number of cyberattacks by state and (state-contracted but technically) non-state actors. This prompts essential questions about the evolving nature of confrontations between the Great Powers—China, Russia, and the United States.

Toward the end of addressing some of these critical questions, the argument within this chapter will proceed as follows. First, an account of *how* cyber breaches often happen is sketched. Second, an ideal type of the 'Cyber Leader,' *who* can help us learn from and respond to such breaches, is briefly outlined. Third, a shortlist is provided that points to *what* virtues sustain the cyber leader in practice. Because the stakes of technical or moral failure are high within cybersecurity, this chapter argues that

G. Potts (✉)
University of South Florida, Tampa, FL 33620, USA
e-mail: garrettpotts@usf.edu

A. Farhadi et al. (eds.), *The Great Power Competition Volume 3*,
https://doi.org/10.1007/978-3-031-04586-8_13

cyber leaders must exhibit technical expertise and practical and theoretical wisdom to protect and extend the internal goods of privacy and security. To do so, I argue that cyber leaders' moral character is also of the utmost importance. Cyber leaders must be resilient in their resistance to various temptations and harms that might otherwise inhibit the flourishing of those impacted by their service. With this roadmap in mind, let us turn first to the question of *how* cyber breaches often happen.

How Critical Infrastructure and Software Breaches Often Happen

Current events around the time this chapter was written are symbolic of the critical infrastructure threats faced by the U.S. at home and in other key places abroad. Today, more than ever before, some of the most pressing privacy and security vulnerabilities appear to be centered around the imprudent handling or lack of secure remote access and identity management technology. In particular, legacy remote access tools and an absence of identity management tools often create vulnerabilities that result in the breach of critical infrastructure controls and servers storing confidential (and often highly classified) data belonging to the U.S. or its allies.

For example, in May of 2021, the U.S. experienced ransomware attacks carried out by Russian criminals who extorted the Colonial Pipeline, which runs from Texas to New York and carries up to 3 million barrels of fuel per day. They did so by installing malware that compromised the online equipment which operates the pipeline. Within a matter of hours, and before the pipeline could return to its regular operations, the FBI had to pay a hefty sum in cryptocurrency to receive a software application that restored network operations (albeit at deliberately slower than desirable networking speeds). The pipeline ceased operations for nearly six full days as cyber leaders in the public and private sector worked together to investigate precisely what had happened and prevent future breaches like this one from retaking place within the Energy sector. In the wake of this event, gasoline shortages were felt for up to a week or more in some areas. The supply chain hurried to catch up, and truck drivers extended their already long days to expedite the delivery of crude oil and refined petroleum to the hardest-hit metropolitan areas. It is difficult to grasp how greatly a vast number of daily life practices in America depend on this pipeline and other critical infrastructure operations to be up and running (i.e., the energy grid, water treatment plants, etc.). That is until any given operation shuts down and the distribution of its associated goods comes to a screeching halt. When disruptions like this happen in the wake of a cyber breach, ripple effects can be dire. For cyber leaders who were paying close attention, this particular ransomware attack sounded the alarm and increased their vigilance.

Events like the Colonial Pipeline incident should cause us to think about the nature of current and future (potentially more prolonged and detrimental) infrastructure breaches that the U.S. will likely face. In the wake of such instances, we must remember that there is no better way to prevent them from happening again than to increase our vigilance around how future attacks on the cyber battlefield could potentially unfold. Simply paying hackers a ransom fee to restore operations may be necessary for the short term, but this makes for an inferior long-term strategy. Adequate prevention measures and additional security and privacy protections must be considered via vigorous deliberations that must persist between cyber leaders working together across the public and private sectors.

As we reflect and seek to learn from cyberattacks, it is helpful to look backward before looking forward to answer *how* breaches to critical infrastructure tend to happen. In so doing, one discovers that many U.S. infrastructure operations were initially built as what former CIA cybersecurity expert, Eric Cole, refers to as closed-loop systems, meaning that they were never intended to be placed online ([4], p. 38). Instead, they were deliberately kept offline, with infrastructure controls and operations protected behind barbed wire and armed guards in many cases (Ibid., p. 177). This was not done out of convenience but instead to heighten security. Careful consideration was given to the known vulnerabilities of legacy remote access platforms that could become a weapon in the hands of U.S. foreign enemies and domestic violent extremists.

In light of Cole's [4] analysis, let us briefly consider another textbook case example that is emblematic of how critical infrastructure breaches tend to happen, which I will return to for an extended discussion later in the chapter. This one occurred at the water treatment plant in Pinellas, FL. Much like the Colonial Pipeline breach, a malicious actor gained remote access to the county's water facility, this time, however, in a failed effort to poison the supply chain. Chemical concentrations were altered within the plant's central control system, forcing operators to manually and quickly shut off the water supply. In this case, infrastructure officials lacked identity access management technology, instead opting to share a commonly known password amongst the plant operators to remotely access critical controls. Furthermore, the remote access tool that they used was a legacy product with known security vulnerabilities that did not meet the latest compliance standards. Together, these failures resulted in a successful breach of the system.

A fundamental problem in this discussion about *how* breaches tend to happen is that remote work has become more normalized (particularly in the wake of the COVID-19 pandemic). Consequently, carrying out critical infrastructure projects online has become more customary while the earlier security concerns around legacy technology have at the same time been taken for granted in many cases. Despite secure remote access tools being available in the marketplace, many organizations have settled for the same outmoded technology deemed unsafe by earlier public and private U.S. critical infrastructure officials. Often, as the cases above indicate, this happens without proper identity or privileged access management technology in place, which unfortunately creates the perfect storm. Now, I should be clear that remote access technology is here to stay, which is wonderful in many cases. Still, public and private U.S. organizations can no longer settle for the cheap and less secure tools that have a repeated history of being hacked. This is especially significant when considering the remote access technology implemented to run our nation's most critical infrastructure operations. Thus, I argue that, when it comes to things like identity access management solutions and securing remote access technology with privileged access management, the U.S. must work together with cyber leaders, sparing no expense to protect and secure its critical infrastructure operations. Remote work conducted without attentiveness to the latest privacy and security standards presents a dangerous threat to flourishing organizations and human lives.

Attacks on critical infrastructure operations are one of two important themes that I wish to highlight within this section about *how* breaches tend to happen. The second theme has to do with how similar patterns of remotely exploiting vulnerabilities in widely-used software tools can and has contributed to breaches of privileged—and in some cases, highly classified–data belonging to the U.S. or its allies in the public and private sectors.

To cite just one example, consider the news that broke in March of 2021, about two months before the Colonial Pipeline incident made national headlines. Since January of that same year, data revealed that the Chinese government had been carrying out attacks for two full months by successfully exploiting a vulnerability in Microsoft Exchange's email servers [5]. This vulnerability allowed Chinese state-sponsored hackers, operating as "Hafnium," to "insert backdoors into systems which they could return to later" (Ibid.). They remotely accessed enterprise networks by posing as internal clients (the clients whose emails had been hacked) to gain control of systems and then pursue privileged escalations to access increasingly sensitive data that was thought to be protected under virtual lock and key.

Prominent targets initially included universities, think tanks, and defense contractors, revealing that Chinese motives principally entailed both the theft of intellectual property as well as the acquisition of any classified information that they felt could potentially weaken ongoing U.S. operations at home or abroad. As if this were not bad enough, the continued backdoor access that Hafnium had to Microsoft Exchange servers caused China to execute what became known as "a massive smash-and-grab raid," where Hafnium shared the exploited vulnerability with other Chinese hackers to advance national interests and intelligence in as many industries as possible before the backdoor was closed. At the same time, Microsoft scrambled to develop a security patch (Ibid.). In the end, this hacking operation affected at least 30,000 organizations worldwide, resulting in the theft of intellectual property and the exposure of many national governments' classified data.

Such instances—particularly the Colonial Pipeline ransomware attack and the Hafnium hack-job—reveal that Russia and China appear to have very few reservations about playing dirty on the cyber battlefield. By way of successful cyberattacks, Russia deliberately attempts to destabilize the practices of American life, while China continues to weaken our innovative advantages in the global arena. Russia appears to prefer threatening America's democratic project by sowing seeds of confusion and pandemonium (via disinformation as well as persistent attempts to breach critical infrastructure). At the same time, China acts as a thief in the night, stealing any information that will increase its industrial prowess while also trying to negatively impact ongoing critical infrastructure operations. In this way, China, too, seeks to threaten America's democratic project via cyber breaches that expand its autocratic knowledge, power, and influence.

Russia and China's deliberately reasoned approaches for threatening democracy in America might differ. In the case of both of these Great Powers, however, the nature of *how* they and domestic violent extremists breach our cyber systems tends to be quite similar, which this section of the chapter briefly attempted to show. Consistently, remote access technology is utilized to gain a foothold in places where hackers do not belong. Then, without adequate identity and privileged access management solutions in place, escalation privileges are sought to access the most sensitive data or the most critical of infrastructure operations. Keeping this brief narrative about *how* cyber breaches tend to happen in mind as we look ahead is essential. We will need to consider it again as we reflect on the nature and responsibilities of cyber leaders and the virtues necessary for them to protect the interests of the U.S., its citizens, and its allies in the months and years ahead.

Who is a Cyber Leader?

Now that we have considered *how* some of the most significant threats associated with cyber warfare tend to happen, this section grapples with questions regarding *who* can help us to learn from the breaches further. I argue that this responsibility falls under the 'cyber leader,' who serves within the practice of cybersecurity. *A 'cyber leader' is someone who demonstrates craft-expertise in the practice of cybersecurity to promote its internal goods. This leader views work as a service first and exhibits virtues that sustain the institution within which the work takes place, the work itself, as well as key relationships of practice.*

In what follows, I will outline what it means for the cyber leader to submit to a prosocial practice that promotes human flourishing and the common good. Understood as a practice, cybersecurity has a tradition-based history and internal goods that are unique. The primary functions, or *teloi*, of cybersecurity, are the internal goods of privacy and security. Following the philosopher MacIntyre [10], this chapter maintains that standards of excellence associated with promoting and delivering these internal goods must continually be refined in light of the ever-evolving nature of cyberspace. As this field changes, so the cyber leader also must adapt to those changes. However, one's work as a cyber leader must always serve the internal goods of privacy and security *first*. Only individuals who remain principally devoted to the goods internal to cybersecurity may genuinely be called 'cyber leaders,' whose work itself is a service to human flourishing and the common good.

Notably, this ideal type of the 'cyber leader' aligns with Robert Greenleaf's (2002) widely-praised account of the 'servant leader' who sees oneself as "a servant *first*" (2002, p. 27). It is "this conscious choice [to serve] which brings one to aspire to lead," and so the cyber leader necessarily views oneself as a leader *second* (Ibid.). As Greenleaf states about all servant leaders, their positive "difference manifests itself in the care taken to make sure that other people's highest priority needs are being met" (Ibid., p. 27). Indeed, cyber leaders certainly can and do ensure that individuals highest priority needs are being met. This happens by extension of the effects that organically ensue via their prioritization of privacy and security in cyberspace.

By promoting the internal good of privacy, I mean to say that the practice of cybersecurity exists to prevent the unethical or dangerous publicization of sensitive or personally owned data. The nature of private data itself, however, is multifaceted. For instance, private data could correspond to personal health records, personal legal documents, forms of identification, patented product information, service blueprints, classified governmental documentation, private customer data and passwords, etc.

Regarding the internal good of security, this too is multifaceted. For example, consider how cyber leaders protect individual and institutional data from being compromised and how their work to engineer and continuously test excellent security tools ensures that many of our basic needs as Americans are met without choking the supply chain or poisoning our natural resources. Today we have become so dependent upon remote access technology that, without cyber leaders, there would be no way to effectively secure remote and privileged access to critical infrastructures such as oil and gas, the energy grid, and water treatment plants, all of which are necessary to meet many of the most basic needs required for a great number of practices in American life to remain ongoing.

Now, it is relatively easy for cyber leaders to become distracted by other pressures that cause them to prioritize something in place of the goods internal to cybersecurity. As Robert Greenleaf says about all leaders, they are likely to face temptations involving giving themselves over to "the need to assuage an unusual power drive or to acquire material possessions" (Ibid., p. 27). Furthermore, leaders may have to make challenging decisions that promote internal goods but endanger their political standing, such as acknowledging that a breach has happened, much to the disdain of their superiors. When these temptations or challenges intoxicate or distract individuals, what often happens is that they exhibit disordered priorities or morally incoherent patterns of self-protection, which deter them from prioritizing the genuine ends of cybersecurity in all relevant cases and contexts. Something like this has happened in instances such as the ones mentioned in the introduction, where Chinese and Russian hackers exercised their technical skills for malicious reasons, prioritizing 'external goods' over 'internal goods.' (i.e., profit in the form of ransom and power in the form of information acquisition). That the hackers mentioned above did so without regard for the goods internal to cybersecurity and for self-gain rather than the common good and human flourishing inherently disqualifies them as cyber leaders by my definition.

Moreover, as is the case with all practice-based work, per MacIntyre [10], it is imperative that cyber leaders situate themselves within a community that engages in *vigilant deliberation* about how to continuously refine standards of excellence associated with the practice of cybersecurity. For example, democratic deliberation should remain ongoing around engineering or reengineering products that meet the latest privacy and security standards in the public and private sectors. Cyber leaders who engage in any aspect of the practice (i.e., customer support, engineering, consulting, etc.) further need to be aware of the latest compliance standards (i.e., HIPPA, PCI,

GDPR, AICPA, FERPA, NERC, etc.) so that they can better serve the needs of their users in industries as diverse as healthcare, retail, finance, education, and critical infrastructure. With so much hanging in the balance when it comes to ransomware and cybercriminal attacks, technical excellence in the practice of cybersecurity becomes a moral responsibility, and it is one that cannot be genuinely fulfilled without sustained deliberation and collaboration between practitioners who are mutually engaged in the work for the right reasons, or ends.

Now that we have clarified *who* the cyber leader is and how that impacts this individual's orientation to work, I next emphasize *what* virtues cyber leaders must exhibit in their day-to-day operations. The following virtues, so I argue, will be necessary for cyber leaders to sustain the institution within which their work takes place, the ends of the work itself, as well as key relationships of practice.

What Virtues are Required of the Cyber Leader?

Turning now to a shortlist of some of the virtues required of our ideal 'cyber leader,' we will find that, as the philosopher MacIntyre [10] says, the virtues are *what* "sustain" practices and practitioners of cybersecurity (p. 129). The virtues do so by enhancing the relationships required for healthy organizations that promote vigilant deliberation about the ends of the work that must be carried out. Furthermore, the virtues are formed in communities of practice which possess extended narratives about what is worth wanting. Through this formation, cyber leaders come to exhibit strengths of character and intellect, which will be necessary to both desire and achieve the goods internal to their practice—in this case, privacy and security (Ibid.). Without the virtues, so MacIntyre and other MacIntyrean business ethicists argue, one would not be able to resist the "various particular harms, dangers, temptations, and distractions which provides every quest with its episodes and incidents" throughout one's life (Ibid.). So, to summarize, the virtues "furnish us with increasing self-knowledge and increasing knowledge of the good," which correctly forms and even transforms our desires and intentions in light of what promotes the common good and human flourishing across all of our given social roles (Ibid.). The virtues, therefore, are necessary to maintain a commitment to excellence in one's work as well as in one's life more broadly.

Given that it is my primary concern in this chapter to understand a virtue approach to cyber leadership in what has become known as "the era of the Great Power Competition," I will limit my discussion of the applicability of the virtues to this particular workplace context ([7], pp. 31–53). To better contextualize this discussion, it should be emphasized again that the virtue approach outlined herein is informed by the ongoing dialogue I have just alluded to within MacIntyrean business ethics circles about the relationship of virtue to business, leadership, and management. This section is framed in light of that conversation. As such, it takes a Neo-Aristotelian approach to consider the intellectual virtues of technical, practical, and theoretical wisdom as well as the character virtues of integrity and constancy, albeit at the expense of much more that can and should be said about them all. Indeed, additional and equally important virtues could also be added to this shortlist of the ones required to sustain cyber leaders in their practice. Still, the following catalog provides us with an essential foundation for future accounts on cybersecurity and cyber leadership to extend. This section begins by discussing the intellectual virtues first before highlighting the character virtues mentioned above.

Technical Wisdom

Regarding the intellectual virtues, let us first consider the role of technical wisdom (*techne*) and its relation to sustaining the internal goods of privacy and security that I have associated with the practice of cybersecurity. This virtue constitutes the practitioner's cultivation of task-specific wisdom, which will be necessary to execute technical expertise as a cyber leader. Without technical wisdom, for example, a cyber leader who works as a product engineer would not be able to tell the difference between an outmoded product versus an excellent remote access or identity access management product. Knowledge of these matters first requires a significant degree of technical 'know-how' that one only garners by acquiring the wisdom associated with tradition-based background narratives about cybersecurity. These conversations have been extended across time and applied to several various workplace contexts.

Given the complexity of cybersecurity concerns and the technicalities of the products that sustain the internal goods of this practice, before a cyber leader who serves as an engineer, for example, can build or update an excellent security product in collaboration with others, this individual must first engage in the disciplined study of cybersecurity under a teacher of the practice. Per MacIntyre [10], all practices require teachers. Therefore, other cyber leaders may find that the essential aspect of their work in the practice of cybersecurity involves teaching others. By engaging in coursework and apprenticeship around matters such as network security, application security, incident response, vulnerability detection, etc., teachers enable aspiring practitioners to acquire the technical wisdom that is a prerequisite to servant leadership in cybersecurity. Furthermore, it would be beneficial for any individual who pursues cyber leadership as a service to engage in some form of apprenticeship with an experienced professional—preferably one who works on the same aspects of the practice that the aspiring practitioner also wishes to do. To be clear, technical wisdom about cybersecurity is required of any individual who finds him or herself working within the practice in some capacity. The increasing expansion of technical knowledge helps to outline what the latest standards for excellence—qua, compliance–ought to be. Submitting to the technical wisdom related to cybersecurity is the first step that one must take to garner the theory (*theoria*) necessary to achieve excellence in this practice (*praxis*).

Unfortunately, several examples already indicate just how high the stakes are for anyone who attempts to engage in this practice without staying up-to-date on the latest technical wisdom that promotes the internal goods of privacy and security. For example, returning to the earlier discussion within the introduction of this chapter about a hack that occurred at the control center of the Pinellas County water treatment plant, we can see from a few different angels how practitioners of cybersecurity who either lacked or ignored the latest technical wisdom associated with this practice indirectly facilitated a breach that could have cost a great number of human lives in a city that is home to 15,000 residents. The successful breach resulted in a failed attempt to poison county drinking water by using remote access technology to increase the concentration of sodium hydroxide by a factor of 100. Fortunately, in this case, operations at the plant were quickly brought to a halt after the supervisory control and data acquisition (SCADA) system was breached.

While the danger to citizens was quickly averted, control operators at the plant nonetheless ignored at least five Critical Infrastructure Protection (CIP) protocols that reflect the latest technical standards for excellence in cybersecurity. First, for however long the water treatment plant had been remotely delivering operational service, it was completely unprotected by any form of firewall that should have been installed. Second, the SCADA system was running an outmoded version of Windows 7, which Microsoft had since updated with multiple security patches that the plant operators had not downloaded. Third, the treatment operators lacked any form of multifactor authentication or password management software that should have been in place for those who accessed remote support technology to control water treatment operations. Fourth, and even worse, all plant operators shared the same internally distributed password for their TeamViewer remote support license. This was in blatant violation of common knowledge about the need for password privacy and security. Fifth and finally, the remote support product itself that plant operators used to control the water treatment plant has had a history of being hacked, particularly by the Chinese state intelligence system, who, for example, installed Winnti trojan malware on TeamViewer servers in 2016 [14]. For this reason, other remote access platforms should have been considered in place of TeamViewer. In these ways, technical standards of excellence in cybersecurity were either unknown or ignored by plant operators, resulting in gross negligence that could have cost human lives. This lesson teaches us that cyber leaders must adhere to the continuously refined standards of technical wisdom in the practice of cybersecurity to carry out excellent operational service that promotes the safety and security of those impacted by their work. When cybersecurity practitioners do not adhere to the relevant technical wisdom associated with their practice, cyberthreats and cyberattacks become inevitable, and threats to human flourishing are likely to ensue.

Practical Wisdom

The next trait that we should consider as we continue to reflect on the Pinellas County water treatment plant incident–and about what it can teach us regarding cyber leaders' need for certain intellectual dispositions–is the virtue of practical wisdom (*phronesis*). Practical wisdom enables us to learn from our past successes and failures in our work and within our lives. Per MacIntyre [11], all of us "discover that the realities of our practical lives are captured by narratives of failure and success as practical reasoners" (p., 74). It is practical wisdom that enables us to see when we must course-correct after something happens that we did not wish for or intend. Sometimes, too, practical wisdom operates as a way of reinforcing the means that did lead to successful achievement of the ends we seek in practice. Perhaps, for example, new means were taken to heighten security and privacy standards in practice that resulted in higher rates of success against warding off a breach of some kind. Much like a shepherd who tends to his flock, the virtue of practical wisdom tends to the ends that we seek in practice and clarifies the means by which we must act to achieve those ends.

I have just argued that practical wisdom is partly responsible for helping us to reflect on the privacy and security concerns that require vigorous deliberation within our work communities, and eventually, any course corrections that may be necessary. Moreover, I have stated that practical wisdom is responsible for reinforcing important course corrections by helping us see that our newfound success via pursuing alternative means should serve as a formative learning experience—one that changes our cybersecurity habits as we look to the future. Next, in the case of the Pinellas County water treatment plant incident, I wish to show how practical wisdom was exercised in these ways by cyber leaders who make up the U.S. Joint Cybersecurity Advisory. The result, as we shall see, was that future means for ensuring privacy and security within the critical infrastructure sector became clarified and enhanced in ways that will better protect U.S. citizens in the years ahead.

In the wake of the threat to the public-water-drinking citizens of Pinellas County, a joint task force comprised of members of the Federal Bureau of Investigation (FBI), the Cybersecurity & Infrastructure Security Agency (CISA), the Environmental Protection Agency (EPA), and the Multi-State Information Sharing & Analysis Center (MS-ISAC) engaged in vigorous deliberation about how to learn from this failure. What ultimately evolved from these discussions was a co-authored summary explaining what had led to the failure and the general cybersecurity threats that the emblematic failure exposed for similar organizations. The team's published "threat overview" detailed how cyber actors often "find entry points into legacy Windows operating systems and leverage Remote Desktop Protocol (RDP) exploits" to adjust system operations ([8], p. 2). It further described standard techniques exercised by hackers, almost all of which derived from reflection on the water plant incident, including the ways that they:

- Use access granted by desktop sharing software to perform fraudulent wire transfers.
- Inject malicious code that allows the cyber actors to:
  - Hide desktop sharing software windows;
  - Protect malicious files from being detected; and,
  - Control desktop sharing software startup parameters to obfuscate their activity.

- Move laterally across a network to increase the scope of activity.[1]

What is not to be missed here is the point that the exercise of practical wisdom, done in such a way that it promoted sustained deliberation across a diverse group of cyber leaders, allowed this advisory team to identify the primary sources of failure that led to this breach. In this way, practical wisdom enabled the advisory team to learn from failure so that future excellence in delivering the internal goods of privacy and security could be enhanced at the Pinellas County plant and across the critical infrastructure sector more broadly through the wider application of this advisory memo.

Theoretical Wisdom

While technical (*techne*) and practical wisdom (*phronesis*) are important virtues in their own rights, both are contingent upon the third and final intellectual virtue of theoretical wisdom (*episteme*) for a complete integration between theory (*theoria*) and practice (*praxis*) to ensue. Theoretical wisdom combines, as Aristotle [1] says, "scientific knowledge" with "intuitive reason," which allows new and revised standards of excellence within a practice to emerge (VI, 1141b). In other words, theoretical wisdom applies the knowledge supplied by practical wisdom to a reflection of our technical know-how. This kind of thinking allows us to see ways that past mistakes and failures can inspire new and creative patterns of specialized thinking and execution that enhance the internal goods of cybersecurity. Theoretical wisdom, then, helps cyber leaders 'raise the bar' of excellence in their craft. It is because of the practical implications stemming from this virtue that MacIntyre [10] acknowledges practices such as cybersecurity can and should always be "revising and extending" their ideals of excellence throughout their tradition-based histories (p. 277).

Indeed, what the Joint Cybersecurity Advisory overseeing the Pinellas County water crisis *did,* upon reflection and deliberation about that failure, is emblematic of the exercise of the virtue of theoretical wisdom. This is evident insofar as the report detailed a comprehensive list of "mitigation strategies" that included general recommendations for "cyber hygiene measures" which "may help protect against similar schemes" as well as more specific recommendations about securing water and wastewater systems and implementing secure remote support software in critical infrastructure environments ([8], p. 3). Notably in particular here is the way that, by exercising the virtue of theoretical wisdom and engaging in vigorous deliberation across four organizations that teamed up to achieve a common end, the advisory team theorized a general list of mitigation strategies including recommendations to:

- Update to the latest version of the operating system (e.g., Windows 10).
- Use multiple-factor authentication.

[1] For the full threat overview, see ([8], p.2). The list mentioned herein is represented as it appears within the report.

- Use strong passwords to protect Remote Desktop Protocol (RDP) credentials.
- Ensure anti-virus, spam filters, and firewalls are up to date, properly configured and secure.
- Audit network configurations and isolate computer systems that cannot be updated.
- Audit your network for systems using RDP, closing unused RDP ports, applying multiple-factor authentication wherever possible, and logging RDP login attempts.
- Audit logs for all remote connection protocols.
- Train users to identify and report attempts at social engineering.
- Identify and suspend access of users exhibiting unusual activity.[2]

In this case, the proof of cyber leaders' need for theoretical wisdom is 'in the pudding.' Without exercising this virtue in a way that also puts technical and practical wisdom to work at the same time, the joint task force would not have been able to reflect on the Pinellas County water treatment plant breach in any productive way that could have resulted in such cogently theorized recommendations for future mitigation. One would be hard-pressed to find any cybersecurity expert who disagrees with the above list of recommendations. This is because the general mitigation statements benefited from the multi-organizational deliberations that enabled the final list to become so comprehensive.

Hopefully, at this point, and in light of this extended example of the Pinellas County water treatment plant breach, we are beginning to see the role and need of the virtues for cyber leaders to protect the goods internal to the practice of cybersecurity. As we transition from a discussion about intellectual virtues, I hope it will become equally apparent that cyber leaders must also possess the character virtues.

Integrity and Constancy

Two of the most prominent character virtues that appear in MacIntyrean business literature are integrity and constancy.[3] Both are critical for sustaining cyber leaders in practice. I include them together in one section here because MacIntyre and other ethicists often discuss these virtues as being closely intertwined. For example, per MacIntyre [9], integrity and constancy are said to compliment one another by enabling cyber leaders to "pursue the same goods through extended periods of time,

[2] For a complete list of the additional and more specific recommendations about securing water and wastewater systems and implementing secure remote support software in critical infrastructure environments, see ([8], pp. 3–4). The general list of mitigation strategies mentioned herein is represented as it appears within the report.

[3] While these English terms lack a clear Greek translation, integrity and constancy are often associated with the highest of the Greek virtues–*philotimo*–which implies the preservation of one's honesty and dignity, and therefore one's honor as an individual who is unified and undeterred in his or her moral endeavors.

not allowing requirements of changing social contexts to distract them from their commitments or redirect them" (p. 318). Working together, integrity and constancy enable cyber leaders to exhibit sound moral character and commitments, regardless of the many various pressures or consequences that they may face. As we shall also see, integrity and constancy are further necessary for the genuine possession of any other character virtues. First, however, let us try to get clearer on the role of each individual virtue.

Beginning with integrity, this virtue enables the cyber leader to exhibit the same character across all contexts. For this reason, integrity is often associated with promoting trust and trustworthiness ([13], p. 124). Thus, to be a person of integrity is to be someone who resists the dangers and moral disunity of what we call "compartmentalization," which is "the idea that a person may adopt one set of standards in one sphere [or moment] of his or her life and a different and conflicting set of standards in another," thereby resulting in a compromised and *disunified character* (Ibid., S119). A unified picture of the virtue of integrity requires that the leader "exhibit the same moral character in different social contexts" ([9], p. 318). So, by embodying the virtue of integrity, we do not simply mean that we trust cyber leaders to do the right thing when no one is looking, but also that we trust them to embody a character that promotes the common good and human flourishing across every social role and relationship, despite the personal (and often political) cost. In this way, integrity involves knowing what we ought to do and drawing on a morally unified character that allows us to rise to the occasion, no matter the circumstances.

While the virtue of integrity may be distinguishable by its function to promote the same character across all contexts, the virtue of constancy may be set apart by its function to promote the same goods throughout extended periods of time ([2, 9],emphasis added). Constancy allows cyber leaders to be undeterred in their commitment to privacy and security and their commitment to the practice-based communities and institutions within which they pursue these internal goods. It is the virtue of constancy that beckons cyber leaders to keep submitting to the final ends of their work in practice, even and especially as other desires or temptations loom. Cyber leaders who find themselves in a management role, for example, will likely have to navigate the tension of promoting institutional sustainability while at the same time ensuring that those external goods derived from the practice of cybersecurity remain subservient to the goods internal to the practice. Challenges from equity investors who wish to accelerate profitable revenue growth pre-acquisition may place undue pressure on such a manager to streamline or restructure engineering and product development teams in a way that no longer sufficiently reinvests financial returns into the development of an increasingly excellent security product. Cyber leaders who possess the virtue of constancy would be inclined to question whether these new organizational pressures have resulted in a first-order priority shift that is harmful to their communities of practice and the excellence of their products and services, which must continue to protect those customers who rely on them to sustain privacy and security. Constancy allows cyber leaders to exhibit stability in their decision making and virtuous intentions, which promotes the subservience of external goods to internal goods in such cases [12].

Because the virtue of constancy equips leaders as enduringly faithful and dependable servants of internal goods that are sought together within their communities of practice, it also promotes one's commitment to work as a prosocial service or calling, as opposed to the view of work as a job [2]. While seeing work as a job inclines one to focus more on external goods like money, per the authors of *Habits of the Heart* [3], having a calling orientation to work entails a commitment to the internal goods of the work *first and foremost*. This prosocial view of calling illustrates the belief that one's "work constitutes a practical ideal of activity and character that makes a person's work morally inseparable from his or her life. It subsumes the self into a community of disciplined practice and sound judgement whose activity has meaning and value in itself, not just in the output or profit that results from it" ([3], p. 66). Furthermore, individuals who possess a sense of calling to their work often articulate an experience of being summoned or drawn to pursue a specific form of work for the unique ends which that work provides ([6], pp. 11–12). In the case of cyber leaders, constancy sustains their belief in the genuine meaning and value of the practice-based work that I have referred to herein as cybersecurity. It is the virtue of constancy that enables cyber leaders to remain devoted to the good and worthy ends of privacy and security as a calling. The constancy that such leaders exhibit as part of their service-oriented work orientation promotes internal goods over extended periods of time and despite the allure of things that might otherwise snatch individuals away from these prior commitments.

In addition to protecting the unity of one's character "in different social contexts" and sustaining the pursuit of internal goods "through extended periods of time" ([9], p. 318), the virtues of integrity and constancy also facilitate other important virtues that will be required for cyber leaders to maintain a commitment to their work as a service, or a calling. In the case of integrity, for example, we know that being a person of differing character from one social role to the next is not only a threat to one's own good, the good of others, and the common good, but also that it diminishes one's ability to practice truthfulness and exhibit trustworthiness, which are required for flourishing relationships in practice. Integrity sets limits on the character that one may exhibit across various roles or relationships, requiring that cyber leaders promote virtues such as truthfulness on all occasions, which leads others to view them as individuals who are worthy of trust. In the case of constancy and its facilitation of other virtues, it enables consistent and flourishing patterns of habit to form. For this reason, it too is believed to be "implicit in every other virtue, for no-one can be accounted virtuous in some particular respect if their acts and dispositions manifest these virtues on some occasions but not others" ([2], p. 681). Consider, for example, constancy's role in facilitating justice [13]. Justice "is only possible if one also possesses the virtue of constancy because on those occasions when justice is placed in jeopardy, constancy is required to remain true to one's obligations" (Ibid., p. S124). Thus, just as water needs a filter, so also the virtues demand integrity and constancy for character to become purified and distilled. Without these important virtues, it would be impossible for cyber leaders to retain the moral strengths of character and commitment that will be necessary to resist the many temptations and distractions they will inherently face.

Conclusion

The primary purposes of this chapter involved addressing three critical questions pertaining to cyber breaches and cyber leadership. First, *how* do breaches tend to happen? Second, *who* can teach us what to learn from and how to respond to such breaches? Third, and lastly, *what* strengths of character and intellect are required to sustain cyber leaders in their work?

Section one highlighted *how* common patterns of vulnerability often result in ransomware attacks and software breaches. It was determined that cyber systems are most commonly breached by legacy remote access solutions and leveraged credentials [4]. Consequently, I called for legacy tools to be replaced by enterprise secure remote access solutions that meet the latest compliance standards. Additionally, I pointed to other necessary layers of secure technology that ought to be in place to promote identity and privileged access management so that the right users can access and manage the right tools and privileges—nothing more and nothing less. In the wake of work that has become increasingly remote, such tools protect against exposure to malware, ransomware, and other forms of malicious hacking by enhancing the user authentication process and promoting secure network connections. Malicious advancements on the cyber battlefield will undoubtedly persist, so ensuring that these tools are in place and maintaining continuous deliberation to refine them is crucial to protecting critical infrastructure operations and data privacy and security in the years ahead.

Within section two, I presented the ideal type of the 'cyber leader' *who* can teach us what to learn from and how to respond to cyberattacks. Cyber leaders are well equipped to carry out these functions. This is so because they have undergone a submission to the practice of cybersecurity, entailing rigorous technical training, apprenticeship, and sustained deliberation within their respective workplace communities. In this section, I also emphasized why it is essential for cyber leaders to fundamentally view their work as a service to the common good and human flourishing via their specialized delivery of privacy and security. Preserving this prosocial orientation to work is critical as many various forms of temptation and pressure loom and must be averted. When someone fails to view this practice-based work as a service, the likelihood of failing to promote its internal goods with constancy and integrity is also high. As we have seen, failures to prioritize these internal goods in cyberspace imperil the practice of cybersecurity and threaten institutions as well as individual lives. With these concerns in mind, we said that the ideal of the 'cyber leader' is representative of *someone who demonstrates craft-expertise in the practice of cybersecurity to promote its internal goods first and foremost.*

Section three highlighted along Neo-Aristotelian lines the role that the intellectual and character virtues play in sustaining the cyber leader as well as the practice of cybersecurity. In particular, the intellectual virtues of technical, practical, and theoretical wisdom were shown to work off each other in a way that applies craft-specific knowledge, allowing cyber leaders to learn from successes and failures in practice and to continuously refine approaches that maximize security and privacy.

The virtues of integrity and constancy were also briefly highlighted for their role in forming cyber leaders and ensuring the facilitation of other important character virtues, which must be practiced habitually in an effort to sustain one's quest for that which is good and noble in and beyond the workplace.

The need for responsible and resilient cyber leaders today is greater than ever before. It should be expected that the temptations and harms cyber leaders now face will only continue to mount as a geopolitical competition between the so-called Great Powers continues to evolve. These powers aim to shape global events as major players and in such a way that advances their interests while corroding the goods internal to cybersecurity. China and Russia present the most serious risk of undermining the flourishing of U.S. citizens and allies at home and abroad by hacking critical infrastructure and other SaaS solutions to manipulate infrastructure controls and acquire sensitive data. In light of these threats, cyber leaders who remain committed to their work as a service play a crucial role in promoting human flourishing and the common good. The defensive approach that they must take to protect our cyberspace will be necessary since China continues to stress whole of government hacking efforts to expand its knowledge, innovation, and influence while Russia, too, attempts to destabilize democracy and drive wedges between the U.S. and its allies via similar malicious cyber operations. Keeping these threats in mind, I have argued that U.S. public and private sector cyber leaders ought to *first* understand their work as a service to the promotion of privacy and security, and that they must rely on the virtues to carry out this noble quest in cyberspace.

References

1. Aristotle. (2000). In T. Irwin (Ed.), *Nicomachean ethics* (2nd ed.). Hackett Publishing.
2. Beadle, R. (2013). Managerial work in a practice-embodying institution: The role of calling, the virtue of constancy. *Journal of Business Ethics, 113*(4), 679–690.
3. Bellah, R. N., et al. (2008). *Habits of the heart: individualism and commitment in American life* (3rd ed.). University of California Press.
4. Cole, E. (2021). *Cyber crisis: Protecting your business from real threats in the virtual world*. BenBella Books Inc.
5. Corera, G. (2021). China accused of cyber-attack on Microsoft Exchange Servers. *BBC News*. Retrieved September 1, 2021, from https://www.bbc.com/news/world-asia-china-57889981.
6. Dik, B. J., & Duffy, R. D. (2012). *Make your job a calling: How the psychology of vocation can change your life at work*. Templeton Press.
7. Farhadi, A. (2021). Conceptualizing the Great Power Competition and U.S. Geoeconomic Strategy for the Central and South Asia Region. In A. Farhadi, & A. J. Masys (Eds.), *The great power competition* (Vol. 1). Springer Nature.
8. Joint Cybersecurity Advisory. (2021). Compromise of U.S. water treatment facility. Retrieved September 1, 2021, from https://www.mass.gov/doc/joint-fbi-cisa-cybersecurity-advisory-on-compromise-of-water-treatment-facility/download.
9. MacIntyre, A. (1999). Social structures and their threats to moral agency. *Philosophy, 74*(3), 311–329.
10. MacIntyre, A. (2007). *After virtue: A study in moral theory* (3rd ed.). University of Notre Dame Press.

11. MacIntyre, A. (2016). *Ethics in the conflicts of modernity: An essay on desire, practical reason, & narrative.* Cambridge University Press.
12. Potts, G. (2020). The calling of the virtuous manager: Politics shepherded by practical wisdom. *Business Ethics: A European Review, 29*(S1), 6–16.
13. Robson, A. (2015). Constancy and integrity: (un)measurable virtues? *Business Ethics: A European Review, 24*(S2), 115–129.
14. Wei, W. (2019). Report reveals TeamViewer was breached by Chinese hackers in 2016. *Hacker News*. Retrieved September 1, 2021, from https://thehackernews.com/2019/05/teamviewer-software-hacked.html.

Dr. Garrett Potts is an Instructor of Religious Studies at USF. His work focuses on religio-cultural competence, particularly in the areas of Health, Business, & Religion. Dr. Potts teaches multiple undergraduate courses around these topics, but primarily he teaches at the intersection of Religion and Healthcare.

Cyberspace Leadership and Management

The Need to Inoculate Military Servicemembers Against Information Threats: The Case for Digital Literacy Training for the Force

Peter W. Singer and Eric Johnson

Every minute of every day, men and women in uniform are attacked by a weapon that threatens them, their services, and the nation. Yet the U.S. military has not trained them to prepare for this onslaught. It is time for this to change. Over the last several years, misinformation and deliberately spread disinformation, pushed by both foreign and domestic sources, have proliferated online. They have shaped not just what people read and believe, but also how they act. This "weaponization of social media" (Like War Book: The Weaponization of Social Media. https://www.likewarbook.com, [1]) has created a formidable challenge in nearly every policy area, from aiding the forces of terrorism and extremism, to being a tool of great-power competition, to damaging the vitality of our democracy. This challenge is not just to our wider national security, but also to the military itself. Every day, millions of servicemembers at every rank use social media. In so doing, they regularly are targeted by and engage with the viral spread of false information online. The resulting effects on them and the military affect operational security, force reputation, and even the physical health of servicemembers.

Introduction

The shameful and illegal Jan. 6 takeover of the U.S. Capitol, fueled by viral lies and largely motivated and coordinated via social media, suggests that the problem may be more dire. Of those members of the mob charged so far in federal court (a subset of the

P. W. Singer
Arizona State University, Tempe, AZ 85281, USA
e-mail: Peter.Singer@asu.edu

E. Johnson (✉)
New America, Washington, DC 20005, USA
e-mail: eric.b.johnson1.mil@mail.mil

A. Farhadi et al. (eds.), *The Great Power Competition Volume 3*,
https://doi.org/10.1007/978-3-031-04586-8_14

total participants), it appears that at least one in five have a military history, approximately three times the rate at which military or veterans are represented in American society.[1] Some data suggest that this problem may be even more widespread: In 2020, the FBI notified the Department of Defense of 143 servicemembers involved in extremism-related probes. Further, screening the social media of forces deployed to protect the inauguration of President Joe Biden reportedly led to a dozen members being "relieved of duty.[2]" More examples will almost certainly emerge as more focus is given to the issue by the military, law enforcement, and researchers, especially as they sift through the terabytes of data leaked from the Parler social network that included posts from multiple military members and geotagged U.S. military bases.[3]

In the wake of the Joint Chiefs' powerful statement [4]condemning these acts and the incoming Secretary of Defense's pledge to act on this new risk during his confirmation hearings, we are at a "Now what do we do?"[5]moment. We believe that a key policy item in this agenda should be to bolster the digital literacy of the force.

The reason for this need comes from not just an emergent threat but a gap in our broader response to a changing world. In facing the rise of digital threats over the last decade, the Department of Defense has launched initiatives ranging from new force structure to altering strategy documents. This focus on the organization itself is also where most of the proposed policy solutions have been focused.[6]

But in sharp contrast to how both the policy (and the policy debate) would address any other area of force effectiveness (as it has done on everything from lethality[7] to physical fitness),[8] the role and training of the individual servicemember in this new online battle has largely been ignored.

The Case for Digital Literacy Training

In a world of claimed "alternative facts," [9]where even the very phrase "fake news" became twisted and weaponized,[10] it is useful to start by defining the problem and its terms. A simple breakdown is offered by the United Nations Educational, Scientific, and Cultural Organization's resource for those who teach journalism[11].

[1] Dreisbach [2].

[2] Twitter [3].

[3] Singer [4].

[4] Gaouette et al. [5].

[5] PBS News Hour [6].

[6] Crane [7].

[7] Cummings et al. [8].

[8] Bond and Reichert [9].

[9] Wedge [10].

[10] Rojas [11].

[11] UNESCO [12].

Disinformation: Information that is false and deliberately created to harm a person, social group, organization or country.

Misinformation: Information that is false but not created with the intention of causing harm.

Mal-information: Information that is based on reality, used to inflict harm on a person, social group, organization or country.

In military terms, information acts like a biological weapon. Like a virus, it can infect the host and spread, both harming the host and also using it a means to reach a wider community. Malicious information thus operates simultaneously at the level of the individual *and* the network, and it should be countered at both levels, too.

Through changing thoughts and beliefs, the viral spread of information can lead to real world actions that can rapidly scale, which is why it has quickly become a key weapon in the arsenal of everyone from terrorist groups to foreign spy agencies [12]to politicians and celebrities.[13] And this weapon can be wielded with both tactical and strategic effect, across any key issue. For example, in the ongoing competition with great powers[14] that particularly plays out in gray zones,[15] Russia has repeatedly targeted NATO forces on deployment in Eastern Europe with such online disinformation campaigns.[16] In many ways, they are modern-day versions of KGB "active measures campaigns[17]" from the Cold War and, just like those campaigns, would also be accelerated to larger effect if there were outright conflict.

Similarly, social media has also altered a host of other age-old defense challenges, and arguably made them far more difficult. American and allied militaries, for instance, have also experienced a rise in "honey pot" entrapment via social networks for use in espionage. So too, "insider threats" can now be cultivated from afar. Both Islamic and far right non-state extremists have already used social media to recruit and mobilize servicemembers to act contrary to the very fabric of the U.S. military's values, policies, and regulations.[18] And, finally, operations security issues are massively magnified in a world where almost anything can be posted online for literally anyone in the world to see, and then mine for data to exploit against that individual servicemember, their unit, or the larger force.

Weaponized information has also struck at the larger society that the U.S. military exists within, and thus here too hits the force itself. A pertinent and ongoing illustration of this challenge, and the high stakes, is what public health professionals call the "infodemic" of false information about COVID-19.[19] The viral spread of false information about everything from the disease's origin to simple best practices to protect against it has been one of the reasons that the pandemic has been so lethal, and

[12] Fyre [13].

[13] Singer and Brooking [14].

[14] Inboden and Fontaine [15].

[15] David and Nora [16].

[16] Brewster [17], Social Sharing. CBC [18].

[17] Avramov et al. [19].

[18] Kanno-Youngs [20].

[19] The Lancet Infectious Diseases [21].

has undermined force protection measures along the way. For example, the spread of various myths about masks have acted to reduce their use and, as a result, put service-members and their families at risk.[20] This challenge may grow even more difficult with the combination of vaccines being a currently voluntary decision and COVID-19 vaccine disinformation campaigns starting to organize—anecdotally,[21] a voluntary sign-up for the vaccine at a U.S. military post this last week saw only a third of the eligible active-duty servicemembers do so.

While it may be less pernicious than lost lives or battles, soldiers' misuse of social media also causes an almost daily drumbeat of embarrassing episodes for the force, which then cause what feels like a never-ending series of headaches for leaders. Dealing with something said or done on social media that should not have been said or done on social media is now a regular challenge not just at the unit command level, but all the way up to the most senior levels of the Department of Defense. These episodes range from simple juvenile antics and mistakes, which become attached to the larger force because they are done by a servicemember online, to more serious examples like the recent case of a junior officer who made grossly inappropriate comments about the Holocaust on social media. One of us interviewed a now retired service chief and he commented that one of the biggest surprises he had in that most senior leadership role was "how much time" he ended up spending on social media-related incidents.

Implementing Digital Literacy: What Would It Look Like?

Getting a better handle on the changing digital threat environment should involve more than shifting organizational charts or creating new units. Just as the online threat has shifted from hacking networks to those on them, so should digital literacy training evolve beyond annual cyber awareness training focused on what link not to click. Fortunately, the United States does not need to re-invent the wheel to develop such training. Sweden, for example, implemented an early training manual for countering influence activities as far back as the 1980s,[22] which was later adopted by Finland.[23] It is no coincidence that the Scandinavian states are now considered among the best nations in the world at dealing with online threats.

There are also many existing educational resources in the U.S. civilian world, which range from books and videos to educational gamifications. Along with the National Association for Media Literacy Education and the Florida education system's CyberFlorida project, our own organization (New America) is building a portal to assemble these tools into one place for educators, and maybe also the force.[24]

[20] Frodl [22].

[21] Dwonskin [23].

[22] Countering information influence activities [24].

[23] Mackintosh [25].

[24] New America [26].

Each of these tools is about providing the skills, knowledge, and abilities that soldiers should possess concerning social media. For example, a RAND/IREX study [25] found that exposure to short media literacy videos made even the most avid and partisan users of social media less likely to engage with and spread false information. Similarly, a Harvard study of an inventive digital literacy video gamification found that such training interventions yield the equivalent of an "inoculation" for the individual, creating "significant and meaningful" reductions in the effectiveness of deliberately manipulated information against them[26] (and the larger network they are within).

Whatever the tool used, the National Association for Media Literacy Education's guide *Building Healthy Relationships with Media* offers a simple set of guiding questions to begin any user's engagement with media: "Why was this made? Who made it? What is missing? How might different people interpret it? How do I know this is true? Who might benefit from this message? Who might be harmed by this message?[27]".

In deploying these tools into the force, the Joint Staff J-7, with its responsibilities for doctrine, education,[28] concept development and experimentation, training, and exercises and lessons learned, would be the appropriate coordinating body, with the implementation then flowing down through each of the service's training commands, such as the U.S. Army Training and Doctrine Command.

The need for these new skills cuts across ranks—much like cybersecurity awareness is handled now, digital literacy should be woven into professional military education at all its levels, from boot camp all the way up to senior leader seminars for new flag officers. To be clear, though, a digital literacy initiative is not about creating new professional military education course loads. Instead, much of it should be via individual self-study lessons, with links to existing digital literacy resources vetted by the force. It should then be supplemented by various forms of hip-pocket training, formerly Noncommissioned Officer Opportunity Training, using "guided discussions" as another means of training that blends digital literacy with a key matter of concern. For example, imagine a sergeant directing all of his or her subordinates to pull out their smartphones and search for COVID-19 stories, then leading a guided discussion about safe behavior on COVID-19 versus disinformation and how it works. Similarly, Leader Professional Development Opportunities for officers and noncommissioned officers can be executed in the same way as the guided discussion, but taken further to address specific force protection and operations security concerns that connect with social media.

The utility of such tools is that they can also be easily deployed outside of active duty and reserve professional military education and training. For Defense Department civilians and contractors, the implementation should be much the same as how cybersecurity awareness training is already baked into contracts and new personnel

[25] IREX [27].

[26] Roozenbeek et al. [28].

[27] The National Association for Media Literacy Education [29].

[28] Joint Chiefs of Staff [30].

assignments. As certain elements of the threat also target military family members (such as the above COVID-19 example), many of the same resources should also be made available through Family Readiness Groups, where they can be voluntarily accessed (we should also expect that the experiences will provoke many to have valuable conversations with spouses, partners, and children, creating a kind of "train the trainer" effect).

As is the practice in civilian digital literacy training (and is already the case in current Department of Defense cyber security training courses), the modules might conclude with quick online skills evaluations. A typical approach is presenting a scenario through which a soldier could demonstrate knowledge of the key features of digital literacy, such as pulling out facts versus opinion statements or identifying altered media. While the individual should be expected to pass the evaluation for completion, any outward reporting of the specific scores should be anonymized (and there should be no career consequence of individual performance on the evaluations). Instead, the overall community scores are about providing a means of feedback and assessment of key areas that might need particular attention in future iterations.

However, it is important to stress that the measurement of success of digital literacy training is not merely about improving one's individual defenses or even the broader resilience of the force. Yes, those who have such skills are better positioned to identify, understand, and, when appropriate, respond to mis/disinformation they encounter in the social media space. But having digital awareness goes beyond just that. It is really about gaining the skills to succeed in a world where information has shifted from being scarce to often overwhelming in its (too) easy availability. We all now should be able not just to access information online but also to evaluate and communicate it in new and better ways.

"Pre–bunking" the Counterarguments

As with any new proposal to face a new challenge, there are potential concerns that have to be answered about the force providing such training.

Some might ask whether digital literacy is even something for the military to teach. One could envision a fear that it might turn the institution into a "nanny state" telling soldiers what to think or even violating their First Amendment rights.

To begin, digital literacy is *not* a policing of behavior or censorship of ideas, but rather a provision of skills. Its training would be about empowering servicemembers to better navigate the digital world that they engage with literally every day. As the RAND Corporation report *Exploring Media Literacy Education as a Tool for Mitigating Truth Decay* sums, the focus is "on teaching participants how to think without dictating what to think." The goal of the force training would be to give servicemembers a better awareness of the environment that they are both constantly in and constantly targeted by and the critical thinking skills to operate more effectively and safely in it.

Secondly, part of why that added training is needed is not just because members of the military are frequent targets, but also because the consequences of their actions can be very different, including for themselves. All Americans have First Amendment rights, but military servicemembers operate under a different legal regime regarding their acts and speech (as confirmed by the Supreme Court from even before the advent of social media, under 1986's *Goldman v. Weinberger*).[29] This also holds within the force's own regulations, such as Army Regulation 600–20, *Army Command Policy*. In cases of criminal behavior online (such as in terms of child pornography and extremist affiliations), there are obvious legal prohibitions, whether someone is in uniform or not. But there are also a wide range of behaviors online that might be legal for civilians under the First Amendment but violate military professional norms and codes and even lead to Uniform Code of Military Justice punishment.

Another potential concern is whether digital literacy training would be a costly or difficult proposition for the joint force. Fortunately, as noted above, there are already multiple resources available, many of them free and open source. In turn, the implementation does not require changing organization charts, new legislation, or massive new line-item budgets. In fact, one might even think of it as a means of indirect cost savings, akin to how we train soldiers in everything from cybersecurity best practices to how to exercise correctly, in order to reduce the number and severity of data breaches or soldiers injured at the gym. Even more, digital literacy fits in with top priorities already identified by the force for its future, such the Army's focus on people and modernization.[30] Gaining digital literacy helps *modernize people* and their ability to navigate a changing world.

Conclusion

There should no longer be any doubt that the spread of misinformation and deliberate disinformation campaigns are a threat to our servicemembers, Defense Department civilians, their families, and, by extension, national security. As such, the Defense Department should alter how it not just organizes, but also trains, all the way down to the individual level. Digital literacy is a necessary capability that servicemembers need to possess in the twenty-first century. If U.S. forces are to defend against new online threats as well as more effectively lead and communicate in this new battlespace, the force needs to fill this key missing part of its training.

References

1. Like War Book: The Weaponization of Social Media. https://www.likewarbook.com.

[29] US Supreme Court [31].

[30] US Army [32].

2. Dreisbach, T. (2021). Nearly 1 in 5 defendants in capitol riot cases served in the military. NPR. https://www.npr.org/2021/01/21/958915267/nearly-one-in-five-defendants-in-capitol-riot-cases-served-in-the-military.
3. Twitter: Lauren Peikoff's tweet. https://twitter.com/laurenpeikoff/status/1351607005136433153?s=20.
4. Singer, P. (2021). Three steps to fight online disinformation and extremism. https://www.defenseone.com/ideas/2021/01/three-steps-fight-online-disinformation-and-extremism/171563/.
5. Gaouette, N., Liebermann, O., Starr, B. (2021). Military Joint Chiefs condemn 'sedition and insurrection' at US Capitol as federal and local officials scramble to ensure security. CNN Politics. https://www.cnn.com/2021/01/12/politics/joint-chiefs-condemn-sedition/index.html.
6. PBS News Hour: Lloyd Austin's Senate confirmation hearing for Secretary of Defense. https://www.youtube.com/watch?v=MgOcMePVlU8
7. Crane, C. (2019). The United States needs an information warfare command: a historical examination. War on the rocks. https://warontherocks.com/2019/06/the-united-states-needs-an-information-warfare-command-a-historical-examination/.
8. Cummings, S., Cummings, J., Kivelin, J., Spencer, J., & Cuomo, S. (2018). Ten ways to fix the U.S. Military's close combat lethality. War on the rock. https://warontherocks.com/2018/03/ten-ways-to-fix-the-u-s-militarys-close-combat-lethality/.
9. Bond, N., & Reichert, B. (2019). It's not about the test: Toward a better fitness training regimen for the army. https://warontherocks.com/2019/02/its-not-about-the-test-toward-a-better-fitness-training-regimen-for-the-army/.
10. Wedge, M. (2017). The historical origins of alternative fact. psychology today. https://www.psychologytoday.com/us/blog/suffer-the-children/201701/the-historical-origin-alternative-facts.
11. Rojas, H., Valenzula, S., & Tong, C. (2020). Fake new is anything they say!. Mass Communication and Society. https://www.tandfonline.com/doi/full/https://doi.org/10.1080/15205436.2020.1789661
12. UNESCO (2018) Journalism, "Fake News' and disinformation: A handbook for journalism education and training. https://en.unesco.org/fightfakenews.
13. Fyre, T. (2020). Inside job: The challenge of foreign online influence in U.S. Elections. War on the Rocks. https://warontherocks.com/2020/10/inside-job-the-challenge-of-foreign-online-influence-in-u-s-elections/.
14. Singer, P. W., & Brooking, E. (2018). What clauswitz can teach us about war on social media. Foreign affairs. https://www.foreignaffairs.com/articles/2018-10-04/what-clausewitz-can-teach-us-about-war-social-media.
15. Inboden, W., & Fontaine, R. (2020). Great power, great responsibilities: global competition in an age of uncertainty. War on the Rock. https://warontherocks.com/2020/03/great-power-great-responsibility-global-competition-in-an-age-of-uncertainty/.
16. Barno, D., & Bensahel, N. (2015). Fighting and winning in the "GRAY ZONE." War on the Rocks. https://warontherocks.com/2015/05/fighting-and-winning-in-the-gray-zone/.
17. Brewster, M. (2020). Canadian-led NATO battlegroup in Latvia targeted by pandemic disinformation campaign.
18. Social Sharing. CBC. https://www.cbc.ca/news/politics/nato-latvia-battle-group-pandemic-covid-coronavirus-disinformation-russia-1.5581248.
19. Avramov, K,, Inboden, W., Pope, P., & Walton, C. (2019). Influence operations and active measures: the history of soviet and Russian political warfare in the west. War on the Rocks. https://warontherocks.com/2019/11/influence-operations-and-active-measures-the-history-of-soviet-and-russian-political-warfare-in-the-west/.
20. Kanno-Youngs, Z. (2020). Delayed homeland security report warns of 'Lethal' white supremacy. New York Post. https://www.nytimes.com/2020/10/06/us/politics/homeland-security-white-supremacists-russia.html.
21. The Lancet Infectious Diseases. (2020). The Covid-19 Infodemic. https://www.thelancet.com/journals/laninf/article/PIIS1473-3099(20)30565-X/fulltext.
22. Frodl, K. (2020). Debunked myths about face masks. https://www.mayoclinichealthsystem.org/hometown-health/speaking-of-health/debunked-myths-about-face-masks.

23. Dwonskin, E. (2020). Vaccine opponents outline online campaign to sow distrust coronavirus vaccine. The Washington Post. https://www.washingtonpost.com/technology/2020/12/23/anti-vaxx-covid-vaccine/?utm_campaign=wp_evening_edition&utm_medium=email&utm_source=newsletter&wpisrc=nl_evening.
24. Countering information influence activities–A handbook for communicatiors Swedish Civil Contingencies Agency (MSB). https://www.msb.se/RibData/Filer/pdf/28698.pdf.
25. Mackintosh, E. (2020). Finland is winning the war on fake news. What it's learned may be crucial to Western democracy. CNN. https://www.cnn.com/interactive/2019/05/europe/finland-fake-news-intl/.
26. New America. Cyber Florida, Florida Center for Instructional Technology and New America Launch New Partnership to Improve "Cyber Citizenship" Skills for K-12 Students. https://www.newamerica.org/international-security/press-releases/cyber-florida-fcit-new-america-partnership-to-improve-cyber-citizenship/.
27. IREX (2020) Randomized control trial finds IREX's media literacy messages to be effective in reducing engagement with disinformation. https://www.irex.org/news/randomized-control-trial-finds-irexs-media-literacy-messages-be-effective-reducing-engagement.
28. Roozenbeek, J., van der Linden, S., & Nygren, T. (2020). Prebunking interventions based on "inoculation" theory can reduce susceptibility to misinformation across cultures. *Harvard Kennedy School (HKS) Misinformation Review*. https://doi.org/10.37016//mr-2020-008.
29. The National Association for Media Literacy Education: Building Healthy Relationships w/ Media: A Parent's Guide to Media Literacy. https://namle.net/wp-content/uploads/2020/09/parent_guide_final.pdf.
30. Joint Chiefs of Staff: J7 Directorate for Joint Force Development. https://www.jcs.mil/Directorates/J7-Joint-Force-Development/.
31. US Supreme Court. Goldman v. Weinberger, 475 U.S. 503 (1986). https://supreme.justia.com/cases/federal/us/475/503/.
32. US Army. Action Plan to Prioritize People and Teams. https://www.army.mil/article/239837/action_plan_to_prioritize_people_and_teams.

Peter W. Singer is Strategist at New America, a Professor of Practice at Arizona State University, and Principal at Useful Fiction LLC. A *New York Times* Bestselling author, described in the *Wall Street Journal* as "the premier futurist in the national-security environment," Dr. Singer has been named by the Smithsonian as one of the nation's 100 leading innovators, by *Defense News* as one of the 100 most influential people in defense issues, by *Foreign Policy* to their Top 100 Global Thinkers List, and as an official "Mad Scientist" for the U.S. Army's Training and Doctrine Command. No author, living or dead, has more books on the professional US military reading lists.

Lieutenant Colonel (LTC) Eric Johnson was a Chief of Staff of the Army Senior Fellow in New America's International Security Program. A Psychological Operations (PSYOP) Officer by trade, his primary area of research is information warfare. Most recently Johnson served as the Director of Information Warfare, Assistant Chief of Staff, G9, 1st Special Forces Command (Airborne), at Fort Bragg, North Carolina. He previously commanded 9th Psychological Operations Battalion, deploying to Afghanistan where he was responsible for establishing the first Information Warfare Task Force, which integrated and synchronized Military Information Support Operations, (MISO), Operations Security (OPSEC), Electronic Warfare (EW), Space, and Cyber capabilities throughout Afghanistan.

The War for Cyber Talent: Can the US Win It?

Ronald Sanders

Abstract Our Nation has become increasingly digitally dependent, and in so doing, it has also become more vulnerable in cyberspace. A significant part of that vulnerability concerns people. For example, users like us—are still the principal cause of cybersecurity breaches. Over three-quarters of them the result from some socially engineered attack vector, with the remaining breaches more technical (and technically sophisticated) in nature. As a consequence, people must also stand guard against those incursions. The latter is the focus of this chapter…the people who are responsible for protecting our data and our networks, including those who engage in what are arguably deterrent offensive cyber operations. In that regard, our Nation's cyber workforce has become a major factor in its ability to protect our critical—and increasingly vulnerable—information and communications technology (ICT) infrastructure…as well as to project power in the cyberspace domain to influence the behavior of other aspiring 'Great Powers' like Russia and China, their proxies, and other regional geopolitical rivals. However, this is not a war in the benign, labor market sense. One involves the competition for talent, although that is part of it. No, the US is in a real 'people' war in the cyberspace domain, with real consequences for national security. To that end, the chapter will begin with a brief introduction that seeks to define the cybersecurity talent gap that is the basis of that war, and then it will then discuss the size of that gap, and/or how the US can attempt to measure it. The chapter will then discuss how the US can begin to close that gap, especially with respect to the pipeline that ultimately produces cyber talent that may be deployed in the US national interest. It will begin with efforts to attract more young people to

---

This chapter is based in part on a congressionally-directed report issued by the National Academy of Public Administration (The Academy) in January 2022 entitled "A Call to Action: The Federal Government's Role in Building a Cybersecurity Workforce for the Nation" (https://napawash.org/academy-studies/dhs-cybersecurity-workforceink); this author, an Academy Fellow himself, served as a subject matter expert (SME) in support of the Academy Panel and Study Team that prepared that report, providing advice through a series of interviews, written submissions, and other interactions with that Panel and the Study Team that have been summarized and incorporated herein.

R. Sanders (✉)
University of South Florida, Tampa, FL 33620, USA
e-mail: rpsanders@usf.edu

A. Farhadi et al. (eds.), *The Great Power Competition Volume 3*,
https://doi.org/10.1007/978-3-031-04586-8_15

cyberspace-related academic disciplines in elementary and secondary schools, post-secondary vocational schools, and colleges and universities (the most obvious ‘raw’ source of cyber talent), before eventually finding their way into cyberspace-related professions. In so doing, the chapter compares and contrasts how our Great Power competitors develop such talent and offers some suggestions on how the US’s ‘supply side’ cyber pipeline may be improved. The chapter will then narrow its focus to the Nation’s largest employer of such talent: The US government, including its armed forces and its civilian and contractor workforce. And as we will see, the Federal government is also potentially the largest producer and supplier of that talent, both directly and indirectly. The chapter will close with a description of the US government’s current efforts in that regard, as well as with some recommendations on how those efforts may be accelerated to help close the national security cyber talent gap. That gap—between the US and its global geopolitical rivals, particularly those that aspire to Great Power status and otherwise—is real and growing.

Keywords Great powers · Cyber talents · National security · Cyber security

Introduction and Overview

As an official with the National Security Agency (NSA) once quipped to this author, cybersecurity is really all about the ‘wetware,’ that is, the people…those who may cause a breach, and those who are responsible for protecting us from them. This is true today more than ever, as our Nation has become increasingly digitally dependent in virtually (pun intended) every aspect of our everyday lives. *We the People* literally live on-line these days, driven in part by COVID 19, as that extraordinary pandemic, along with other technological and cultural factors, have forced us to rely on the Internet for school, work, shopping, and even social interaction. Bottom line up front: Our economy, our culture and society, and even our national security now depend on information and communications technology.

And in this author’s view, we have passed an inflection point in that regard, especially in the US. We will never go back to ‘the way things were’ pre-pandemic. Indeed, one can argue that living more and more of our lives on-line was inevitable, and COVID merely accelerated that trajectory.

Unfortunately, as our Nation and its citizenry have become increasingly digitally dependent, we have also become more vulnerable in cyberspace, and that too comes down to people. Unfortunately, users like us are still the principal cause of cybersecurity breaches, over three-quarters of them the result of some socially engineered attack vector. That is just a fancy way of saying that most breaches result from someone being duped into clicking on a malware-laden email attachment, plugging an infected device into their computer, or inadvertently disclosing login or personally identifiable information (PII).

So long as people populate cyberspace, ‘stuff’ like this will happen, and it is the people who must guard against that stuff that is the focus of this chapter…the people who are responsible for defending our data and our networks—including those who engage in what are arguably offensive cyber operations.[1] That is where we will spend our time. Why? Because our Nation's cyber workforce has become a major factor in its ability to protect its critical—and increasingly vulnerable—information and communications technology (ICT) infrastructure…as well as to project power in that domain to influence the behavior of other aspiring 'Great Powers' like Russia, China, and their proxies.

In other words, when pundits describe a 'war' for cyber talent, they are not kidding. To that end, we will begin with a brief, stage-setting introduction that seeks to define the cybersecurity talent gap that is the basis of that war. We will then discuss the size of that gap or alternatively, how we can attempt to measure it. But we will spend most of our time and attention discussing how the US can begin to close that gap, especially with respect to the talent pipeline that ultimately produces cyber talent that may be deployed in the national interest. It will start with efforts to attract more young people to that profession in elementary and secondary school and eventually, post-secondary vocational schools, as well as colleges and universities, the most obvious 'raw' source of cyber talent. In so doing, we will compare and contrast how our Great Power competitors develop such talent and offer some suggestions on how the US's 'supply side' cyber pipeline may be improved.

In so doing, we will narrow our focus to the Nation's largest employer (and as it turns out, developer) of such talent: The US government, its armed forces as well as its civilian workforce. As we will see, the Federal government is not only a major cybersecurity employer, but it is also potentially the largest producer and supplier of that talent, both directly and indirectly. Accordingly, we will close with a description of its current efforts in that regard and offer some recommendations on how they may be accelerated to help close the cyber talent gap from a national security standpoint. For that gap—between the US and its global geopolitical rivals, particularly those that aspire to Great Power status and otherwise—is real and growing.

The Price of Digital Dependence

Why has cyberspace become a such battleground? Part of it is literally us…as in those of us who live in the US. The US is the most digitally dependent nation on the planet, and that dependence can be exploited asymmetrically. In other words, it does not take massive armies, navies, or air forces (fueled by enormous defense budgets) to take advantage of that vulnerability; instead, it can be exploited by a few. That too is just a fancy way of saying that David can much more easily slay Goliath in cyberspace than in the 'real' world of kinetic warfare. Thus, it doesn't take much to penetrate a system or network and exploit it—it no longer even takes

[1] Interestingly, policymakers seem to avoid that phrase in favor of less controversial descriptors like 'defend forward' to describe them.

an evil computer genius to do so. All sorts of cyberattack weapons are for sale on the Dark Web—especially when people are the most accessible attack vector.

However, on the 'good' side of the ledger, people also protect us in cyberspace. Cybersecurity professionals, some in uniform but many not, stand between us users and cyber threats of all kinds, from nation-states and their proxies to common cyber 'street criminals' who exploit the gullibility of our fellow citizens of the Web. It just takes one of those to let a bad actor in to steal our PII or our intellectual property. Nowhere is this digital dependence more prevalent than in our national security establishment, broadly defined.[2] For whether it is for better or worse, we Americans have embraced technology as the ultimate 'force multiplier' when it comes to weapons systems and other tools of national and homeland security.

Thus, we are in a war for talent. That phrase has become overused, and it too often (and too glibly) refers to the hyper-competition in the US labor market. Among private companies and/or between them and the public sector (especially the Federal government, including the US military and its civilian workforce) for those who can fill cybersecurity jobs. But this makes the 'war' word seem like just another hyperbolic noun. When it comes to our national and homeland security, it really is a war…in Baron von Clausewitz's classic sense of that word.

Yes, we are at war…a war with our 'near peer' nation-state competitors.[3] And because cyberattacks are also the classic definition of asymmetric warfare, we are also at war with smaller but no less antagonistic nation-states like Iran and North Korea, as well as all of their cybercriminal proxies, many of whom act as their agents in this war, in other words, the cyber version of Russia's infamous 'little green men.' That war, somewhere between hot and cold, is going on as we speak, and ultimately, it starts with people…in this case, those who are responsible for attacking us on one hand, and those who defend us from those bad actors on the other.

To be sure, we have our proxies too, although as a matter of law, they are strictly defensive. Thus, the civilian cybersecurity professionals that man the virtual watchtowers in the US private sector—especially those that comprise our nation's critical infrastructure, like electric and other utilities or the financial service industry[4]—are part of this war. Their employers and their supply chains are often the subject of a cyberattack, or as an attack vector to get at part of our national security establishment. And unlike many of those who are doing the attacking, those defenders are not subject to the command and control of our national military command apparatus;

---

[2] For purposes of this chapter, we will exclude cybercrime from this broad definition, at least that cybercrime that is perpetrated by 'common' criminals who are pursuing purely criminal ends. Obviously, that line becomes less clear when it comes to criminal organizations that have some state sponsorship, even tacitly, and thus, as a practical matter, we must choose to include the latter in our focus here.

[3] Indeed, as explained below, this author believes that they are no longer near the US when it comes to peers; rather, they are at least our equals, if not potentially our betters.

[4] The Department of Homeland Security's official definition of critical infrastructure can be found at the following link: https://www.dhs.gov/archive/coronavirus/cybersecurity-and-critical-infrastructure#.

indeed, they are not even required to report an attack, much less act in the national interest in responding.

Defining the Cyber Talent Gap

For purposes of this chapter, the 'war for talent' that we will refer to is the one the US is actively engaged in with those nation-states and their proxies who threaten our national security, directly and otherwise. To be sure, there is a far less malign war for talent in the US (and to somewhat lesser extent, the international) cybersecurity labor market, but that 'war' is classically over such mundane things like pay and benefits and working conditions. Those bread-and-butter issues are not unimportant—indeed, they are also at the root of our national security establishment's (in)ability to compete with the US private sector for the cybersecurity talent to defend us, at least in part—but their impact on our national security is less direct.

Here is what we do know: Our Nation's so-called 'near peer' competitors in cyberspace, Russia and China in particular, produce far more cybersecurity talent than we do—that is, more graduates with training in STEM and cybersecurity-related fields—and they employ (and deploy) that talent in shady ways, and in some cases legally prohibited, in the US. This has become conventional wisdom in the US, but at the risk of stating the obvious, it is a matter strategic importance to our country. However, that conventional wisdom is more anecdotal than empirical, and becoming more precise in that latter regard is problematic, on both sides of the talent ledger.

However, as in all things, defining the gap is easier said than done, particularly since cybersecurity is an applied discipline that includes individuals with formal training in it—as manifested by a college or university degree that says 'Cybersecurity' on the diploma, or an industry cybersecurity certification like a CISSP—but in the view of many, including this author, it is much broader than that. For example, the field's 'supply side' also includes individuals with STEM skills, like computer scientists or electrical and electronic engineers, but it may also encompass mathematicians (for cryptography) and even individuals with 'soft' skills like foreign languages or intelligence analysis, the latter with formal training in psychology, sociology, criminology, and the like.

As a result, in Florida, due at least in part to the urging of this author's organization, the state's higher education Board of Governors (BOG) has expanded their definition of 'cybersecurity degree' to include some dozen-and-a-half academic majors that prepare a student for a job (or a career) in cybersecurity. In the past, they only 'counted' degrees that actually had the word 'cybersecurity' in their title. Thus, today, when the State of Florida attempts to gauge the size of its cybersecurity workforce, it counts those earning degrees in Electrical/Electronic Engineering, Computer Science and Engineering, and Mathematics (the latter with a focus on Cryptography), as well as less technical degrees—at least from a STEM standpoint—such as cybercrime, intelligence studies, and digital forensics.

Note that none of these academic majors says ‘cybersecurity’ in its title. Still, they all contribute to the state’s (and the Nation’s) labor pool of college-trained graduates who *could* enter the cybersecurity profession and thus reduce our digital vulnerability. How many of them actually do so is another matter altogether? Thus, cybersecurity’s supply side is no longer the exclusive denizen of ‘wetware’ with training in that narrow field, or even those with STEM expertise [1].

Accordingly, it may be useful to take a page from the way the Bureau of Labor Statistics defines the labor force. It uses a ‘tiered’ approach that starts with all of those who are employed or actively seeking same—that is, the civilian labor force (CLF)—and necks that down to individuals who are in the so-called Relevant Civilian Labor Force (RCLF)…in other words, those who are actually qualified to fill jobs in a particular occupation.[5] Using that definition, the ‘relevant’ US cybersecurity workforce is broader than just those who graduate (or who are trained) in cybersecurity.

Measuring the Cyber Talent Gap: What Should We Count?

That also begs another critical question: What is the size of this gap? Today, there is not a single, official estimate of the cybersecurity labor force, either on the demand side (in terms of overall employment levels, including vacancies) or on the supply side. Measuring the gap too is far easier said than done.

Take the latter for example. As we have posited, one cannot just count the number of college students graduating with a degree in cybersecurity, as that surely results in undercounting those that may qualify for a cybersecurity position, especially at the entry level…graduates in everything from Computer Science and Electrical/Electronic Engineering to Mathematics, Mass Communications, and Criminal Psychology. And while many of those graduates do not go on to fill cybersecurity positions, we know that some do, so counting the ‘supply side’ of this equation becomes problematic.

But here Is what we do know: Russia and China produce far more STEM and related graduates than the US, and while they and we may define their skills differently, especially in a qualitative sense, not all of them will become (or at least start out as) cybersecurity professionals, they represent the most basic of talent pools: those who can do cybersecurity work…and not just for legitimate employers in the cybersecurity market. Thus, in the case of more centralized, command-and-control economies (like Russia and China), they are being deployed and directed to address the security interests and objectives of their nation-states, both defensively and offensively.

In that regard, in this author’s view (as well as that of many other pundits and policymakers[6]), we are losing the war for talent, at least in a quantitative sense.

[5] The definition of relevant civilian labor force (RCLF) can be found at the following link: Frequently Asked Questions on Equal Employment Opportunity Tabulation (census.gov).

[6] Sampling of articles.

For example, we know that Russia and China produce more cybersecurity-relevant talent than the US does. Still, even this suffers from a definitional problem as in 'what exactly counts' in the talent war?

In the US, there are various counts—more accurately, just guesstimates—conducted by various organizations, some private and some governmental, but there is no one 'official' way of quantifying both supply and demand.[7] And that is mainly with respect to STEM-focused efforts, to the potential neglect of those emerging 'softer,' non-STEM skills and programs—like analytic tradecraft and critical thinking, as well as advanced education in the social sciences, strategic languages,[8] and regional/international affairs—that are becoming so important to cybersecurity [1].

Note that these estimates do not include other sources of cybersecurity talent, many of which are classified. Thus, even while we know more-or-less empirically that Russia and China are producing more college and university graduates in STEM and cybersecurity-related disciplines, we can only estimate those individuals that their armed forces train and deploy in cyber-focused military units, or those in paramilitary 'reserves' or even semi-official militias that have overt or covert nation-state support…including those who conduct what can only be called information operations against the US and its allies. Nor can we even begin to count those in our peer and near-peer competitors who have the skills—sometimes deliberately developed, sometimes self-taught—to engage in black- and gray-market activities, many extra-legal, that have some official or tacit state sanction.

However, whatever those 'shadow' numbers may be, we can only speculate that they are considerable, particularly given the asymmetric cyber threats they pose. But should we count and compare these to the US and its allies? This author believes so (although that's easier said than done). Just as we try to estimate the number of infantry battalions and armored units or aircraft squadrons that Russia and China can field compared to our own, so must we begin to estimate cyberspace capabilities and capacity, even if only in a quantitative sense.

Those counts and comparisons will be difficult, as the 'talent pool' available to those nation-states to pursue their national cyberspace interests (defensive, exploitive, and even offensive) differs from ours even in the most definitional sense—for example, for the most part, our cybersecurity professionals employed in the US private sector cannot readily be deployed to support national aims, whereas their Russian and Chinese counterparts can, and in many cases, are—but in this author's opinion, we have to start somewhere.

To make it even more problematic, we also know that both of those nations (as well as others) employ what those of us in the cyber talent business refer to as the 'Russian

[7] The National Academy's report can be found at https://napawash.org/academy-studies/dhs-cybersecurity-workforceNAPACISAreport.

[8] NSA's STARTALK grant program offers a possible model for cybersecurity; established in 2010, it provides grants to local schools and school districts to develop and deliver K-12 curriculum in certain 'strategic' foreign languages like Mandarin or Dari (see https://www.startalk.info/). The University of Maryland's National Foreign Language Center administers the program for NSA (see https://nflc.umd.edu/projects/startalk).

Gymnast' method of talent development: That is, their governments officially search for and identify youngsters with an aptitude for STEM/cybersecurity and place them in 'special schools' that prepare them for service in their country's cybersecurity apparatus, on both the offensive and defensive sides of the battle. How many does that model produce? We do not know (again, in an unclassified sense), but we do know that those nation-states employ that method…indeed, they have for years.

We do not do this in the US…or at least we like to say we don't. To be sure, while we do not officially identify, conscript, and develop talented young people for cybersecurity (or some other national security related) service, we let the labor market do its thing. Many communities offer public, non-profit, and even for-profit instruction (in forms ranging from after-hours classes to charter and full-curriculum magnet schools) in cyber- and IT-related skills like coding. And in many instances, those opportunities are limited to families with the resources to take advantage of them.

To be sure, some variations are more accessible to disadvantaged but otherwise high potential youth, but these are even more limited. And while these efforts may not have the direct support of the US government, it does offer many millions of dollars in related grant programs that are specifically designed to subsidize such development.

Thus, as far as this chapter is concerned, our hypothesis is that that a significant cyber talent gap exists between the US and its various geopolitical rivals, and even though we cannot measure it as precisely as we can a nation-state's kinetic military capabilities, our main focus will be on how our country can acquire, develop, and deploy the talent it needs to protect itself (and the rest of us) from those asymmetric cyber threats that emanate from beyond our borders.

Closing the Gap, Part 1: Start with the K-12 Pipeline

The 'count and compare' problems enumerated above notwithstanding, let us assume that we are behind our Great Power competitors—and at least on a per capita basis, other regional geopolitical rivals—in the production of cybersecurity professionals, at least in a quantitative sense.[9] If that is true, is there anything we can do about it?

Most agree that it must start at the beginning of the talent pipeline, in K-12 curricula…by adopting a more market-oriented US version of the 'Russian Gymnast' strategy that attempts to increase the supply of cybersecurity professionals by focusing on the early years of their development. And that means giving parents, teachers, and school systems the ability to identify, attract, and most importantly, *incentivize* young people (especially women and underrepresented minorities) with the aptitude and attitude to enter the cybersecurity profession. Thus, as depicted in Fig. 1, we believe that all K-12 students should be schooled in basic cyber hygiene,

[9] We like to think that we have superior quality—that is, better cyber 'ninjas'—than our geopolitical adversaries, but given the incidence of successful cyberattacks, that may be debatable.

Fig. 1 Cybersecurity curriculum content

Cybersecurity Curriculum Content	Student Audience
Cyber Hygiene	All K-12 students
Cyber Citizenship	Middle/High School (HS)
Cyber Careers	HS Juniors/Seniors
Cyber Competencies	HS Seniors/HIPO Juniors

and that most should also be taught how to be responsible 'cyber citizens' in today's electronic public square. However, while those efforts would reach virtually every student, a far smaller number would learn about cybersecurity (and related) careers. From that cohort, a smaller number still—those with the interest and aptitude in cybersecurity and its related disciplines—would be afforded an opportunity to take more targeted courses of study, in their later years in high school and early years in college.

Think of it as a funnel that teaches K-12 students about cybersecurity and gradually concentrates on the relative few that want to become elite cyber warriors. The key word here is 'want,' as American students are never forced—at least by their government—to do so, but rather have the free will (obviously influenced by parents, teachers, peers, etc.) to follow a more academically rigorous path.

However, with some relatively small-scale Federal exceptions—like NSA's GenCyber and DHS's Cybersecurity Education and Training Assistance Program (CETAP)—national efforts to provide those opportunities by making more targeted K-12 cybersecurity education available to K-12 students are virtually non-existent. Moreover, again with some small-scale exceptions, the top of that funnel is left mainly to state, local, and community imagination and innovation and resources. And these efforts, while more narrowly concerned with cybersecurity, do 'profit' from national efforts to encourage more STEM interest (and curricula) amongst K-12 students, as their products contribute to the potential pool of available labor…if only because those students in more general K-12 STEM programs may acquire the basic skills and interest to go on for more advanced cybersecurity-related training in college and/or in vocational school.[10]

NSA's K-12 STARTALK program may serve as a model here.[11] Established by the Congress in 2009, STARTALK provides millions in funds to NSA to support foreign language training for K-12 students. NSA funnels those funds through the University of Maryland's Center for the Advanced Study of Languages (CASL)

[10] The US missed a real opportunity with the bipartisan K-12 Cybersecurity Act of 2021. Signed into law October 8, 2021 by President Biden, the Act has a laudable but relatively narrow purpose: to give resources to local school districts to help them protect themselves against cyberattacks, based on a 'deep dive' study on the cybersecurity vulnerabilities in school districts across the country, to be conducted by the Department of Homeland Security's Cybersecurity and Infrastructure Security Agency (CISA). In this author's view, the Nation's needs are far greater.

[11] See Note 9.

via a competitive grant program that supports innovative pilots for elementary and secondary school students to learn certain strategic foreign languages like Chinese or Pashto.[12] CASL evaluates the performance of those grants and makes recommendations to the NSA, DOD, and the Congress, regarding more permanent funding. The effort, which complements Boren Scholarships for college—level undergraduate and graduate students in foreign languages, as part of the National Security Education Program (or NSEP, discussed further below), has been quite successful to date, but it should be emphasized that it only provides 'start-up' funding, with the prospect of longer term support up to the recipient—for example, a school or school district—based on the efficacy of the recipient's pilot effort…and the hope that local communities or states will provide sustainment funding.

While the US Department of Education already has similar grant programs in place to support STEM and related pilots, in this author's view, STARTALK could easily be replicated and scaled to support K-12 cybersecurity education as well. However, in the meantime, it has been primarily left to states to do so on their own, with Maryland and Virginia, and more recently Florida, leading the way.[13]

For example, this author's organization, the Florida Center for Cybersecurity is the recipient of a \$20 M grant from the Florida Department of Education (FLDOE) to fund a cybersecurity 'Pathways' program that would provide one-year funding to K-12 schools, as well as state colleges and universities, to expand, evaluate, and eventually export innovative programs in cybersecurity education across the state. Such programs would, as depicted in Fig. 1, include basic/foundational cyber 'hygiene' and citizenship curricula, and for those select students who demonstrate the interest and aptitude, more targeted instruction in cybersecurity competencies that will prepare them for a job or a more advanced course of instruction in a cybersecurity-related academic discipline. The program would also fund developmental programs to increase the supply of frontline educators equipped to teach these subjects.

There will be more publicity around Florida's Cybersecurity Pathways program as it unfolds, but it is but one example of what schools, school districts, states, and institutions of higher learning are doing, largely on their own, to increase the supply of cybersecurity talent. However, in this author's view, if the Nation is truly serious about closing the cybersecurity talent gap—and reducing our vulnerability to cyber mischief on the part of our geopolitical rivals—that is where it must begin, and more Federal support will be necessary perhaps as part of DHS CISA's newly-expanded \$100 M budget for cybersecurity workforce development [5].

---

[12] Ibid.

[13] The EARN Maryland Grant Program (http://dllr.state.md.us/earn/earnwhatisearn.shtml) and the Virginia Cybersecurity Public Service Grant (https://www.schev.edu/index/tuition-aid/financialaid/state-student-aid/cybersecurity-public-service-grant) are great examples of state-led support for cybersecurity education.

Closing the Gap, Part 2: Cybersecurity Scholarships

Another way of improving the production of the Nation's cybersecurity talent pipeline, especially for undergraduate and graduate students—is by helping them cover the cost of their increasingly high tuition through full and partial scholarships. The Federal government has used this tool for many years with its cyber-focused Scholarship for Service program administered by the National Science Foundation since 2000.[14] The Defense Department has offered a its own improved version since 2001,[15] and while these programs have enjoyed moderate success, much more could be done in this area.

In concept, the NSF scholarship program is great as far as it goes. It provides substantial financial incentives—in the form of tuition assistance, books, even a stipend for living expenses—to deserving cybersecurity students, and it conditions all of that support on a commitment to post-graduation public service…specifically, to go to work for a Federal agency after graduation for a period of up to three years.[16] Graduating students even get special 'excepted' appointing authority to expedite their hiring by a Federal agency. And if they don't make good on that post-graduation commitment (for example, if they try to get out of it), the scholarship winner would have to pay back the money he or she received.

DOD's more recent cybersecurity scholarship program (aptly titled just that, or 392 DoD CySP for short), involves its cybersecurity officials from the outset. In effect, a DOD hiring manager (presumably one with a vacancy in the queue) has to sign off on the award of a DoD CySP scholarship to a student. In so doing, they not only commit to providing the student with an internship, but they also promise to consider that same student for a permanent position upon graduation…in other words, they make a pre-graduation decision—albeit a tentative one—to hire that student. So, they're ahead of the game.

So, where does that leave us? There is no question that student scholarships, especially those that come with a difficult-to-escape government service commitment, can help close the cybersecurity talent gap. Still, existing programs can be revised and expanded to improve their 'yield' dramatically. Both NSF and DOD programs expand the supply of college-trained cybersecurity talent, albeit on a relatively small scale, and they both offer programmatic advantages and disadvantages. In addition, both require a service commitment on the part of awardees as a quid pro quo for their government-funded scholarship, although as noted, awardees can also 'buy out' that commitment.

Another option worth considering is a variation of the US military's Reserve Officer Training Corps (ROTC). Note that NSA and this author discussed piloting

[14] A description of the National Science Foundation's Scholarship for Service Program can be found at https://www.nsf.gov/pubs/2021/nsf21580/nsf21580.htm#summary.

[15] A description of the Department of Defense Cybersecurity Scholarship Program can be found at: https://dl.dod.cyber.mil/wp-content/uploads/cysp/pdf/unclass-attachment-c-scholarship-and-vacancy-announcement.pdf.

[16] Ibid.

this model back in 2009. Like its uniformed counterpart, that pilot would have sponsored the creation of ROTC-like cybersecurity detachments at select participating colleges and universities, staff them with experienced practitioners (perhaps retirees) from NSA and DHS to serve as faculty and mentors, screen and admit students in cybersecurityrelated degree programs, and use those detachments as a way of developing even more graduates in academic disciplines related to cybersecurity…but at a scale that would satisfy the growing cyber workforce needs of the Nation. And while these 'civilian ROTC' graduates would owe a post-graduation Federal service commitment just like their NSF and DOD colleagues, once that commitment was fulfilled, they could become part of a larger cadre of cyber militia or Reserves and subject to recall in a cyber emergency (more on a cyber militia below).[17]

Closing the Gap, Part 3: Adapting to Changing Career Paths

Policymakers must also take changing career patterns into account. Here again, the evidence is more anecdotal than empirical. It begs for more longitudinal study, but this author (along with many others) believes that the career patterns for the post-Millennial generation will look far different from those of the past, when both public and private organizations could expect their new hires to want steady, career-long employment. They offered a 'social contract'—in the form of vested pensions, seniority-based compensation, and other rewards for loyalty and tenure—that comported with that premise. The objective (on both the supply and demand sides of the labor market) was to keep talent once acquired.[18]

That may no longer be the case, especially when it comes to hypercompetitive—and thus, far more mobile—professional occupations like cybersecurity, where an expert can offer his or her general expertise to almost any organization. To be sure, each organization has its own data and systems, as well as its own policies and protocols. Still, they must comply or comport with externally established cybersecurity standards, like those promulgated by NIST and/or the Center for Internet Security (CIS), so practitioners can be much more mobile, both organizationally and geographically. And when coupled with the COVID-driven 'work wherever you want to' trend in professional employment—something that is not likely to disappear—the

[17] While this program is not classified, it was authorized in the classified annex of the FY 2009 NDAA and cannot be reported here.

[18] For example, it has become a truism, supported by both empirical and anecdotal evidence, that those newly hired into Federal service will not likely 'sign up' for a three-decade career…in part because the Federal government's current retirement system no longer acts as a set of vesting 'golden handcuffs' like the old CSRS used to do; rather, the current version is far more portable (albeit less generous as well), thus allowing Federal employees in high-demand occupations like cybersecurity to leave, perhaps after they've been trained and/or fulfilled their service commitment to the government (see https://ourpublicservice.org/publications/beneath-the-surface-understanding-attrition-at-your-agency-and-why-it-matters/).

net result is a more episodic and eclectic career path, with cybersecurity practitioners moving from job to job, organization to organization, sector to sector, etc.

This has significant implications for cybersecurity employers. New hires may join to take advantage of a high starting salary, a great training and development program, or other educational opportunities (including professional certification) and then move on. That is something that the US military, as well as some of the Federal government's 'premier' cybersecurity employers—like NSA, FBI, and DHS CISA—are beginning to see, as young employees choose them not for the pay (as discussed below, all of those agencies are salary-constrained) but for the training they receive. Still, once their service obligation is fulfilled, they then move on to bigger and better opportunities. It truly is a 'sellers' market' when it comes to cybersecurity, at least insofar as more-or-less 'free' labor markets are concerned.[19]

Moreover, this is not just an entry-level phenomenon. For example, while NSA's non-retirement attrition has remained at an historically acceptable level (indeed, some would argue that it is too low), the agency has begun tracking so-called 'regrettable losses' in its more senior ranks, as those relatively few—and relatively scarce—extraordinarily talented cyber experts receive six-figure salary offers from private companies and even NSA contractors that they just cannot refuse. So, they leave, and they take unrivaled expertise and experience with them. And Draconian efforts to stop that exodus by force, through service agreements and other restrictions, may have unintended consequences.

So, if 'indentured servitude' is not the answer, what is? If increased mobility in today's hypercompetitive cybersecurity labor market is a fact of life, how can organizations adapt to the inevitable exodus of scarce talent?

One way is to just live with it. In other words, expect that an organization's 'best and brightest' cybersecurity talent will likely move on and like the US military, plan accordingly…by 'over hiring' at the entry level. That is not to say that organizations, particularly in the public sector, should not maximize other personnel flexibilities (such as 'work wherever or whenever you want to' policies),[20] and above all, by focusing on the intangibles that may attend public service. With all of that, agencies should expect—and thus plan for—much greater turnover amongst their ranks of cybersecurity professionals.

NSA has already done some of this, and two of its sister Federal agencies (and sometime-competitors for talent), FBI and DHS CISA, have also started down this path, albeit far more slowly than this author would advocate. But at the end of the day, changing career patterns may very well force an acceleration of this trend, as more and more organizations—including but not limited to the military and the US government—begin to see themselves as both employers <u>and</u> developers of cyber talent for the Nation.

---

[19] Of course, that is not necessarily the case with command economies like China and to a lesser extent, Russia.

[20] For example, salary-constrained Federal agencies could adopt internal promotion rules that favor capability over seniority and experience, or assignment schemes that vary someone's work experiences. Information on the Department of Homeland Security's Cyber Talent Management System can be found at https://www.govinfo.gov/content/pkg/FR-2021-08-26/pdf/2021-17824.pdf.

Closing the Gap, Part 4: Searching for (and Developing) Cyber Unicorns

Even if the US is able to increase the number of college-trained graduates in STEM/Cybersecurity-related academic disciplines, most will have only a three to four-month internship to show for it, and the question remains "will anyone hire them?" One need only look at the cybersecurity vacancy announcements in such popular job sites as Indeed, Zip Recruiter or Linked In—most of which require multiple years on 'hands on' work experience—to wonder.

However, the evidence here is also anecdotal, and this author's current organization, in partnership with the SANS Institute, has commissioned a research project to try to prove that hypothesis empirically.[21] Euphemistically entitled 'The Search for Cyber Unicorns' because it hypothesizes that most employers set their talent sights unrealistically high, it seeks to determine the experience that cyber employers are actually looking for in the labor market, compared to what most cyber students—representing the 'raw material' at the entry level of that market—are bringing to the supply side. The research is using a combination of surveys, focus groups, and case studies to ascertain whether that gap is real, and more importantly if it is, how best to close it. with results to be published later in 2022.

Another aspect that this research project will attempt to address; that is, whether many (or any!) cybersecurity employers have strategies in place to close the experiential gap between supply and demand…through entry-level training and developmental programs and the like. Here again, the Cyber Unicorn project's hypothesis is that most cyber employers, particularly those in the small and medium-size range, do not offer such 'bridge' programs, in part because they are unaffordable at any scale, especially when compared to the 'price' of finding and hiring a few Cyber Unicorns. As a result, they search for those scarce Unicorns in a cybersecurity labor market that is already hypercompetitive…and in so doing, they avoid the expense of incurring the cost of fielding their own gap-closing strategy.

The net result is a bifurcated, two-tiered labor market, where very large employers (and the military) invest in bridge developmental programs to meet their immediate needs, only to have a significant fraction of that developed talent siphoned away in what is a hypercompetitive secondary labor market for more experienced, 'second-job' cybersecurity talent. Thus, the majority of small and medium-sized organizations—those that cannot afford their own bridge programs, including many state, local, and smaller Federal, government agencies—look to 'feed' off of those larger organizations that have incurred the expense of taking raw cybersecurity talent and converting it into more fully developed and experienced professionals.

However, an unfortunate byproduct of that tiered structure is even keener competition for that scarce talent in the secondary labor market. This drives the price of that talent even higher to levels that most government agencies cannot afford. But more on this later. The Cyber Unicorns research project will see what the empirical

[21] The project is being led by noted cyber workforce scholar Dr. Diana Burley on behalf of the Florida Center for Cybersecurity and the SANS Institute.

evidence is. Still, its secondary hypothesis is that internal developmental strategies are the exception rather than the rule. As a result, the entry-level experience gap—or alternatively, the never-ending search for Cyber Unicorns—will persist.

Assuming that that experience gap exists, there are several proven ways to close it. One is the US military and its All-Volunteer Force (AVF). Thus, when it comes to cybersecurity talent, it deliberately seeks recruits that have a basic aptitude or potential for that kind of work through its primary 'feeder' pipelines like ROTC and the military academies…in part because it simply cannot offer salaries that are competitive for top-trained talent. In other words, with some exceptions, the US military stops short of searching in vain for cyber talent at the full-performance level, at least for their uniformed ranks; instead, like so many of its career fields. It looks more for potential—typically in the form of a college degree or relevant technical training in a cybersecurity-related field—with the expectation that it will have to train and develop that potential talent until it is deployable…for example, to a US Cyber Command Cyber Mission Team.[22] And because that is its staffing model, not just in cybersecurity but across the board, it is willing to invest in such 'bridge' training at scale.

Of course, the military expects a return on that investment, in the form of an ability to retain and deploy that now-developed talent over the course of a multi-year career (although, as discussed above, that assumption too is becoming debatable). Thus, that military 'force management' model is based on the premise that the attraction of advanced training, a commitment to its mission, and the promise of a longer-term military career with all of its attendant tangible and intangible benefits, will overcome the lure of higher salaries that its uniformed talent could command in the cybersecurity labor market. And while this author believes that premise to be true, at least in a general sense (and the US military's retention statistics tend to bear this out), that too needs a 'reality check' against emerging career patterns, especially in this particular career field.

That reality check (and the policy adjustments that may come with it) notwithstanding, the assumption remains: that is, that US military's developmental strategy effectively provides a bridge between the raw potential of a graduate in a cybersecurity-related discipline and the hands-on experience required by the military—indeed, the country!—to accomplish its cyber missions. And as a result, it may serve as a model for other employers, not just the largest ones, but coalitions of smaller ones who must otherwise compete in the hypercompetitive secondary market for cyber talent.

Another model has been perfected by major cybersecurity contractors, especially those that support the US government, albeit one that comes at greater expense. For example, this author has first-hand experience with the consulting firm Booz Allen Hamilton's internal cyber talent development strategy, and without providing any 'insider' information, can attest to its efficacy. But those internal efforts, largely matched by its major competitors in the contract cybersecurity domain, were scaled

[22] Of course, it also 'rents' that talent from contractors as well; however, much of the military's cybersecurity work is inherently governmental and cannot be outsourced.

on the premise that the firm had more than enough business, mostly in support of the US government and major private sector employers, to warrant the investment in an internal staff development program that turned raw college-trained recruits into cleared cybersecurity professionals who could be deployed to meet the full-performance requirements of its multiple clients. But the key word here is 'multiple' clients, for without the aggregated demand of those clients, Booz Allen and its competitors could not (and likely would not) invest in their own bridge programs.

> Another example, this one quite successful but still on a relatively small scale, is the Cyber 'Boot Camp' operated by the University of South Florida's Muma College of Business. Funded in part by a generous donation from the Tampa-based cybersecurity firm ReliaQuest, it was based on failure...that is, ReliaQuest's conclusion that it could not hire USF cybersecurity graduates because they just did not have the skills it (and its competitors) was looking for. So, the Muma College of Business, working in collaboration with that company and others, developed what amounts to a 'finishing school' for college students that prepares them for the actual world of cybersecurity work. ReliaQuest does not gain exclusive access to Boot Camp graduates; rather, any organization can compete for those graduates. But the point is that much of the cost and the risk of producing truly job-ready graduates in that 'finishing school' (regardless of their academic major), is shifted to USF.
>
> Other 'bridge' programs undoubtedly exist—indeed, finding them is one of the objectives of the Florida Center's Cyber Unicorns research project—but these may serve as models...ways of closing the hypothesized experience gap in the cybersecurity labor market. And if the US ignores them, it does so at its peril, and the shortage of US cybersecurity talent will persist.

Closing the Gap, Part 5: The Federal Government as Cyber Talent Source

As we have suggested, the various parts of the Federal government—both individual agencies and their subordinate elements, as well as DHS's CISA and DOD's NSA and CYBERCOM, all of which have cross-government and even national responsibilities—are a principal source of demand for cyber talent...indeed, collectively, they represent the single largest employer in that regard, and when you add cybersecurity-related student scholarships, school-to-work programs (like apprenticeships and internships),[23] and other more general grant programs supporting cyber workforce development,[24] it becomes apparent that Federal government has a significant influence on cyber talent development, both directly and otherwise.

---

[23] Although if one examines cybersecurity-related vacancy announcements on the Federal government's central job Web site, *USA Jobs*, it is apparent that it too is guilty of a sometimes-fruitless search for fully qualified Cyber Unicorns: https://www.usajobs.gov/.

[24] Grants issued by NSA's Centers for Academic Excellence (CAE) program office are just as much focused on addressing the Nation's cyber talent needs as they are NSA's. For example, NSA has issued a series of grants supporting cybersecurity academic programs to some of the over 350 CAE colleges and universities (see https://search.usa.gov/search?affiliate=nsa_css&sort_by=&query=CAE+grants&commit=Search).

Thus, on the one hand, the Federal government offers more employment opportunities, entry-level and otherwise, than any other single employer, especially when you consider the fact that once an individual earns 'competitive' civil service status with any one agency, he or she can move laterally to any other Federal agency with relative ease, and in many cases, will likely also receive first consideration when it comes to internal promotion opportunities (within their 'home' agency or another one) as well. And in addition to the sheer number of employment opportunities, the Federal government also offers the chance to serve an important mission—after all, who wouldn't be attracted to a job that involved protecting the networks of the Nation or going after bad cyber actors—and in the process, to acquire a top-level security clearance, as most Federal cybersecurity jobs require them.[25]

And finally, at least some agencies—most notably NSA—offer intensive, multi-year training programs that will mold raw, entry-level cyber talent into 'full performance' professionals. Indeed, as we have noted, this is the US military's primary cybersecurity talent development strategy. That promise of advanced training that attracts many to the US armed forces, especially those looking for hands-on cybersecurity work experience to complement their academic training. So, what's not to like?

Despite these advantages, the Federal government simply cannot compete when it comes to pay not so much in terms of average salaries for average talent, as the Federal government does not do too badly in competing for those that have basic qualifications. However, when it comes to competing for top graduates, Federal agencies are often second or third choice (or worse!). And when you couple that deficit with the often arcane and lengthy Federal hiring and security clearance processes, which in some cases can take months (especially when a clearance is involved), it is no wonder that top entry-level cyber talent looks elsewhere.

It is not like Federal agencies, or the Office of Personnel Management (which makes most of the Federal government's hiring and compensation rules) do not know this. Still, change has come hard to the US civil service in general. Nevertheless, a number of initiatives—many originating at the agency level, but some more recently emanating from OPM itself—have helped the Federal government's cybersecurity employers be more competitive in that regard, and these are worth noting.

Perhaps the most competitive of Federal cybersecurity employers—as well as the most concerned about the development of cyber talent as a national security imperative—has been the National Security Agency (NSA). The good news is that it has been able to leverage its one-of-a-kind (and super-secret) mission, as well as world class training, to recruit and retain some of the country's most talented cyber warriors. And it has done so since at least 1993, with the advent of something called the Defense Civilian Intelligence Personnel System (DCIPS), which gave NSA and its sister DOD intelligence agencies unprecedented personnel flexibilities, especially with regard to hiring and pay. Its details are beyond the scope of this chapter, but

[25] According to many government contractors, a job candidate with a Top-Secret clearance can command an additional $15,000 or more in salary (see https://news.clearancejobs.com/2020/07/10/do-different-clearance-levels-translate-to-more-money-in-your-bank-account/).

suffice it to say that NSA has fared quite well in the 'war' for cyber talent, especially when it comes to talent that gives the US an edge in comparison to the other Great (and regional) Powers and their proxies.

Indeed, this author knows from first-hand experience that NSA has some of the best cyber 'ninjas' on the planet. While the Nation is fortunate for this, it is in at least some part attributable to the extraordinary flexibilities provided by DCIPS.

> That has not gone unnoticed, by the rest of DOD on one hand, and agencies like FBI and DHS CISA on the other. Indeed, the latter convinced the Congress to give it DCIPS-like personnel flexibilities back in 2013, largely in order to 'level the (personnel) playing field' with NSA.[26] At long last, DHS finally issued rules operationalizing those flexibilities in late 2021, for immediate implementation, so that is a positive development. Also positive was the fact that at the end of the day, the 'new' system does provide DHS CISA with the ability (among other things) to be more competitive for cyber talent, giving it expedited hiring authority and the ability to pay above government-wide salary rates for that talent.
>
> However, on the negative side, it took DHS far too long to exercise that statutory authority, all the while putting the networks and data that DHS was charged with protecting at risk.[27] And contrary to press reports [2], it was not because those authorities were so complex as to require that extended 'gestation' period; indeed, it is the view of this author (as well as those on his Subject Matter Expert team that supported the DHS effort) that that system could have been designed, developed, and implemented in 18-24 months. To be sure, DHS's Cybersecurity Talent Management System (CTMS) goes beyond the flexibilities provided by DCIPS and other alternative Federal personnel systems, but mindful of the old adage that "perfection should not be the enemy of the 'good enough,'" one has to ask whether those additional flexibilities were worth the wait.
>
> However, in this author's opinion, the most important drawback to the approach taken by DHS is the planned scope of CTMS. Its original design called for the establishment of a governmentwide Cybersecurity Service comprised of cyber warriors numbering in the hundreds (if not thousands), hired and paid by DHS using the flexibilities afforded by CTMS and then deployed via rotational tours to other Federal agencies as the 'guardians' of their assigned networks and systems. Unfortunately, that vision does not seem to comport with the reality of CTMS implementation, which envisions—at least for the foreseeable future—and initial cadre of less than 150 employees, all focused on CISA's more centralized responsibilities, with other Federal agencies left to their own devices, hamstrung by existing civil service rules to try to hire cybersecurity talent.
>
> The good news here is that under its new Director, the Office of Personnel Management is in the process of revising those existing rules to make the Federal government more competitive overall for talent including cybersecurity talent, but more on that below.

[26] This author led a team of Subject Matter Experts (SMEs) on contract to DHS to help them design, develop, and implement those flexibilities, and as noted above, it is the opinion of those SMEs that it took far too long to do so.

[27] Although it should be noted that DHS CISA managed to use a number of existing civil service flexibilities, like recruiting bonuses and retention allowances of up to 25%, to remain competitive even as it waited far too patiently for the new system to replace them. Information on the Department of Homeland Security's Cyber Talent Management System can be found at https://www.govinfo.gov/content/pkg/FR-2021-08-26/pdf/2021-17824.pdf.

DOD took a decidedly different approach than DHS. As a consequence, in just over two years it was able to implement comparable cybersecurity personnel flexibilities provided by Congress a year after the latter did so for DHS—at least initially for civilian personnel of NSA's counterpart, US CYBERCOM, as well as the various military components (like the Navy's 10th Fleet and the 23rd Air Force) that supported it. It did so in part by simply renaming NSA's variation of DCIPS as its Cybersecurity Excepted Service (CES) and extending it, albeit with some minor tweaking, to CYBERCOM and those various service elements. And even now, it is in the process of hiring hundreds of additional civilian cyber warriors under its auspices.[28] So, where it may count the most—that is, the protection of our Nation's military cybersecurity data, systems, and networks—we seem to be making progress.

And even as DHS begins to implement its scaled-down Cybersecurity Talent Management System for DHS CISA, the Biden Administration's Office of Personnel Management is making incremental reforms to the US civil service system that, while not directly targeted at the Federal government's cybersecurity workforce needs per se, will make it far more competitive in so doing. For example, OPM has issued new rules that will enable Federal agencies to hire newly minted college cybersecurity graduates 'on the spot' and without regard to veterans preferences that while laudable as a general matter, have impeded the US government's ability to bring in college talent, especially in technical disciplines. That 'on the spot' authority has some limitations, but if a Federal agency needs to recruit cyber talent, it can do so with far greater speed.

In addition, OPM has also made it far easier to bring back cybersecurity talent—especially previously cleared talent—that may have left US government employ for greener pastures in the private sector. As noted above, that trend is almost certainly part of changing post-Millennial career/work-life patterns. New graduates may take advantage of the military or Federal government's superior entry-level training but then once trained, depart for a more lucrative cybersecurity job with a contractor or company. However, at least anecdotally, there is some evidence that these so-called 'regrettable losses' may eventually want to return to the fold. OPM has changed its reinstatement rules to permit former civil servants to come back…not at their previous (low!) pay grade as before, but at a level that takes into account the experience they may have earned in the private sector.

For example, a graduate in a cybersecurity-related academic discipline can be hired relatively quickly by NSA or DHS CISA[29] (perhaps as part of a DOD or NSF scholarship pay-back commitment), attracted by their mission, the prospect of a security clearance, and the intensive training and work experience that those agencies may make available to them. However, as we have suggested, once they have completed the training—and whatever post-graduation service commitment they may have made upon initial appointment—they may be lured away by a Google

[28] See https://public.cyber.mil/cw/dod-cyber-excepted-service-ces/.

[29] Although security clearances continue to delay their actual EOD (entry on duty) date, this author's organization has proposed reforms in this area to DOD and OPM, but they remain pending their review.

or Hewlett-Packard, or even a Crowd Strike or Mandiant. Are they gone forever? No; there is increasing evidence that some of those who leave will eventually come back, bringing a wealth of experience with them, and if and when they do, the Federal government should—and now can—make it easy for them to do so, by simple reinstating them…but at a higher pay level commensurate with their non-government experience.

To be sure, agencies will have to be prepared to take advantage of this 'passport' authority, as well as all of the other flexibilities (in some cases, agency-specific) that we have cataloged above, and in so doing, accept more comings-and-goings amongst employees in certain occupational categories (like cybersecurity). Still, at least now, OPM rules permit them to do so. And if they do, they can begin to win the war for cyber talent.

Winning the War for Cyber Talent: Some Final Issues and Solutions

> So, here is what we know…or at least strongly suspect. There is a cybersecurity talent gap, not just when it comes to the public and private sector labor market, but more importantly, when it comes to our peer and near-peer geopolitical rivals in cyberspace. In sum, they are beating us when it comes to that particular war for talent. And to the extent that the cyberspace domain becomes contested—overtly in a hot war or more covertly (and more likely) in a cold or lukewarm one—this has significant national security implications for the US, especially given our country's disproportionate digital dependence.
>
> However, by the same token, the US government may also be the way to ultimately close that gap, via direct action (scholarships, hiring) and indirect intervention, as a deliberate developmental 'factory' that has the potential to satisfy national cybersecurity workforce needs beyond those of the government's *per se*. However, there are certain antecedent actions it must take if it wants to do so.
>
> In that regard, it is clear that the Federal government not only serves as a major recruiter of cybersecurity talent, but also as a significant developer of that talent…as a sponsor for a number of formal development programs, as well as an informal—indeed, so far largely inadvertent—source of individuals once trained by its agencies and now employed elsewhere. Most notable among the latter is the 'joint'[30] NSA/DHS Centers for Academic Excellence (CAE) program, which among other things, certifies that the curriculum of a growing number of colleges and universities—at both the undergraduate and graduate levels—meets certain cybersecurity content standards,[31] and that those courses of study can be traced back to the

[30] The reason this author has put 'joint' in quotation marks is that most observers (including this author) would argue that at least to date, DHS has contributed little to the effort, relying on NSA's funding and small staff to sponsor efforts that only incidentally support NSA's internal recruiting needs.

[31] NSA calls them Knowledge Units (KUs): https://www.nsa.gov/Academics/Centers-of-Academic-Excellence/CAE-CO-Advanced/.

National Cybersecurity Workforce Framework (NCWF) issued and periodically updated by NIST.[32]

The CAE program now boasts over 350 'certified'[33] schools and produces far more cybersecurity talent than NSA needs, and while DHS's Cybersecurity and Infrastructure Security Agency (CISA) has largely been absent in direct support of the CAE program, one would hope that this will change, given the interest (and impetus) of the US Congress in this area. Thus, Congress gave CISA upwards of $100M in the FY 2021 National Defense Authorization Act to fund national cybersecurity workforce development programs. As noted at the very beginning of this chapter, it also commissioned the National Academy of Public Administration to make recommendations to CISA in that regard.[34]

This chapter is not the place for a detailed description of the current or proposed cybersecurity workforce development efforts planned or undertaken on the part of NSA and DHS.[35] Still, the two agencies remain central to the Federal government's role in closing the cyber talent gap. Accordingly, in this final section, we will focus on the ability of the Federal government to fulfill two of its three roles in the cybersecurity labor market. That is, as a recruiter and developer of its own cybersecurity talent, and secondarily, as a deliberate developer of that talent for others in the US public and private sectors.[36] The latter is of particular interest here, given the extraordinary mobility of cybersecurity professionals, particularly those trained by such 'tip of the cybersecurity spear' three-letter Federal agencies as NSA, FBI, and CIA.

Thus, we posit that like it or not, the Federal government—through turnover (non-retirement and retirement) alone—will continue to supply a major fraction of trained and cleared cybersecurity talent to the rest of the Nation, whether it does so advertently or otherwise, so its ability to recruit and develop those individuals, whether or not they ultimately stay with the Federal government, is of significant interest here. Indeed, this author believes that those Federal agencies should accept and even embrace the role of 'cybersecurity talent developer' for the Nation and do more to take advantage of that phenomenon by sponsoring a true Cybersecurity Reserve Corps[37] or 'cyber 'militia' that can be mobilized in times of National need.

In that regard, as we have previously noted, the Federal government's ability to help the Nation close its cyber talent gap is actually aided by changing work-life and career patterns, as more and more students deliberately take advantage of government-sponsored training programs, like those offered by NSA and more

---

[32] NSA is in the process of updating its KU standards as this chapter goes to press: https://www.nsa.gov/Academics/Centers-of-Academic-Excellence/CAE-CO-Advanced/.

[33] NSA/DHS certification is not as rigorous as the typical academic accreditation process—for example, see SACS, AACSB or ABET accreditation standards for comparison—but they are nevertheless quite rigorous: https://sacscoc.org/accrediting-standards/.

[34] See Note 1 at the beginning of this chapter. As described therein, the US Congress asked the National Academy of Public Administration (NAPA) to study the matter, and the latter recently issued a report that includes such a description (link); as stated in Note 1, this author served as a Subject Matter Expert (SME) to the NAPA Panel and Study Team that completed that report, and his views are reflected therein.

[35] Ibid.

[36] The third role, that of regulator, is beyond the scope of this chapter.

[37] Such a cadre already exists, but its focus has been almost exclusively on recruiting US government retirees by eliminating some of these disincentives—like the annuity offset—associated with re-employing Federal annuitants (https://www.esd.whs.mil/Portals/54/Documents/DD/issuances/140025/140025v3005_dodi_2017.pdf?ver=2017-08-15-121335-967).

recently, DHS CISA (which by all accounts are world class) but then move on to other jobs, often in the private sector, after they complete them…taking their training, frontline work experience, and a top security clearance with them. And in this author's opinion, that is a good thing, as there is just not enough of that cyber talent to go around, especially if employers remain nostalgically rooted in the now outmoded 'once we hire them, they will stay' career path model. Like the *Ronin* of ancient Japan, many new cybersecurity employees—almost always the best of them—will still leave after they have been trained, as will more senior employees (again, almost always the best of them) when they have acquired the expertise and experience that will make them more marketable.

Thus, through programs like CAE, as well as more cybersecurity-related scholarships, the US can increase production of college-trained (and less-than-college-trained) graduates in STEM/Cybersecurity-related discipline. It can adopt a military model to develop and field 'bridge' programs that will take raw cyber grads and 'finish them' to meet the full-performance needs of cybersecurity employers, that is one solution to that talent gap challenge.

However, that also means that US cybersecurity professionals—especially those that are developed under US government auspices—must also be inculcated to act in the national interest…even as they may endeavor to protect the interests of their particular employer.

That is obviously easier said than done, but as detailed in the final section, models can be found in American history that may be relevant here. Think about the 'Minutemen' and other militias who were called upon to supplement our country's fledging regular armed forces during the Revolutionary War, or the citizens (many of them women) who provided civil defense during World War II and the Cold War. Civilians all, they were trained not only in the technical skills they needed to support the country, but the values and *ethos* necessary to do so as well.

And as noted, this model already exists, at least in part, in the US military's Reserve and National Guard components, in which enlisted and commissioned personnel serve part-time (typically after a period of full-time service in uniform) as the needs of the Nation dictate. Indeed, they are most often called upon to augment active-duty personnel in times of national emergency. In this author's view, a debilitating cyber-attack on our critical infrastructure, especially one sponsored or perpetrated by a rival nation-state, would certainly qualify.

So how do the US military's Guard and Reserves ensure that their personnel—all of whom have their own civilian 'day jobs' in the public and private sectors—act in the national interest when recalled? Through intensive annual training (for example, 'drill' weekends) and short periods of military service that not only keeps their skills current but also serve as a constant reminder of the oath they take…and the *ethos* that comes with it.

While this chapter is not the place for detail, suffice it to say that that model could easily be adapted to the many cyber *ninjas* who were first developed by the US government but who now work for other US employers…without the need for the rigors that are perceived to come with uniformed service. Thus, those cyber professionals, former public servants initially trained and developed by a three-letter

Federal agency, could be 'tethered' back to the US government through some sort of Reserve-like commission that included periodic training not only designed to keep their skills (and perhaps their security clearances) up to date, but also deliberately intended to remind them that they are part of something bigger…that is, that they serve the Nation in time of need.[38]

Unfortunately, we are not there yet. This author talked to an IT director for a West Coast patent law firm that had been attacked by ransomware, and he told me that his firm had had its data encrypted by a cybercriminal; however, because it had 'good' backups, the law firm refused to pay and had its systems restored in a matter of days. That is all well and good, but when this author suggested that the cybercriminals in his case may not have actually been after a ransom, but rather may have used their attack as a diversion to penetrate his firm's networks and just 'watch' for new intellectual property that they could exfil—a favorite tactic of certain nation-states—he said, "Why do we care? Our clients aren't losing any money directly, so if cybercriminals get advanced notice of a new product, no big deal!".

Well, aside from the potential theft of IP, and the loss of millions in R&D leading up to it, the fact is that it is in the national interest to care about this, even while it may not bother the interests of a single firm.

In this author's view, the US (and its workforce of cybersecurity professionals) need to acknowledge that the 'boundary' between public and private sectors has become blurred, not just from a labor market standpoint but also from a mission standpoint as well. Thus, national and private interests converge and conjoin. US cybersecurity professionals—especially those who have been trained by the military or NSA who protect our Nation's critical infrastructure—are 'dual hatted' as public servants, just as much as their uniformed and civilian counterparts in the US government. In other words, they must be loyal not only to (and protect the interests of) the employer they work for, but also to the US and its national interests. And if that is the case, we must rely on the judgments of individual private employers in those critical industries, or the individual skills and personal motivations of their cybersecurity professionals, to do the right thing for the Nation (vs. their company) in any particular cyber incident.

Bottom line: It's no secret that the United States faces advanced, persistent threats to the security of the networks and data that drive our daily lives—from state, state-sponsored, and non-state actors, to criminal organizations, hacktivists, and 'lone wolf' cyber terrorists, even insiders who work amongst us. How do we begin to defend ourselves against those threats? To take a lesson from the war on terror, this author (and many others) think it will take a network to defend our networks, but in this case, a human one: a network of citizen cyber-soldiers developed and deployed across our public and private sectors, all connected in support of our common defense.[39]

The Constitution's preamble empowers the federal government to provide for that common defense. In our republic's earliest days, that meant relying on those same

[38] See this author's views at https://www.federaltimes.com/opinions/2015/11/09/successful-cyber-defense-relies-on-collaboration/.

[39] Ibid.

citizen soldiers, drawn from among the 13 original colonies. When a threat arose, they came together as one, to fight for the greater good. This author believes that this simple and profound model can be adapted to address the cyber wars (declared and otherwise) that lay ahead…to look back to our origins and apply the citizen militia model to the cyber-defenders of tomorrow.[40]

In that regard, it is clear that our geopolitical rivals in cyberspace respect no boundaries, and those who defend us against them must be able to operate in kind, no matter where they work. The members of a boundary-spanning 'militia' of defenders would be in a position to respond quickly and collaboratively to a potentially crippling cyber-attack. And the foundation for that network of cyber-defenders has already been laid. President Barack Obama issued several executive orders during his Administration to promote the sharing of cyber threat information, and organizations and institutions are beginning to form together to do just that. However, sharing information is just the first step. That information needs to be translated into insights and actionable intelligence, something only a boundary-spanning network of cybersecurity experts—trained to the highest standards, vetted for trustworthiness, and connected by a shared *ethos*—can do.

Thus, authority and precedent exist to do this. Executive Order 13,636[41] instructs the Secretary of Homeland Security to bring private sector cyber experts into government on a temporary basis. Likewise, the intelligence community is authorized to create the civilian equivalent of the US military's vaunted ROTC program to develop civilian intelligence officers and analysts, including those who work the cyber beat.[42] However, these authorities have been woefully underutilized, and we believe they can be better leveraged to develop the sort of connected cadre of cyber defenders—think of them as a virtual Cyber National Guard—who can be 'called up' to bring their expertise to bear whenever and wherever needed.

Something like this has been done before. The nation created a similar network of intelligence operatives and analysts ten years ago, in part by requiring the sixteen different agencies that comprised the US Intelligence Community (IC) to share their best and brightest professionals, and in so doing, create a cadre of 'joint' intelligence officers that could "connect the dots."[43] Patterned after a similar requirement for military officers mandated by the Goldwater-Nichols Act of 1986,[44] the IC's initiative has produced a network of over 12,000 intelligence professionals, connected and collaborating in support of our common defense. HON Mike McConnell,[45] the former

[40] Ibid.

[41] https://obamawhitehouse.archives.gov/the-press-office/2013/02/12/executive-order-improving-critical-infrastructure-cybersecurity.

[42] See Note 21. See also Ref. [3].

[43] See https://9-11commission.gov/report/. This author was responsible for designing and implementing the IC's civilian 'joint duty' program when he served in the Intelligence Community from 2005 to 2010.

[44] See https://history.defense.gov/Portals/70/Documents/dod_reforms/Goldwater-NicholsDoDReordAct1986.pdf.

[45] VADM McConnell is now the Executive Director of the Florida Center for Cybersecurity, a co-sponsor of the Great Powers conference series, and this author's supervisor.

Director of National Intelligence, credits the program with changing the mindset and culture of the U.S. intelligence community, and recently told us that it was "one of the most fundamental and profound mechanisms to achieve a culture of collaboration among the organizations charged with identifying and thwarting our adversaries."

We need that same 'jointness' mentality in our cyber defense, with a militia-like network of cybersecurity professionals fundamentally bound together by common cause and collaboration, especially if we are to overcome the quantitative and doctrinal disadvantages that we may face in the global war for cyber talent.

References

1. Borum, R., & Sanders, R. (2020). Preparing America's Cyber Intelligence Workforce. *IEEE Security & Privacy*, 18(5), 67–73. Preparing America's Cyber Intelligence Workforce | IEEE Journals & Magazine | IEEE Xplore.
2. Ogrysko, N. (2021). Why the new DHS cyber talent management system was nearly 7 years in the making. *Federal News Network*.https://federalnewsnetwork.com/workforce/2021/11/why-the-new-dhs-cyber-talent-management-system-was-nearly-seven-years-in-the-making/.
3. Sanders, R. (2015). Successful cyber-defense relies on collaboration. *Federal Times*.https://www.federaltimes.com/opinions/2015/11/09/successful-cyber-defense-relies-on-collaboration/.
4. The Defense Civilian Intelligence Personnel System. An Independent Assessment of Design, Implementation, and Impact. (2010). *The National Academy of Public Administration*.https://s3.us-west-2.amazonaws.com/napa-2021/studies/the-defense-civilian-intelligence-personnel-system-an-assessment-of-design/FINAL-DCIPS-REPORT-June-2010.pdf.
5. William, M. (2021). (Mac) Thornberry National Defense Authorization Act (NDAA) for Fiscal Year 2021, Public Law No. 116–283. https://www.congress.gov/116/plaws/publ283/PLAW-116publ283.pdf.
6. Wagner, E. (2021). OPM Rule Makes It Easier to Hire Recent College Grads. *Government Executive*. OPM Rule Makes It Easier to Hire Recent College Grads-Government Executive (govexec.com).

Dr. Ronald Sanders is a Fellow of the National Academy of Public Administration and a decorated, twenty-year member of the US government's Senior Executive Service. He served as Staff Director for the Florida Center for Cybersecurity, where he helped coordinate cybersecurity research, education, and outreach efforts among Florida's 12 State University System (SUS) institutions. From 2017 to 2020, he served as the Director of the University of South Florida's School of Public Affairs, where he oversaw various graduate education and research programs in public administration and urban and regional planning, as well as its Florida Institute of Government. In addition, until his unintentionally wellpublicized resignation in October of 2020, he chaired the US Federal Salary Council (a Presidential appointment), which makes recommendations on pay raises for the Federal government's 2M+ white collar employees. He also served on the Advisory Board of the National Security Agency and Vice Chair of the National Intelligence University Foundation.

Leveraging Talent to Dominate in Cyber War—An Army Perspective

Colonel Chad Bates and Major Charlene Rose

Abstract In order to dominate in a persistent engagement with the nation's adversaries in cyber warfare, the United States Government must posture a skilled and experienced workforce through a strategic, nationwide partnership for a whole-of-society defense against adversarial threats. Recruiting and maintaining such a high performing workforce requires an innovative talent management plan that will prevail in an expanding competition on acquiring these highly skilled cyber experts. This plan will assist the United States Cyber Command (USCYBERCOM) to effectively enforce the nation's cybersecurity objectives in the growing great-power competition within cyberspace. In this domain, adversarial cyber strategies range from malicious and sporadic, to calculating and cohesive. Requiring an agile national U.S. cybersecurity posture to engage against these threats through innovative technological solutions, maximizing cyber human capital, and continuously evaluating adversarial cyber strategies. The ingenuity of the human brain is critical in meeting the nation's strategic goals in the cyber domain. While also leveraging technological advancements in order to enable the pursuit for dominance in this great-power competition. Across the continuum of engagement within cyberspace, the country must leverage its cyber talent across all sectors in order to more effectively meet cybersecurity demands through streamlined lateral linkages as a whole-of-society defense. Nationwide partnerships will support the response rate necessary for the evolving cyber domain, leading to an agile cyber workforce reinforced by the matrix of U.S. cyber agencies and organizations. The cyber posture strategy must consider talent recruitment, development and retention to successfully meet talent management goals. This paper will review the U.S. Army Cyber Command's perspective of these key talent management components as it relates to a national, whole-of-society cyber security concept.

C. C. Bates
U.S. Army War College, Carlisle Barracks, PA, USA
e-mail: chad.bates.mil@armywarcollege.edu

M. C. Rose (✉)
United States Military Academy, West Point, NY 10996, USA
e-mail: c.a.rose828@gmail.com

A. Farhadi et al. (eds.), *The Great Power Competition Volume 3*,
https://doi.org/10.1007/978-3-031-04586-8_16

Keywords Cyber war · Cyber warfare · Russia · China · Iran · North Korea · United States

Introduction

In order to protect national interests, the country requires a foundational cybersecurity workforce, trained to improve base cyber defenses and to protect critical infrastructure against existential threats. The nation also needs a formidable force of highly skilled, technical forces specialized to conduct classified operations, utilizing cutting edge technologies, to act in the interests of the United States against its adversaries. These specialized cyber warriors are some of the most technically skilled in the government. Shrouded cyber warriors can impose debilitating costs upon the nation's adversaries to provide credible deterrence, backed by the power and resources of the U.S. government. They are the strategic weapons of tomorrow that can hold critical infrastructure or adversarial offensive cyber capabilities at risk, analogous to the nuclear weapons of the Cold War. While the U.S. government, the academic community, and the collective cyber industry can train personnel on technical cyber skills, specialized cyber warriors must have the unique attributes, mental acuity, and grit that makes them indispensable. Similar to the high-caliber missions carried out by the U.S. Army Special Forces relative to other ground forces, highly skilled cyber warriors operate on an advanced level relative to the rest of the cyber workforce. Additionally, operational feedback from this workforce can inform the training curriculum to remain nimble within the continuously changing cyber domain. This adaptation will provide the United States government the required edge to remain dominant in this environment now and in the future. As Peter Senge wrote in his book *The Fifth Discipline* and echoed by Arie de Gues in his book, *The Living Company*, "The ability to learn faster than your competitors may be the only sustainable competitive advantage." The Department of Defense must invest data-driven strategies into its human capital to maintain a competitive edge in cyberspace.

U.S. Goal to Dominate in Cyber Warfare

The U.S. aims to posture a resilient force that reduces the magnitude and capabilities of adversaries in cyberspace. USCYBERCOM established its vision to focus on "defending forward" to increase resiliency and maintain continuous engagement in the cyber domain [32]. *Defending forward* shifts the Army Cyber Command posture away from "be prepared," the Command's defensive posture focus in 2015; instead, *defending forward* moves the point of conflict closer to adversarial networks. Doing so meets the intent of USCYBERCOM's Commanding General, General Paul Nakasone, who envisions a persistent cyber force that imposes greater costs on adversaries and their decision calculus [19].

To streamline the integration of the operational strategy into the USCYBERCOM's talent management blueprint, the command intends to collaborate across combatant commands, services, departmental agencies, and the industry. The collaborative effort supports operational goals and intends to attract new critical mission partners, such as the Defense Information Systems Agency (DISA), the National Security Agency (NSA), and the broader Intelligence Community (IC). NSA already bridges the gap between the DOD and the IC across all battlespace domains—especially cyber. In order to accomplish this work, the NSA's Center for Cybersecurity Standards streamlines collaborative efforts between the U.S. government and the industry [22]. Following these efforts, cross communicating insights will enable Cyber Command to inform commanders and civilian government leaders in the policymaking and national security decision-making process [31]. Additionally, a rigorous information sharing system strengthens USCYBERCOM's predictability of cyberattack effects. Furthermore, a streamlined system elevates operators' understanding of how to respond to indicators of discreet cyberattacks and associated behaviors of processes and systems. However, operational partnerships must take careful considerations in making adjustments to accessibility barriers, such as security clearances, to mitigate threats by rogue operators within the bureaucracy or other malicious enterprises seizing newfound connections [18]. Collaborative insight, as a result of integrated operations, will allow the U.S. to achieve sustainable cyberspace superiority.

Cyberspace Superiority

USCYBERCOM requires dominance in security and conduct of operations, by working as a joint force with anticipatory persistence, to contest adversarial actions (JP 1-02 2016). Due to the constantly evolving threats within the cyberspace, USCYBERCOM requires an evolving doctrine to assist in developing policies and capture processes that maintain a decisive advantage through integrated operations. The 2018 USCYBERCOM Vision breaks down its guiding criteria for superiority through persistence by describing an operational philosophy: USCYBERCOM's continuous defensive and offensive operations must maneuver seamlessly to create an operational advantage for us, while denying the same to adversaries by keeping operations at close proximity to adversarial networks.

Achieving effective operations through agile forces, operational partnerships, and persistent engagement requires a multi-faceted talent management strategy. Developing an agile force requires retention of talent to fill key roles to support Cyber Command missions and to build an experienced and mature cyber workforce. Placement and retention within the broader Cyber Command formation is a top priority for the talent management blueprint. However, establishing operational partnerships can offset voids in capabilities as a result of attrition into other cyber agencies or sectors. Operational partnerships are a major factor for strategizing the management of the Nation's cyber talent pool. Streamlining public–private partnerships in the U.S. by reevaluating career pipeline opportunities and expanding current partnership programs is a proactive measure to counter adversarial cyber threats to the United States. The global cyber talent shortage implies a competitive labor market that expands beyond domestic sectors.This market competition creates an impetus for a whole-of-society defense approach to preserve national security interests. Integrated operations will therefore stabilize the foundation for persistent engagement and cyber strategies moving forward.

Shifting Cyber Command's emphasis from a secure defense posture to active engagement will need a sustainable talent management blueprint that withstands the cybersecurity dilemma. The security dilemma arises as forces develop or deploy due to the notion that states induce fear in other states as they secure themselves [7]. Acknowledging the implications of talent management reform on the motives and momentum of the nation's adversaries should not call for a heedful approach; rather, Cyber Command should take a prescient approach in strategizing talent management adaptations. Ben Buchanan's six-step defensive model in *The Cybersecurity Dilemma* (preparation; detection; data collection; analysis; containment; and decontamination) is useful for framing the combination of technological and human capital needs in outpacing adversaries in light of the security dilemma. Integrating universal defensive models, like that of Buchanan, into the talent management strategy can shape a more effective human capital model and cyber operation strategy. USCYBERCOM's talent management blueprint should fulfill key components of a defensive model. Using a defensive model as a basis for both a human capital strategy and an operational strategy can create a strong bridge between mission critical and enabling components within the cyber domain. While the talent management strategy should have the capacity to evolve with changes in cyber war, a blueprint should still consider universal defense models to establish a foundational philosophy with longevity. A seamless integration of an effective operational model into the human capital model can allow Cyber Command to posture superior cyber defenses against the nation's adversaries while minimizing risks of adversarial retaliation.

U.S. Cyber Capabilities

The U.S. military aims to dominate in three key components of cyber warfare: exploitation, attack, and defense [4]. U.S. Cyberspace operations contribute to multiple levers of national power through its role in establishing sanctions, communicating to adversaries, combatting adversarial information operations, facilitating overmatch of adversary military capabilities in all domains, and intelligence operations. These dynamic roles within the cyber domain requires the U.S. to maintain a whole-of-society approach to keep pace. A key component for establishing a strong cyber talent pool in the U.S. is through education and development programs. Academic communites within the U.S. continue to develop education curriculums to prepare students for information security roles. While bachelor degrees in computer engineering, computer science, software engineering, and information systems historically lead to cyber roles, an assessment by the Joint Task Force for Computing Curricula found shortfalls within these programs [24]. Their work identified shortfalls based on a particular set of qualifications and knowledge areas for different information security careers [16]. Therefore, as cybersecurity operations demand more from information security roles, program curriculums must continuously evolve to maximize the strength of the U.S. cyber workforce.

The Security Agency Act of 2018 elevated the mission of the former National Protection and Program Directorate (NPPD) within the Department of Homeland Security (DHS), establishing the Cybersecurity and Infrastructure Security Agency (CISA) to build the national cybersecurity capacity. CISA and USCYBERCOM, maintain the strategic-level cyber operations domestically and militarily, respectively.Both agencies achieve and maintain cyberspace superiority through the management of critical cyber tools, talent, infrastructure, and partnerships as part of a whole-of-society defense. CISA's 2020 Year in Review highlighted its priorities on "election security, supply chain risks from cyber threats, protecting the.gov domain, soft target security, and critical infrastructure protection," and the COVID-19 pandemic and election events served as guiding frameworks for the CISA cyber campaign that year. During the same timeframe, USCYBERCOM continued to evolve its workforce across the supported combatant commands and agency partnerships by identifying cyber domain initiatives to address the retention of critical work roles, monitoring and outpacing adversarial cyber operations, and building the Cyber formation.

The Department of Defense (DOD) is responsible for defending the U.S. homeland and interest from attacks, including those that occur in cyberspace. DOD's cyber domain includes cybersecurity and operations to defend DOD networks, systems and information, which includes support for operational and contingency plans. Specific cybersecurity activities include information sharing, interagency coordination, partnerships with the private sector, and expanding to partnerships abroad. It also acknowledges that national strategic deterrence in cyberspace requires the collective effort of multiple government departments, agencies, and partnerships. In 2015, the DOD Cyber Strategy established a baseline for the DOD's order of cyber domain priorities: (1) to defend DOD networks, systems, and information; (2) to defend the U.S. and its interests; (3) provide integrated cyber capabilities for military operations or contingency plans at the direction of the President or the Secretary of Defense. In order to execute these priorities, the Cyber Mission Force (CMF) is comprised of highly skilled cyber operators organized into sub-groups to support the national cyber strategy [5]:

Cyber Protection Forces: an augment to traditional defense measures to defend priority DOD networks and systems against priority threats; serves the authority of combatant commands.

National Mission Forces: teams that defend the U.S. and its interests against cyberattacks of significant consequence; serves the authority of USCYBERCOM.

Combat Mission Forces: teams that support combatant commands by integrating cyberspace effects into operational and contingency plans; serves the authority of combatant commands.

As a Service Cyber Command under USCYBERCOM, the U.S. Army Cyber Command (ARCYBER) continues to evolve its workforce to meet the USCYBERCOM intent. ARCYBER develops its active duty, Army national guard, Army reserve components, and Department of the Army Civilians to achieve cyberspace superiority and contribute to national strategic deterrence. U.S. Cyber Command aims to achieve the commander's intent through five imperatives that mutually support and enhance success (Fig. 1). The five imperatives serve as a framework for cyber operations and the management of cyber personnel.

Overall, USCYBERCOM coordinates and executes cyberspace operations for its 133 cyber teams through four Joint Forces Headquarters for Cyber (JFHQ-C), operated by each Service Cyber Command, and the National Cyber Mission Force

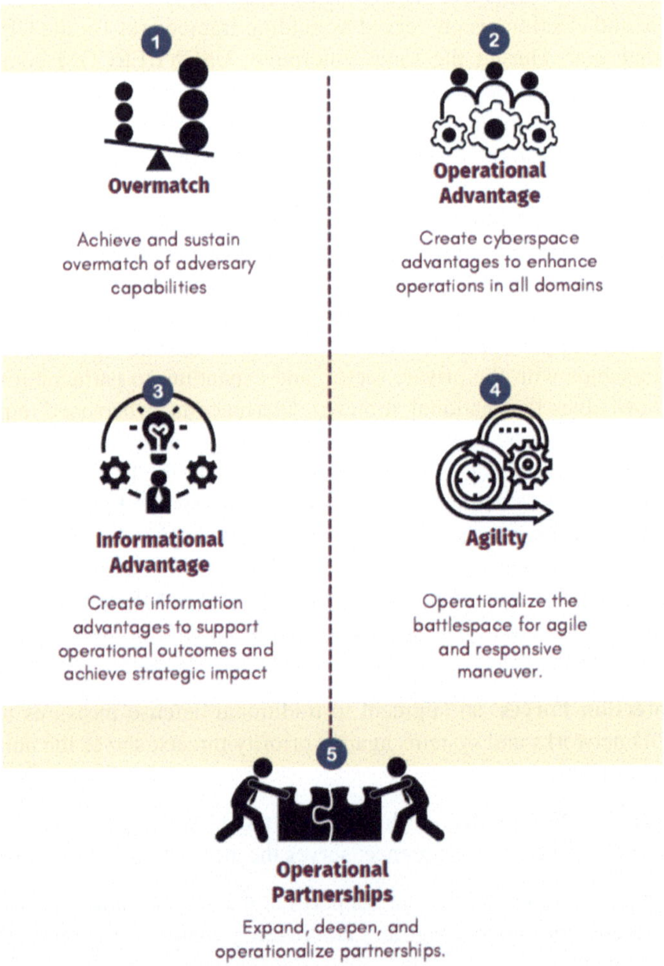

Fig. 1 US Cyber Command's five imperatives for mission success

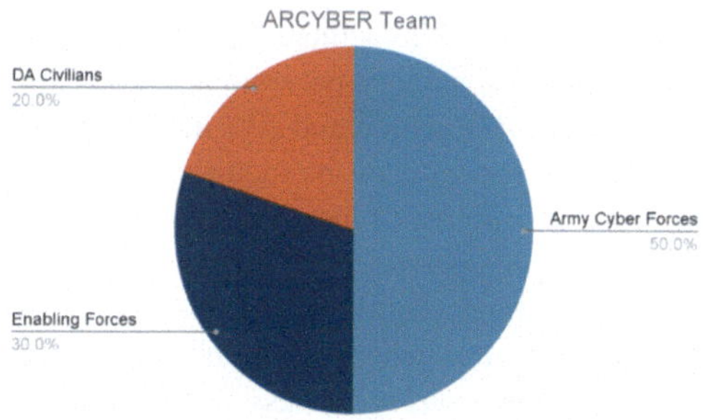

ARCYBER Teams——Who are in these Cyber Teams?
ARCYBER teams are composed of officers, warrant officers, and enlisted forces from the Army's Cyber branch. Additionally, there are warrant officers and enlisted personnel from the Military Intelligence, Fires, and Signal Corps branches that form the enabling forces of the team. Department of the Army Civilians are the enduring members of the team and execute enabler and core cyber functions.

Fig. 2 Distribution of cyber professionals in team structure

Command. These commands also collaborate and coordinate with different U.S. Government and Intelligence Community organizations. This collaboration enables unity of effort, appropriate visibility of threats and operations, and improves the responsiveness and effectiveness of these missions.

ARCYBER is responsible for manning, training, and equipping the 41 active duty cyber teams (about 1800 personnel) of the 133 teams within USCYBERCOM's CMF. Additionally, the U.S. Army National Guard and the U.S. Army Reserve are building an additional 10 and 11 cyber teams respectively, for a total of 62 Army teams (around 2600 total personnel) by the end of 2024. Personnel on these ARCYBER teams are comprised of professionals from the Army's cyber branch, enabling forces, and DA Civilians in distributions annotated in Fig. 2 [5].

Of the 1800 positions on the 41 active duty teams, sixteen different military occupational service (MOS) types and two civilian operational series comprise the team structures. Additionally, these positions are categorized into 26 different work roles on the teams which can be unique to specific MOS types or filled by different ones depending on individual aptitudes and talents. CMF work roles serve specialized jobs on different cyber teams, requiring specific training and qualification requirements to enable special access to unique equipment and information within the intelligence community. These work roles are defined by USCYBERCOM and prescribed the appropriate training, qualification standards, and certifications in order to execute their specific role to support operations.

Within each work role there are three proficiency levels, each requiring additional training, exams, and years of experience in order to qualify. These proficiency levels start at the basic level, requiring initial training and qualifications to meet. Depending on the work role, this requires an additional three to 30 months to complete. Senior proficiency requires additional training and two to four years of experience. Master proficiency requires a significant demonstration of skills and another additional 3–4 years of experience above senior proficiency. The longest qualification timeline takes around 10 years from entry into the operational force to achieve Master proficiency. Overall, a healthy cyber force should have a good balance between new and experienced personnel, with an average of 40% basic, 40% senior, and 20% master qualified personnel. Higher proficiency levels of the force indicate its maturity, experience, and skill level of the force, and quantifying the number of these personnel with these proficiencies allows the force to sustain itself in a persistent engagement.

Critical work roles, considered mission critical, dictate the number of missions assigned. An analogous example is seen in aviation, where the number of available pilots determine the capacity to complete a mission, regardless of the number of aircraft available. Cyber critical work roles take the most time to fully qualify for missions, require unique attributes and abilities, and their training is also the most expensive. A *defend forward* posture requires *network operators* to enter adversarial systems to execute the operation, as directed by *exploitation analysts* who guide them through the labyrinth of barriers to access. *Capability developers* serve as the enabling component to these offensive operations, creating the tools or code to gain entry and impose cyber effects. Our *increased resilience* operations require *network and host analysts* as well as *analytical support officers* to to fortify cyber defenses by hunting for variants in our own networks. They then remove any trace of adversarial code, harden the networks against similar actions, and continue hunting within the larger enterprise. However, these key work roles can only execute their missions based on the efforts of those serving in other work roles within the high capacity CMF. These team members conduct all of the analysis and planning required before, during, and after operations to make these missions successful. Each requires their own unique operational and technical skills and a sophisticated knowledge of how to integrate themselves into cyber operations. Individual skills and talents are valuable in cyberspace operations but the Army relies upon cohesive teams to be effective and to dominate in warfighting domains.

Commercial Competition

Global: The global shortage of cyber professionals currently stands at a 3.12 million gap between workers and jobs available, according to $(ISC)^2$. While the current standing is an improvement compared to the 2019 report, uncertainties from the pandemic contributed to this trajectory shift. For example, cyber professionals reported losing their jobs due to the pandemic. Additionally, organizations reported downsizing efforts to save costs on personnel expenses in order to redirect spending to technological expenses. The global cyber talent shortage continually increased each year, peaking at 4.07 million in 2019 prior to the first drop in 2020. While the cyber labor market saw losses as a result of the pandemic's impact on many businesses, organizational leadership across industries recognize the impetus of having strong cybersecurity for their businesses. The industry illustrated their recognition of the importance of strong cybersecurity through contingency measures such as budgeting cyber ransomware into fiscal year plans. This effect may lead to the cyber workforce gap reverting back to its preceding pattern of growth due to a heightened demand for effective cybersecurity. Furthermore, the 2020 $(ISC)^2$ Cyber Workforce

Estimate suggests that the global cyber workforce must grow by 89% to meet its cybersecurity demands. Meanwhile, unemployment among cyber professionals is about 2% globally. The competitive cyber labor market therefore continually inflates wages across all sectors.

United States: The following analysis by Colonel Chad Bates and Major Charlene Rose for this publication reviews the cyber labor competition within the United States:

The U.S. contains the largest population and gap of cybersecurity professionals (ISC^2 Cybersecurity Workforce Study 2020). Emsi, a labor data analytics firm, estimates that the national cybersecurity demand of 370,295 jobs is twice as great as the supply of 178,443 qualified candidates (Emsi 2020). Figure [2] illustrates the nearly two to one ratio within the United States cyber labor market.

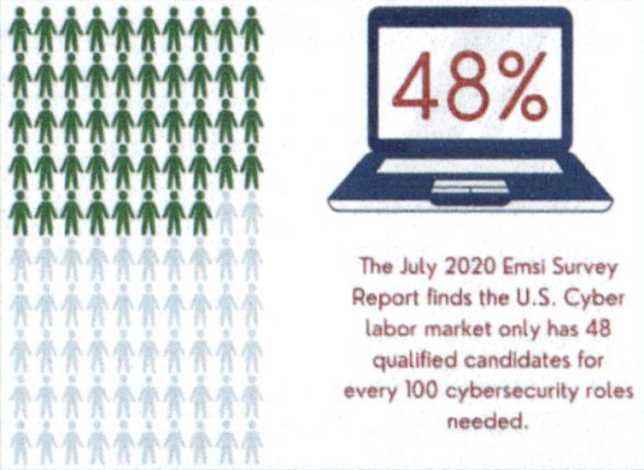

This shortage and demand has created an extremely competitive market for cybersecurity talent. Recruiters and employers offer top benefits and compensation packages to skilled and experienced workers, usually trying to entice them from their current employers. According to the Enterprise Strategy Group-Information Systems Security Association International (ESG-ISSA) survey of cybersecurity professionals, 40% of respondents stated that they are solicited by recruiters at least once per week while 72% are recruited at least once per month. Over the four years of the project and surveys, these recruitment activities have been consistent and will likely grow in the future. (ESG-ISSA Survey Report, 2020 - pg. 41). Additionally, 38% of respondents responded that the skills shortage has led to cybersecurity teams being unable to learn new techniques or fully implement current procedures due to heavy workloads (ESG-ISSA Survey Report, 2020).

Adversaries in the Great-Power Competition in Cyberspace

The nation's adversaries in the cyber domain consist of state actors as part of the great-power competition, criminal organizations, and individual malicious actors [5].

Cyberspace threats continue to grow and expand trans-regionally as a means for other states to offset their comparative disadvantages in armed conflict, compromising U.S. national security. The 2018 DOD Cyber Strategy identifies threats to include intellectual property theft, meddling in democratic processes, and other disruptions to the U.S. government and commerce. USCYBERCOM has identified serious investments in cyber military capabilities by Russia, China, Iran, and North Korea, reducing the U.S. competitive advantage. The Department of Homeland Security (DHS) identified Russia and China as the most capable nation-state cyber adversaries; however, they also determined that Iranian and North Korean cyber actors pose serious threats to U.S. interests. Technological advances, while beneficial to USCYBERCOM and national cyber interests, can also accelerate the demonstrated capabilities of other states who have undertaken sophisticated cyber campaigns.

The 2018 USCYBERCOM Command Vision recognizes that adversaries acknowledge the United States' traditionally high threshold for response to adversarial cyber activity. As a result, adversaries exploit U.S. vulnerabilities to offset their inferior physical warfare capabilities through advancement of their cyber operations. The 2018 USCYBERCOM vision document stated: "adversaries operate continuously below the threshold of armed conflict to weaken the nation's institutions and gain strategic advantages." Based on these near-peer or other great powers intentions, we can expect these adversaries' continued investment in the talent that enables these activities to meet their objectives within cyberspace. Instead of building more physical or kinetic military forces, they are building large organizations of military and quasi-civilian cyber personnel motivated to work for their governments. The United States government needs to have a cyber workforce that can compete against these adversaries. A competitive US force is mature in making decisions, experienced in cyberspace activities, and highly skilled to outmaneuver these competing forces. The success of cohesive adversarial operations should imply a greater impetus for a whole-of-government US cyber defense and eventually a whole-of-society defense.

RUSSIA

The October 2020 threat assessment by DHS identified that Russia possesses sophisticated cyber capabilities and can use cyber attacks to cause damage to critical infrastructure in the United States. They have demonstrated a capacity to execute cyber-enabled influence campaigns that disrupted other state internal affairs. Russia accomplished these sophisticated cyber effects through disinformation campaigns, espionage, and destructive cyberattacks. Further analysis on these operations has indicated a fragmented command and control structure since Russian cyber units appear to not synchronize their operations and oftentimes compete with each other on the same targets without apparent awareness of each other (Bowen, 2021). Original recruitment and co-optation strategies for Russian cyber operations relied heavily on criminal and civilian hackers, but Russia illustrates constraints to its cyber capacity as a consequence of recruitment and resourcing issues. Overall, Russia's cyber operations remain disjointed, leading to competition among its domestic agencies for personnel, resources, and influence (Ertan et al. 2021).

CHINA

DHS expects China to focus its cyber operations targeting the U.S. in critical manufacturing, defense industrial base, energy, healthcare, and transportation sectors. The FBI associates the Chinese threat with its talent plans, which are talent recruitment programs that bring outside talent into the country. Further analysis indicates that China's talent plans are malicious in nature, incentivizing participants to break laws and steal trade secrets and technologies to advance Chinese interests. Talent recruits enter into contracts with Chinese organizations, that are oftentimes affiliated with Chinese government, that require them to follow Chinese laws, contribute to Chinese-exclusive information sharing, and to assist in the recruitment of more talent (Heginbotham et al. 2015). In contrast to Russia, China is approaching cybersecurity in a seemingly unified way and constitutes a greater cybersecurity threat to the U.S., based on talent plans identified by the Intelligence Community.

IRAN 

Iran's intelligence managers bid out its cyber tasks to multiple contractors to execute reactionary cyber operations that lack control and cohesion. The bidding system utilized for Iran's cyber operations creates competition between contractors vying for influence in the Iranian government (Gundert et al. 2018). This tiered and mixed cyber operation system staffs a combination of government and private cyber operators leading to some successful attacks, but an overall lack of trust within the Iran cyber community. To date, Iran's cyber attacks on the U.S. have been retaliatory in nature with operations usually based on political responses to actions such as economic sanctions. While Iran's historical pattern of cyber retaliation enables US cyber operators to anticipate Iranian cyber attack plans, Iranian cyber attacks on foreign targets have grown increasingly aggressive (Theohary, 2020).

 **NORTH KOREA**

The North Korean government commands a centralized, controlled cyber operations force that predominantly uses criminal hacking for monetary gain (DHS 2020). Monetary benefits support the North Korean political goal as a self-purported socialist autarky (Fisher 2016). Despite the isolation and lack of internet access for the North Korean citizenry, the state has produced some of the most proficient hackers (King 2019). North Korea illustrates its capabilities through criminal hacking that targets banks, networks of major corporations, South Korean military networks, and through concerted cyber attacks - with some including partnerships with China (HHS Cybersecurity Program 2021). Despite economic sanctions, North Korea continues to fund its military programs by stealing billions of dollars using cybercrime (DOJ 2021). While North Korea's cyber strategy does not present the biggest threat to the U.S., they've illustrated the power derived from controlled and deliberate cyber attacks that may enable them to gain more of a competitive advantage through the continued building of strategic programs such as their nuclear weapons program.

Summary

In order to stay competitive in this war for cyber talent, US Cyber Command and ARCYBER must not only continue to monitor and analyze the cyber labor market but also ensure its workforce is prepared for the country's shift toward a cohesive approach to persistent engagement in the cyber domain. A whole-of-society defense, composed of a mature and persistent force, which leverages cyber talent is a critical aspect for meeting the five imperatives from the 2018 USCYBERCOM Vision. ARCYBER should consider the national approach as it relates to the shift towards defending forward. In contrast, a quasi-capitalist approach to talent management as a symptom of the global cyber talent shortage will prevent the U.S. from effectively executing any of the imperatives. While market competition can arguably lead to more technological innovation, streamlined partnerships can achieve similar or better innovative progress while strengthening national strategic deterrence of adversaries in the cyber domain. Integrating these contextual considerations into the ARCYBER operational and people strategy will allow the CMF to effectively streamline cyber operations, keep the point of conflict closer to adversarial networks, and in turn, reduce the attack surface. Understanding the complexities in the national cyber strategy shift to defending forward must consider implications related to the competitive cyber labor market. However strategists must bridge the gap of analysis by ensuring the human capital model aligns with the operational model to maximize operational effectiveness. Otherwise, the U.S. Cyber strategy may experience issues related to disjointed national operations between the public and private sector, leading ultimately to a greater national security issue.

Talent Recruitment: Acquiring the Most Talented

In 2014, the Army created the Cyber branch—creating the military occupational specialty (MOS) code of 17A for cyber operations officers, 170A for cyber warfare technicians—the warrant officers of the cyber branch, and 17C for cyber operations specialists for enlisted personnel. The Cyber branch established the United States Army Cyber School within the Cyber Center of Excellence at Fort Gordon, Georgia in 2014, which began to teach courses in 2015. Since this establishment, the Army has successfully recruited some of the brightest cyber professionals to serve either in uniform or as a Department of the Army Civilian.

In order to qualify for the enlisted Cyber Operations Specialist (MOS 17C), recruits must receive qualifying scores on the Armed Services Vocational Aptitude Battery (ASVAB). The minimum required scores for 17Cs are 110 on the general technical exam and 112 on the skilled technical exam (80th percentile of all Army recruits). Candidates must also score a minimum of 60 on the Information Communication Technology Literacy (ICTL) test before they are accepted for accessions. Another incentive for candidates is the $60,000 signing bonus for a six-year contract of service. This contract obligates individuals to serve five years following Advanced Individual Training (AIT). Historically, ARCYBER fills 17C positions within 48 hours of opening, indicating a competitive offering for 17C careers.

For those personnel already within the Army and holding the rank of Sergeant (E5) or higher ranking can apply to be a Warrant Officer and serve as a Cyber Warfare Technician (MOS 170A). Warrant officer candidates must hold the appropriate clearances and specified certifications (like SEC + ; CASP; or CISSP), nine semester hours of college, and three years of experience in cybersecurity operations. As the technical experts of the Army, Warrant officers have an equally demanding selection process to be eligible to begin the Warrant Officer Basic Course and to become a 170A. Candidates will be evaluated by a board after meeting the minimal qualifications stated above. On average, the Office of Chief of Cyber (OCC) evaluates 900 packets every year to fill the available 120 Cyber Warrant Officer slots in the school—a 13% selection rate. Based on the qualification process and selection rate, 170A candidates face a competitive career pipeline.

Cyber Commissioned Officer Candidates need to obtain and maintain the appropriate clearances, hold a Bachelor's of Science (or higher) degree in specific STEM disciplines, and pass the Cyber Branch assessment. The assessment includes an interview, aptitude assessment, written questionnaire, and requires other supporting documents as part of a packet review. In 2019, about 3,000 Reserve Officer Training Corps (ROTC) cadets competed specifically for Cyber Branch for their commission, two percent of which were selected (filling almost 60 positions). The Office of Chief of Cyber (OCC), at the Cyber Center of Excellence at Fort Gordon, tests officer applicants on foundational skills required to assess Cyber Officer (17A) minimum requirements. The top 400 scores of the exam are picked for further interviews to assess their cyber knowledge, interests, and aptitude to be a successful Cyber Officer. Of the 400 cadets, less than 60 were selected for the limited cyber branch slots. The extremely low selection rate relative to the volume of 17A candidates indicates a rigorous qualification process. Furthermore, ARCYBER may need to investigate the preparatory pipelines for a 17A career to enhance candidate eligibility for the branch. Lastly, ARCYBER should consider reevaluating the limitations on 17A slots to ensure it minimizes the amount of talent missed each year.

Another way to enter into the Cyber branch is through the Cyber Direct Commission Program. Since early 2018, the OCC has received and evaluated almost 1,000 Direct Commission application packets. As of early 2021, 16 officers have been appointed (2 Majors, 9 Captains, 5 First Lieutenants) with 11 additional candidates currently processing through requirements (1 Lieutenant Colonel, 2 Majors, 5 Captains, 3 First Lieutenants). Most are for active duty appointments, but appointments in the Army National Guard and Army Reserve have also been accomplished. These officers have proven skills and experience in many fields, to include data science, machine learning, software developers, access developers, hardware and radio frequency engineers, DevSecOps Engineers, and data architects. In March of 2020, this program was opened to active duty enlisted and warrant officers to provide opportunities for direct commission into the officer corps.

Department of the Army (DA) Civilians may join the Cyber Corps and often arrive already qualified for their work roles based on their prior military service. The high quality of these civilian candidates results in less training compared to new personnel to qualify as mission ready. However, competition with the commercial cyber sector, security clearance requirements, and opportunities in other government agencies results in about twenty percent consistent vacancy rate with high turn-over.

Cyber Warriors by the Numbers

Overall, the Army has a recruitment and assessment program that is bringing in top-tier talent from the U.S. population. There are many applicants for the limited positions and those interested score in the highest percentiles of eligible candidates. They see serving in Cyber as an exciting and rewarding career, where they will receive premier training and education, tied with experiences that can only be conducted within the authorities of the U.S. Government.

A comparative analysis of Army enlisted (17C) recruit data from 2015 to 2019 shows stronger candidates pursuing Army Cyber branch compared to candidates for other Army branches. Twenty-three percent of Army Cyber recruits already possessed a bachelor's degree when they enlisted. To compare, only five percent of the annual, total recruit population holds this degree or higher. Thirty-five percent of

Cyber recruits scored in category 1 on the Armed Forces Qualification Test (AFQT) compared to only five percent of the total, annual recruits population. The AFQT uses a combination of a candidate's ASVAB subtest scores to calculate an overall score to rank all recruits against each other and category 1 score accounts for the top 93rd percentile of all recruits. Lastly, the average age for Army Cyber candidates is 23, compared to the Army average age of 20 for candidates across the Department. The higher average candidate age implies varying levels of experience in different fields before choosing to join the Army Cyber branch. Fifty percent of Cyber candidates are ages 18–23, forty percent are ages 24–29, and ten percent are ages 30–42. Overall, Army Cyber candidates hold comparatively higher credentials, evaluation metrics, and talent potential.

Given these outstanding candidates, the Army has designed a program to provide them the skills and tools to develop into highly skilled cyber warriors. In 2021, the Army easily accessed enough candidates to meet its quota of 530 enlisted candidates, 60 warrant officers, and 130 officers. Of those enlisted recruits that arrive at the US Army Cyber School every year, almost a fourth fail to meet course standards. Most of the failures are due to academic reasons—related to the difficulty of the curriculum, however security clearances and other reasons also contribute to this rate. Individuals that can't meet these requirements are usually re-classified to a different Army branch. Since replacements can't be placed into the training cycle for these failures, this loss of a seat in a class cohort means less personnel for future missions. This discrepancy in the enlisted screening process requires an improvement in initial screening processes to better identify candidates that can handle the rigor of initial training courses in the U.S. Army Cyber School. A task the Army Talent Management Task Force and other organizations responsible for recruiting are addressing.

Understanding the complexity of cyberspace, it will take individuals from many different backgrounds and experiences to remain dominant in this competition. Over the years, the Army has been improving the diversity of the candidates it accepts into the force. For commissioned officers, the number of females that branched Cyber has increased from 11% in 2017 to 20% in 2021. This has been largely due to more female cadets pursuing STEM degrees at the United States Military Academy and in other universities with ROTC programs. Ethnic diversity has also been steadily increasing, with non-white officers making up 43% of the new class cohort in 2021, a significant increase from 29% in 2017. An area of improvement in racial diversity is the number of black or African-American officers, which saw major increases in 2020, but reduced back to only 6–8% of the incoming class in 2021. In order to gain more candidates in these roles, the nation needs to create more opportunities for women and other racial minorities in secondary schooling across the United States. Students' time in college or the academies invigorates desires to pursue cyber educational opportunities.

Warrant officers have also made significant increases in recruiting female non-commissioned officers to pursue a cyber warrant career. In 2017, only 3% of the new class were female but this has increased to 20% of the class in 2021—a significant increase. Additionally, each class has also been very ethnically diverse, with 40–50% of the new classes being non-white. A large success in diversity is shown in the 2021 classes where 30% of the class were black or African-American. The success with warrant officers is that larger portions of the technical branches are becoming more diverse, which has been seen in the pool of candidates that have applied to the cyber corps. A great trend for the Army and for the cyber branch.

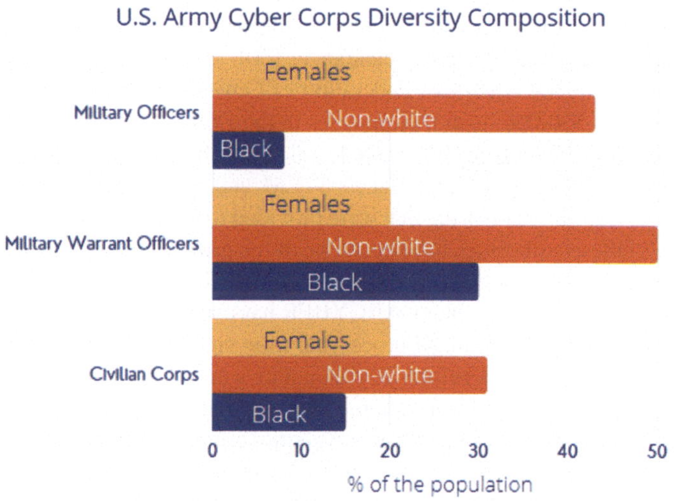

Given these statistics, the Army Cyber workforce is comparable to the commercial sector's diversity makeup for people in technically advanced jobs. Nelson Frank compared a portion of their 27,000 ServiceNow professionals in their 2020–2021 salary survey. Of those that responded, 81% identified themselves as male and 16% female with the remaining 3% not specified. The diversity composition of survey respondents were comparable at 45% non-white, while only 3% of respondents were Black. While the Army Cyber workforce is comparable to the commercial sector, it continues to strive for greater diversity to reap the positive benefits of diversified human capital.

Entering as a Department of the Army Civilian

Even though there have been improvements on hiring DA civilians into cyber positions, there are certain barriers that make it difficult to hire top talent into these positions. To assist in explaining the 20% vacancy rate, new hires have to wait an average of 9 months to receive a final offer of employment from their first initial temporary offer. The main reason for this delay is the time it requires to fully investigate and adjudicate new personnel through all the security clearances procedures and steps.

Some highly skilled applicants turned down job offers solely due to compensation, with high-ranking positions (GS-12, Step 2) starting at about $65 K after taxes (see breakdown in Appendix A). However, the greater hiring discrepancy is related to delays in hiring due to security clearance and administrative processing, resulting in low acceptance rates of final job offers; these requirements take an average of 9 months before sending candidates their final job offer. The delayed hiring process further substantiates the significant number of veteran hires (72%), who usually hold necessary certifications and security clearances [30].

In May of 2021, the American Federation of Government Employees (AFGE), representing over 300,000 federal employees in the DOD, asserted in a letter to the Deputy Secretary of Defense, Honorable Kathleen Hicks, that skill gaps still persist in the DoD for jobs like cyber, intelligence, STEM, and other highly skilled positions even with the flexibilities in hiring that were provided to the DOD. AFGE asserts that the mis-use of these authorities have been one of the primary factors in the skill gap. This is leading to reduced retention and difficulty in hiring long-term employees [3].

In order to improve this problem, ARCYBER has centralized its hiring authorities for all locations. This has reduced processing time significantly, but it still takes a month to process personnel before they can start work. Additionally, ARCYBER has implemented the Career Exploration, Advancement, and Development (CEAD) program to create internships with local universities around Fort Gordon, GA. During these internships, students are exposed to the mission of the headquarters and the security clearance process has begun as they are juniors in college. Upon graduates, students are then able to quickly transition into a civilian role in a short period of time. ARCYBER is also re-looking the grade and level of its cyber positions to be more competitive with other DOD organizations that require the same skill level and expertise.

Future Trends

Authorizations for cyber personnel are increasing every year, since 2017 there has been an almost 300% increase in cyber personnel authorizations within the Army. Growth will continue until at least 2026 but only in small increases over the next several years. This leads to an increase in acquisition programs and expanded training facilities at the U.S. Army Cyber School. Due to the small numbers in growth, this is not a difficult mission for the Army.

The U.S. Army continues to develop personnel management initiatives that align with the Army People Strategy (APS). The momentum from Army-wide personnel reform efforts informs the ARCYBER vision for meeting the USCYBERCOM intent to shift toward defending forward. Army strategists continue to discern relevant factors such as civil-military relations and market trends. U.S. Army Talent Management supports the Army's People First philosophy and the APS *acquire and develop* line of effort by refining relevant programs, including: direct commissioning, the Army Coaching Program, and integrating the GRE requirement at Captains Career

Course [27]. Such programs enable effective pipelines for talented recruits to pursue and grow in an Army career. In addition, the Army established the Army Enterprise Marketing Office, an agency fulfilled by its newly-formed functional area (FA58) to develop a data-drive national marketing campaign in support of the Army's recruiting mission [28]. The multi-faceted campaign aims to improve recruitment rates of talented candidates, to include those in the cyber field.

Talent Development: Creating a Dominant Workforce

ARCYBER Development Pipeline

Every new cyber candidate conducts their initial training at the United States Army Cyber School, within the Cyber Center of Excellence, at Fort Gordon, Georgia. These courses are very challenging and prepare personnel with the basic skills to succeed in their assigned units. For very talented and skilled personnel, they can "test-out" of the training, or portions of the training. However, in an interview with the Commandant of the Cyber School, out of over 600 personnel that tried to "test-out" and get constructive credit, only six have successfully passed all the exams.

For recruits selected to become cyber operations specialists (MOS 17C), they undergo a rigorous 39-week course at Fort Gordon, Georgia after they complete basic training. This is a very difficult course for a large percentage. Currently, about 22% of candidates are unable to meet the requirements of the program, for either academic reasons (76% of failures) or failure to meet the standards of the Army. For those that can not meet the academic requirements, they are assigned into a different MOS within the Army. According to the U.S. Army Cyber Center of Excellence [29], 17Cs are trained to execute offensive and defensive operations in cyberspace to enable freedom of maneuver within the cyber domain and deny the same to adversaries. Such operations include data and network protection, net-centric capabilities, detecting variants, responding to attacks, employing devices, employing computer programs or techniques that combine software, firmware, or hardware to produce cyber effects across all domains [29].

New officers undergo a 37-week program training curriculum at the US Army Cyber School at Fort Gordon, GA. Since all officer candidates have a STEM degree, their pass rates are much higher and only a small fraction are removed from the program. Upon completion of this course, these cyber officers (MOS 17A) serve experts in projecting power in and through cyberspace and the electromagnetic spectrum. They understand maneuver operations to ensure synchronized, relevant, and integrated effects are employed within an ever-changing strategic and operational environment. These officers are subject matter experts in cyberspace operations and employment of Cyber Mission Forces (CMF), including offensive and defensive capabilities at all echelons [11].

Selected warrant officers undergo the twenty-one week Cyber Warfare Technician Warrant Officer Basic Course where they build on their already technical background to become the technical experts within the Cyber Corps. Upon completing the basic course, warrant officers become Cyber Operations Technicians (MOS 170A). 170As serve as subject matter experts and advisors to their units for offensive and defensive cyber operations [33]. Units rely upon the expertise of 170As to develop policy recommendations and technical guidance to optimize combat effectiveness with regard to cyberspect effects. Cyber Operations Technicians directly affect the security of DOD information and networks against adversaries [33].

Upon earning MOS classification, graduates move to their operational unit to fill a vacancy in one of the approximately 2,600 slots within the Cyber Mission Force (CMF) teams (active duty, army reserves, and army national guard). Upon arrival at their unit, another selection process begins to identify personnel for USCYBERCOM's most critical work roles. These are work roles that require extensive rigorous training, demanding qualifications, and very talented individuals to pass the requirements. Follow-on training to fulfill one of these critical work roles can add up to an additional 36 months of training and evaluations. This additional training in these critical work roles can exceed $300,000. A historical pass rate below fifty percent indicates the level of difficulty of this training curriculum, where students are usually recycled through several iterations or dropped for other jobs. However, this course is also continuously changing due to the evolving cyberspace operating environment. The comprehensive selection process for these critical work roles results in the selection of the most capable and dedicated personnel. However, once trained and qualified, this very select group of personnel are the most capable and well-trained cyber personnel within the Department of Defense.

Commercial Sector

The U.S. commercial sector consists of major employers who have the capacity to pay employees well, train for the skills they want, and recruit through their large employee network. Cyber industry leaders invest in the development and training of their employees despite high personnel turnover rates in the private sector. Based on the pattern of significant investments in employee training across the industry, organizations can observe higher personnel costs by choosing not to invest in their workers' growth and development.

Google established a tracked learnings system called Googler-to-Googler (G2G), which is an employee network of volunteers to become teachers and students [25]. G2G lowers costs associated with general training, therefore allowing Google to allocate more of its training budget to specialized training and niche training that is identified to address shortfalls.

Apple credits its low employee turnover rates to its scrupulous onboarding process [17]. By hiring people already committed to the Apple brand and its workplace culture, Apple addresses retention issues by focusing on curating their hiring process to "pick the right apples" [17].

Amazon heavily invests in upskilling its current employees in addition to providing them with competitive wages and benefits [1]. Amazon's vice president of human resources, Beth Galetti, stated that the organization's approach to developing talent values the growth opportunities for its workers, even if those employees will maximize their growth by pursuing careers elsewhere [2]. Examples of Amazon development programs are the Amazon Technical Academy, Associate2Tech program, and the Machine Learning University [25]. Amazon has the capacity to accept a high rate of attrition to maintain its philosophy of investing in its workers.

Summary

The evolving cyber domain demands adaptable training curriculums for ARCYBER warriors to remain nimble in cyberspace. ARCYBER development pipelines must prepare cyber warriors for their roles as Army personnel as well as technical experts. Understanding how their field expertise enhances national security will make them better cyber professionals. ARCYBER can encourage and ensure this by investing in development opportunities for its CMF. Utilizing how the commercial sector trains its workforce can lead to streamlining and better preparing ARCYBER warriors for their next cyber job. Leveraging these insights, ARCYBER can lower training costs and improve its development programs. As cyber warriors move from different cyberspace operational fields (defensive to offensive) they can go through shorter re-training courses that build off their experience and demonstrated skills—better preparing them for their moves within the cyber mission teams and workforce. The competitive labor market in the U.S. cyber commercial sector has forced industry leaders to curate their cyber human capital strategies. ARCYBER can utilize best practices and lessons learned to emulate the commercial sector's high caliber development strategies for its cyber workforce.

Talent Retention: Keeping the Best

Retention by the Numbers

In order to build a mature and experienced workforce, it requires a level of longevity for personnel to continuously serve in cyber positions. Over the past three years, the Army has exceeded its goal in retaining Cyber professionals. In fiscal year 2020, 41% of enlisted cyber warriors eligible for reenlistment, decided to pursue a continued contract with the Army.

Retention rates for ARCYBER critical work roles are 20% on average, over the past three years—meaning that the organization is retaining 20% of those eligible to leave the force when their contracts or service obligations are complete. These personnel are highly qualified and have many opportunities outside of serving on Army teams. Additionally, their higher qualifications and experience lead to very competitive offers from other agencies and the private sector. Personnel with a high opportunity cost to stay in the Army are therefore defined as regrettable attrition if they pursue such offers. Consideration for a Cyber warrior's opportunity cost as a metric in developing strategies to reduce regrettable attrition supports the Army People Strategy's aim to use data-driven talent management practices [20].

Military retention tools include education benefits, retention bonuses, and retirement contributions gained in addition to non-taxable income and free healthcare. Altogether, a mid-grade Cyber Soldier (Non-Commissioned Officer in the rank of E-6) with about six years of experience and trained for high-caliber cyber jobs can earn upwards of $100,000 in compensation. Comparatively, a DOD Cyber civilian employee in certain DOD organizations can earn about $105,000 with the same skills; DOD Civilians can also earn up to $60,000 in loan repayment, a 25% (performance or relocation) bonus, and 8 h of vacation for every pay period. Appendix A outlines of each of these compensation packages for comparison.

Utilizing these retention numbers and putting it into a more recognized business term, the Army's Cyber Mission Force experiences a 15% annual "turnover rate" (# people leaving ÷ # people in the organization) for purely Cyber personnel.

Compensation

Enlisted cyber personnel that choose to extend their contract by re-enlisting are eligible for a selective retention bonus of up to $72,000 for an additional six-year contract or, if they are in a critical work role, they can receive up to $81,000 for the same time commitment. Senior Warrant Officers are eligible for retention bonuses of $20,000 per year for a maximum of four years. To date, this has been very successful with the retention of 19 senior warrants, with 12 of them rescinding their retirement requests to stay on active duty. There is also a written bonus agreement for senior enlisted bonus that provides up to $100,000 for four years of continued service. Since 2017, 37 senior enlisted non-commissioned officers have accepted the agreement and stayed on active duty.

Cyber Assignment Incentive Pay (CAIP), while not a retention measure, is designed to incentivize personnel to complete their required training for their assigned work role as soon as possible. Depending on where an individual is assigned in the organization, the difficulty of the training for their work role, and the overall fill-rate within the command—they can earn from $150 to $1000 per month for basic proficiency levels up to $300–$1500 per month for those that hold a master proficiency level. For masters in the critical work roles, this is an extra $18,000 per year.

Special Duty Assignment Pay (SDAP)—for personnel assigned to the cyber mission force, they are eligible for an additional $150-$300 per month based on their proficiency level (basic to master).

ARCYBER personnel electing to separate receive a compensation comparison during separation counseling to make informed decisions. Taking into account tax incentives for being on active duty, free health insurance, and other incentives—a staff sergeant with six years of service needs to earn an equivalent of $100,000 per year in the area around Washington, DC area in civilian compensation—details are located in Appendix A.

Civilian Comparison

Cyber industry leaders in the U.S. incorporate focused strategies to retain their cyber talent by gaining insight on the desires and needs of its people, analyzing the information gathered, and addressing major issues to reduce regrettable attrition. The global shortage of cyber professionals currently stands at a 3.12 million gap between workers and jobs available, according to $(ISC)^2$. While the current standing is an improvement compared to the 2019 report, uncertainties from the pandemic contributed to this trajectory shift. For example, cyber professionals reported losing their jobs due to the pandemic, which paired with organizations reporting downsizing efforts to save costs on personnel expenses in order to redirect spending to technological expenses. Prior to 2020, the global shortage was increasing each year, peaking at 4.07 million in 2019. While the cyber labor market saw losses as a result of the pandemic's impact on many businesses, organizational leadership across industries recognize the impetus of having strong cybersecurity for their businesses. This effect may lead to the cyber workforce gap to revert back to its preceding pattern of growth. Furthermore, the 2020 $(ISC)^2$ Cyber Workforce Estimate suggests that the global cyber workforce must grow by 89% to meet its cybersecurity demands. Another tech survey states about 1/3 of technical employees are planning on leaving within the year with 21% actively looking [14]. Meanwhile, unemployment among cyber professionals is about 2% globally. The competitive cyber labor market therefore continually inflates wages across all sectors.

By comparison with commercial technology companies, who have a turn-over rate of 13.2% world wide for voluntary and involuntary combined, the Army experiences a higher turnover rate. Involuntary is an additional 7%—or 20.2% overall turn-over rate). This means 13.2% of the entire organization voluntarily leaves each year in these technology companies [6]. Even if companies do manage to attract and onboard a great candidate, chances are they won't stick around for long. Commercial sector job roles that are comparable to that of ARCYBER skill sets include: data analysts with a turn-over rate of 21.7% and Embedded Software Engineer at 21.7% [21].

While Google is often hyped for its comprehensive benefits package and quirky offices, the average length of service for Google employees is just 1.1 years. Amazon's tech workers tend to stay for only one year, and Apple fares little better, with a typical tenure of 24 months [21]. Interestingly, other sources report that Google has great success in its talent retention efforts—which is attributed to their people analytics model [26]. It isn't surprising that these major technical companies don't readily share its retention data due to the impact on its image—or the information that other sources are reporting contain errors which they aren't upfront about. Overall, people analytics focus talent management strategies on data-driven solutions—which lead to better outcomes. As opposed to hunch-driven solutions that prevail in HR practices throughout the industry to this date [12].

Work Force Feedback—ARCYBER

In order to retain the best, ARCYBER has to understand the needs and desires of its workforce. First, the average age of the entire enlisted population when they officially signed their enlistment contract is 23 years old. Fifty percent of ARCBYER personnel were ages 18–23 years, 40% were 24–29 years old, and 10% were 30–42 years old. Comparatively, the average age of the officers when they entered active-duty service is also 23. For a 6-year contract, Army Cyber personnel face the decision to continue serving in the Army around 28 years of age. Personnel who begin serving at the age of 18 would face the same decision at around 23 years of age. The ages of the respective decision points represent different phases of life, where life goals and requirements factor in considerable differences. This requires multiple retention options to meet the requirements of the Army's diverse and unique workforce; however, a majority of the force that is eligible will still leave each year.

To better determine how ARCYBER can improve retention, it conducted a survey of the key work roles in early 2019. 483 personnel received the survey and 333 responded. Of the 333, 287 provided full responses with detailed responses to the questions. After a semantic analysis, the work force's top response was they "loved" the mission. They wanted to execute more missions and wanted the administrative distractions to be minimized. The work force wanted more time to concentrate on the mission, and that most of the Army mandatory requirements were distracting them from the mission and was a factor for them to leave the service after their contract or service obligation expired. Some wanted to just focus on their tradecraft (keyboard time) and not have to worry about the Army promotion requirements either, so there was concern on career progression that is against traditional Army career paths. This was supported by a USCYBERCOM survey in 2020 that asked similar questions of the entire force from all services.

Second, the work force wanted more opportunities and time to build their tradecraft, to learn more skills, educate themselves, tinker on projects, or just attend conferences. They wanted better career paths or development plans that demonstrated a continuous learning plan in which the Army was investing in their development. Also, due to the high operational tempo of those that are fully qualified and experts in their tradecraft, a balance between mission and self improvement is critical to build a better workforce but to also prevent burn-out.

A distant third was related to compensation and recognition. Some wanted to be paid a lot more (those that were likely to leave the force anyway), but the majority just didn't want to worry about money (living paycheck to paycheck), and wanted some other forms of recognition for discoveries, missions executed very well, or higher skill levels.

In Daniel Pink's Book DRIVE, he hypothesized, employees are much more likely to stay with their employer based on three factors: Autonomy, Mastery, and Purpose. This is especially true with highly talented employees that work cognitively challenging jobs. They want autonomy over their work day and schedule. Ability to become masters in their field of expertise, and to be involved with a purpose that resonates with their core values. In comparison with the surveys conducted within ARCYBER, the results match what Daniel Pink proposed to improve retention and productivity in an organization.

In addition, concerning retention, most wanted stability in locations, better quality of life for family, improved employment opportunities for spouses, and other family centric issues that usually relate to an older Army workforce. Families are a very high priority in the decision framework of personnel that are considering staying or leaving the Army cyber workforce.

Civilian Comparison

Cyber professionals in the commercial sector cited career advancement, competitive compensation, and leadership's commitment to cybersecurity as the top three reasons that affect job satisfaction and their decision to leave an organization [23]. The 2019 Nelson Frank Survey also found the top two reasons for job dissatisfaction among cyber professionals to be career development and compensation. A prior survey, LinkedIn's 2018 Workforce Learning Report, found 94 percent of personnel would stay at a company longer if it continuously invested in their career development. The 2020 Nelson Frank Survey found that personnel considered leaving their current employer based on: (1) a lack of career and promotional prospects; (2) low salary or earnings potential; and (3) a desire for new challenges. When these desires of the employees were met, it led to increased motivation, job satisfaction, and feeling valued by the organization—leading to lower turn-over rates and better productivity.

The figure below illustrates a comparative view of grievances from ARCYBER workforce surveys and commercial sector cyber workforce surveys; there is a correlation between commercial sector and Army on what personnel want.

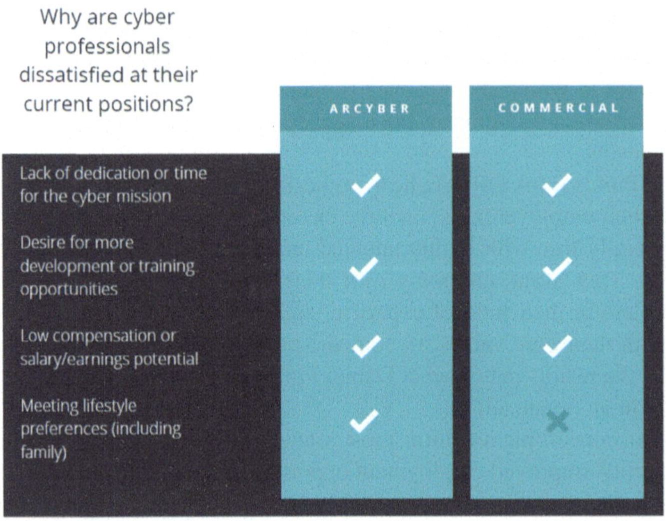

Cyber professionals in both groups reiterated job dissatisfaction based on a dedication to the cyber mission, development opportunities, and competition. ARCYBER surveys found its personnel cited dissatisfaction based on lifestyle preferences as well. Aside from grievances related to compensation, each of these factors are tied to the organizational culture. As the evolving cyber domain becomes a more visible priority to leadership at various organizations, the culture-based concerns may improve. ARCYBER has the organizational structure to instill appropriate workplace culture through its chain of leadership to address grievances specific to its specialized cyber warriors. Furthermore, ARCYBER should recognize the opportunity to address personnel grievances and present them to key leaders, based on the momentum from personnel reform initiatives such as the Army People Strategy. Lastly, ARCYBER should consider the grievances related to Permanent Change of Station (PCS) cycles and duty assignments that affect quality of life and family preferences. These factors distinguish all military personnel from their civilian counterparts, but can be addressed with the available infrastructure, the nature of cyber work roles, and with the momentum of the latest people-focused management strategies.

The DOD Cyber Strategy requires the establishment of a cyber top talent management program that provides its most skilled cyber personnel with focused resources and opportunities to develop their skills over the course of their careers. In this program, the Department needs to use competitive processes, including individual and team competitions to identify the most capable DOD cyber personnel—in order to appropriately apply limited resources to retaining the best of the cyber workforce. This is required because the DOD needs to empower these identified personnel to solve the Department's toughest cyber challenges [9].

Conclusion

The U.S. Government's goal for achieving and maintaining superiority in the Cyber domain is not feasible without a data-driven national strategy that is postured as a whole-of-society defense. Establishing this foundation will enable a talent management blueprint to operationalize partnerships across agencies and sectors to dominate in the continuum of engagement against the nation's adversaries in cyber warfare. The U.S. can optimize cyber technology innovation, both in developing and managing it, by streamlining the management of cyber talent through proper development and nationwide partnerships. A curated people strategy that focuses on development and partnerships can improve the experience within Army Cyber Command (ARCYBER) careers and attract more cyber talent. Investing in cyber professional career opportunities can enhance ARCYBER human capital, despite easing transitions to careers outside of ARCYBER, because it may also ease transitions into the cyber force. Strategizing the ARCYBER posture and personnel management within the context of a whole-of-society defense can elevate the appeal for ARCYBER careers. ARCYBER should continue to gather and analyze information to develop solutions that address the evolving factors of the cyber labor market competition while striving for combat effectiveness in cyber warfare. A talent management blueprint that sets parameters for talent recruitment, development, and retention, can enable ARCYBER to evolve with the landscape without sacrificing its talent management goals.

Appendix A

ARCYBER Compensation Comparison

	SSG Unmarried 6 years experience		Civilian GS 12 Step 2 (NSA)		SSG Married 6 years experience		NSAICEIP TLMS GGE-13	
Salary	$	70.222.80	$	90,106.00	$	76,126.80	$	126,676.00
Taxable	$	41,464.80	$	86,838.58	$	41,464.80	$	123,408.58
Non-taxable	$	28,758.00	$	3,267.42	$	34,662.00	$	3,267.42
Health insurance	$	–	$	3,267.42	$	–	$	3,267.42
Federal tax	$	4,302.78	$	8,938.00	$	3,113.68	$	10,461.00
State tax	$	1,857.95	$	4,163.38	$	1,857.92	$	5,572.50
RCA	$	3,172.00	$	6,487.73	$	3,172.00	$	8,185.50
Take home pav	$	60,890.07	$	67,249.47	$	67,983.20	$	99,189.58
Incentives								

(continued)

(continued)

	SSG Unmarried 6 years experience		Civilian GS 12 Step 2 (NSA)		SSG Married 6 years experience		NSAICEIP TLMS GGE-13	
CAIP entitlement	$	15,000.00	$		$	15,000.00	$	
SDAP	$	2,400.00	$	–	$	2,400.00	$	–
CIV retention incentive	$	–	$	22,526.50	$	–	$	–
Selective reenlistment bonus	$	16,200.00	$		$	16,200.00	$	–
Incentive tax burden	$	6,252.00	$	3,439.00	$	3,795.00	$	
Take home dav with cvber entitlements (after taxes)	$	83,232.00	$	86,336.00	$	95,388.00	$	99,189.58
Future Entitlements								
Retirement contribution	*$*	*10,000 00*	*$*	*4,505.30*	*$*	*10,000.00*	*$*	*6,333.80*
Take home pav with future entitlements	*$*	*93,232,00*	*$*	*90,841,30*	*$*	*105,388,00*	*$*	*105,523,38*

References

1. Amazon. U.S. Benefits and Stocks. Retrieved June 15, 2021, from https://amazon.jobs/en/benefits.
2. Amazon Staff. (2020, October 2). Amazon will invest over $700 million to provide upskilling training programs. https://www.aboutamazon.com/news/workplace/upskilling-2025.
3. American Federation of Government Employees. (2021). Letter to Secretary Kathleen H. Hicks. Retrieved June 16, 2021, from https://www.afge.org/globalassets/documents/generalreports/2021/letter-to-sec.-hicks-from-afge-np-kelley.pdf.
4. Andress, J., & Winterfeld, S. (2011). *Cyber warfare: techniques, tactics and tools for security practitioners*. Waltham, MA: Syngress - Elsevier, Inc.
5. Bates, C., & Hoadwonic, C. (2021). Talent management series: A roadmap to defending forward. *Cyber Defense Review*, 6(4).
6. Booz, M. (2018). These 3 Industries Have the Highest Talent Turnover Rates. LinkedIn Talent Blog. Retrieved June 17, 2021, from https://www.linkedin.com/business/talent/blog/talent-strategy/industries-with-the-highest-turnover-rates.
7. Buchanan, B. (2017). *The Cybersecurity Dilemma*. New York: Oxford University Press.

8. Cybersecurity and Infrastructure Security Agency. (2020). CISA 2020 Year in Review. Department of Homeland Security. https://www.cisa.gov/sites/default/files/publications/CISA_Year_in_Review_2020_Final.pdf.
9. Department of Defense (DOD). (2018). Department of Defense Cyber Strategy 2018. Retrieved June 16, 2021, from https://media.defense.gov/2018/Sep/18/2002041658/-1/-1/1/CYBER_STRATEGY_SUMMARY_FINAL.PDF.
10. Department of Defense. (2015). Department of Defense Cyber Strategy 2015. Department of Homeland Security. https://archive.defense.gov/home/features/2015/0415_cyber-strategy/final_2015_dod_cyber_strategy_for_web.pdf.
11. Department of the Army, Headquarters. (2018). Department of the Army Pamphlet 600-3: Commissioned Officer Professional Development and Career Management. Secretary of the Army.
12. Donaldson, C. (2017, October 12). The 3 most common people analytics problems (and how to fix them). *Inside HR*. https://www.insidehr.com.au/3-most-common-people-analytics-problems/.
13. Ertan, A., Floyd, K., Pernik, P., & Stevens, T. (2020). Cyber Threats and NATO 2030: Horizon Scanning and Analysis. NATO Cooperative Cyber Defence Centre of Excellence. https://ccdcoe.org/uploads/2020/12/Cyber-Threats-and-NATO-2030_Horizon-Scanning-and-Analysis.pdf.
14. Frank, N. (2019). Nelson Frank Salary Survey 2019/20. Frank Recruitment Group.
15. Joint Publication 1-02. (2016). Department of Defense Dictionary of Military and Associated Terms. U.S. Department of Defense. https://fas.org/irp/doddir/dod/jp1_02.pdf.
16. Joint Task Force on Computing Curricula. (2013). Computer Science Curricula 2013. Association for Computing Machinery (ACM), IEEE Computer Society. https://www.acm.org/binaries/content/assets/education/cs2013_web_final.pdf.
17. Khan, H. (2016, April 28). The apple store guide to insanely great customer service. *Shopify Retail Blog*. https://www.shopify.co.uk/retail/119535811-the-apple-store-guide-to-insanely-great-customer-service.
18. Libicki, M. C. (2009). *Cyberdeterrence and cyberwar*. Rand Corporation.
19. Lopez, T. C. (2019). Persistent Engagement, Partnerships, Top Cybercom's Priorities. U.S. Department of Defense. https://www.defense.gov/Explore/News/Article/Article/1847823/persistent-engagement-partnerships-top-cybercoms-priorities/.
20. McCarthy, R., McConville, J., & Grinston, M. (2019). Army People Strategy. U.S. Army. Retrieved March 15, 2021, from https://www.army.mil/e2/downloads/rv7/the_army_people_strategy_2019_10_11_signed_final.pdf.
21. Morris, Z. (2019, June 14). What's driving the tech sector's extreme turnover rate? *Information Week*.https://www.informationweek.com/strategic-cio/team-building-and-staffing/whats-driving-the-tech-sectors-extreme-turnover-rate/a/d-id/1334920.
22. National Security Agency (NSA). (2021). Center for Cyber Security Standards. Retrieved June 11, from https://www.nsa.gov/What-We-Do/Cybersecurity/Partnership/Standards/.
23. Oltsik, J. (2020, July). The Life and Times of Cybersecurity Professionals 2020. Enterprise Strategy Group (ESG)–Information Systems Security Association International (ISSA).
24. Schmidt, L., O'Connell, C., Miyake, H., Shah, A. R., Baron, J., Nieboer, G., Jourdan, R., Senty, D., Winkelman, Z., Taggart, L., Sondergaard, S., & Robinson, N. (2015). *Cyber practices: What can the U.S. air force learn from the commercial sector?* Santa Monica, CA: RAND Corporation. https://www.rand.org/pubs/research_reports/RR847.html. Also available in print form.
25. Stringer, G. (2020). How innovative companies train their employees (Google, Amazon, Uber & more). How Now. Retrieved June 17, 2021, from https://gethownow.com/blog/how-innovative-companies-train-their-employees-google-amazon-uber-more/.
26. Sullivan, J. (2013). How Google Reinvented HR and Drives Success Through People Analytics. InsideHR. Retrieved June 17, 2021, from https://www.insidehr.com.au/how-google-reinvented-hr/.

27. U.S. Army. (2016). U.S. Army Talent Management. Retrieved June 16, 2021, from https://talent.army.mil/.
28. U.S. Army. (2019). U.S. Army Enterprise Marketing Office. Retrieved June 16, 2021, from https://www.army.mil/aemo.
29. U.S. Army Cyber Center of Excellence. (2015, October 1). 17C Cyber Operations Specialist. *U.S. Army.* Retrieved June 17, 2021, from https://pdf4pro.com/fullscreen/17c-cyber-operations-specialist-17ee20.html.
30. U.S. Army Cyber Command. (2021). The Nation's Army in Cyberspace. U.S. Army Cyber Command.
31. U.S. Cyber Command. (2018). Achieve and Maintain Cyberspace Superiority–Command Vision for US Cyber Command. U.S. Department of Defense.
32. U.S. Cyber Command (2018). Achieve and Maintain Cyberspace Superiority–Command Vision for US Cyber Command. US Cyber Command. U.S. Army.
33. U.S. Army Cyber School (2022, April 8). The Official Homepage of the U.S. Army Cyber School. U.S. Army. https://cybercoe.army.mil/CYBERSCH/

Colonel Chad Bates, PhD is an instructor at the U.S. Army War College in Carlisle, PA, where he teaches Cyber and Information Power in the Center for Strategic Leadership.

Major Charlene Rose is an Assistant Professor of American Politics at the United States Military Academy at West Point, where she teaches the American Politics survey course and Political Philosophy in the Department of Social Sciences.

The manufacturer's authorised representative in the EU is Springer Nature Customer Service Centre GmbH, Europaplatz 3, 69115 Heidelberg, Germany. If you have any concerns regarding our products, please contact ProductSafety@springernature.com

Printed and bound by CPI Group (UK) Ltd, Croydon, CR0 4YY
30/09/2026
02225493-0013